Springer Proceedings in Mathematics & Statistics

Volume 115

More information about this series at http://www.springer.com/series/10533

Springer Proceedings in Mathematics & Statistics

This book series features volumes composed of select contributions from workshops and conferences in all areas of current research in mathematics and statistics, including operation research and optimization. In addition to an overall evaluation of the interest, scientific quality, and timeliness of each proposal at the hands of the publisher, individual contributions are all refereed to the high quality standards of leading journals in the field. Thus, this series provides the research community with well-edited, authoritative reports on developments in the most exciting areas of mathematical and statistical research today.

Bernhard Heim • Mehiddin Al-Baali
Tomoyoshi Ibukiyama • Florian Rupp
Editors

Automorphic Forms

Research in Number Theory from Oman

Springer

Editors
Bernhard Heim
Department of Mathematics and Science
German University of Technology in Oman
Muscat, Oman

Mehiddin Al-Baali
Department of Mathematics and Statistics
Sultan Qaboos University
Muscat, Oman

Tomoyoshi Ibukiyama
Department of Mathematics Graduate
School of Science
Osaka University
Osaka, Japan

Florian Rupp
Department of Mathematics and Science
German University of Technology in Oman
Muscat, Oman

ISSN 2194-1009 ISSN 2194-1017 (electronic)
ISBN 978-3-319-11351-7 ISBN 978-3-319-11352-4 (eBook)
DOI 10.1007/978-3-319-11352-4
Springer Cham Heidelberg New York Dordrecht London

Library of Congress Control Number: 2014953955

Mathematics Subject Classifications: 11-XX, 11Fxx, 11F46, 11F50

Printed on acid-free paper

Springer is part of Springer Science+Business Media (www.springer.com)

Preface

Without doubt, one of the most important areas of modern science in mathematics is number theory. Consequently, on 18–22 February 2012, the "International Conference on Automorphic Forms and Number Theory" was held in Muscat, Oman; (see http://www.mathoman.gutech.edu.om/index.html for more information).

The conference took place at the German University of Technology in Oman for 3 days and for 2 days at the Sultan Qaboos University and had the main purpose of bringing together students, postdocs, and experts in the field of modern number theory to share their ideas and to establish new collaborations. Besides talks on number theory, presentations on e-learning were also shared.

The event was opened by welcome speeches by the Dean of Science of the Sultan Qaboos University, Prof. Dr. Saif bin Nasser Al-Bahri; the rector of the German University of Technology in Oman, Prof. Dr. Burkhard Rauhut; and the German Ambassador in Oman, H.E. Angelika Storz-Chakarji. The ceremony opened with a live Oud (a traditional Omani musical instrument) performance. For most participants, it was the first visit to an Arabic country. Special talks were included in the multifaceted program of the conference by students from the local universities who spoke on "science in the Orient" and "the beauty of Oman" to show the close relation of Oman to science. The participants were invited to visit the Grand Mosque, the Muttrah Souq, and the Corniche as well.

Recently, the Sultanate of Oman became a member of the International Mathematical Union. In view of this development, the conference provided a platform to encourage collaboration between scientists of different countries. A special feature of the conference was to open the doors for establishing close international exchanges between scientists and students in Germany, Oman, and Japan to help to develop and shape their scientific career. The workshop gave students and researchers in Oman the unique opportunity to attend high-quality scientific talks.

This volume presents a collection of carefully refereed papers that for the most part have been presented at the conference. This includes surveys, outlines of research projects, and research articles on the following topics:

- Borcherds products
- Congruences and codes
- Jacobi forms
- Siegel and Hermitian modular forms
- Special values of L-series

We would like to thank the Graduiertenkolleg “Experimentelle und konstruktive Algebra” and Prof. Dr. Krieg from the RWTH Aachen for supporting the conference and enabling the participation of several young researchers from the RWTH Aachen. Moreover, we also would like to thank Prof. Dr. Ibukiyama and Prof. Dr. Oda for supporting the researchers from Japan. The conference was generously supported by the Sultan Qaboos University, the German University of Technology in Oman, The Research Council of Oman (TRC), and Springer Science+Business Media. We would like to thank all these organizations.

Muscat, Oman — Bernhard Heim
Muscat, Oman — Mehiddin Al-Baali
Osaka, Japan — Tomoyoshi Ibukiyama
Muscat, Oman — Florian Rupp
August 2014

Conference Photo

International Workshop in
Mathematics in Oman 2012
GUtech
International Workshop in
Oman 2012
GUtech

كلية العلوم

Participants

Masood Alam	Sultan Qaboos University, Oman
Dilnawaz Anwar	Sultan Qaboos University, Oman
Hiroki Aoki	Tokyo University of Science, Japan
Mehiddin Al-Baali	Sultan Qaboos University, Oman
Noora Al-Balushi	GUtech, Oman
Humaid Al-Fulaiti	Middle East College of Information Technology, Oman
Nada Al-Harthi	Ministry of Education, Oman
Fatma Al Kharousi	Sultan Qaboos University, Oman
Mohamed Al-Lawati	Sultan Qaboos University, Oman
Sami Al-Rwahi	Ministry of Education, Oman
Elena Berdysheva	GUtech, Oman
Thomas Bier	Sultan Qaboos University, Oman
Siegfried Boecherer	University of Mannheim, Germany
Mohamed S. Boudellioua	Sultan Qaboos University, Oman
Hatice Boylan	Universität Siegen, Germany
Pallath Chandran	Sultan Qaboos University, Oman
Lucia Cloth	GUtech, Oman
Till Dieckmann	RWTH Aachen, Germany
Volker Enss	GUtech, Oman
Rudolf Fleischer	GUtech, Oman
Andreas Freh	RWTH Aachen, Germany
Valery Gritsenko	University Lille I, France
Sanjiv Kumar Gupta	Sultan Qaboos University, Oman
Tamadher Hamdan Al-Rubaie	Sultan Qaboos University, Oman
Bernhard Heim	GUtech, Oman
Tomoyoshi Ibukiyama	Osaka University, Japan

Hidenori Katsurada	Muroran Institute of Technology, Japan
Sebti Kerbal	Sultan Qaboos University, Oman
Hidetaka Kitayama	Osaka University, Japan
Judith Kreuzer	RWTH Aachen, Germany
Aloys Krieg	RWTH Aachen, Germany
Kamal Makdisi	American University of Beirut, Lebanon
Shamil Makhmutov	Sultan Qaboos University, Oman
Marina Makhmutova	Sultan Qaboos University, Oman
Kenji Makiyama	Kyoto Sangyo University, Japan
Pieter Moree	MPIM Bonn, Germany
Atsushi Murase	Kyoto Sangyo University, Japan
Shoyu Nagaoka	Kinki University, Japan
Jibulal B. Nair	Modern College of Business and Science, Oman
R. K. Nair	GUtech, Oman
Hiro-aki Narita	Kumamoto University, Japan
Lovelyn Nelisha	GUtech, Oman
Paul Nevai	King Abdulaziz University Jeddah, Kingdom of Saudi Arabia
Takayuki Oda	University of Tokyo, Japan
Conjeeveram S. Rajan	Tata Institute of Fundamental Research, India
Balakrishnan Ramakrishnan	Harish Chandra Research Institute, India
Burkhard Rauhut	GUtech, Oman
Nirmal C. Sacheti	Sultan Qaboos University, Oman
Hashim Sadiq	Sultan Qaboos University, Oman
Huda Saleh Al-Sharji	Sultan Qaboos University, Oman
Irina Skhomenko	Mazoon College, Oman
Nils Skoruppa	Universität Siegen, Germany
Patrick Solé	Telecom Paristech, France
Chefi Triki	Sultan Qaboos University, Oman
Abdullahi Umar	Sultan Qaboos University, Oman
Cornelia Wirtz	RWTH Aachen, Germany
Martin Woitalla	RWTH Aachen, Germany
Tsukasa Yashiro	Sultan Qaboos University, Oman

Contents

On Formal Series of Jacobi Forms and Borcherds Products

Hiroki Aoki

Mathematics Subject Classification 2000: Primary 11F50; Secondary 11F46

1 Formal Series of Jacobi Forms

1.1 Siegel Modular Forms of Degree 2

First, we review Siegel modular forms of degree 2. We denote the Siegel upper half space of degree 2 by

$$\mathbb{H}_2 := \left\{ Z = {}^tZ = \begin{pmatrix} \tau & z \\ z & \omega \end{pmatrix} \in \mathrm{M}_2(\mathbb{C}) \;\middle|\; \mathrm{Im}\, Z > 0 \right\}.$$

The symplectic group

$$G := \mathrm{Sp}(2,\mathbb{R}) = \left\{ M = \begin{pmatrix} A & B \\ C & D \end{pmatrix} \in \mathrm{M}_4(\mathbb{R}) \;\middle|\; {}^tMJM = J := \begin{pmatrix} O_2 & -E_2 \\ E_2 & O_2 \end{pmatrix} \right\}$$

acts on $\mathbb{H}_2$ transitively by

$$\mathbb{H}_2 \ni Z \longmapsto M\langle Z\rangle := (AZ+B)(CZ+D)^{-1} \in \mathbb{H}_2.$$

H. Aoki (✉)
Department of Mathematics, Faculty of Science and Technology,
Tokyo University of Science, Noda, Chiba 278-8510, Japan
e-mail: aoki_hiroki@ma.noda.tus.ac.jp

B. Heim et al. (eds.), *Automorphic Forms*, Springer Proceedings
in Mathematics & Statistics 115, DOI 10.1007/978-3-319-11352-4_1

For a function $F(Z)$ on $\mathbb{H}_2$ and $k \in \mathbb{Z}$, define the action of G by

$$F(Z) \longmapsto (F|_k M)(Z) := \det(CZ + D)^{-k} F(M\langle Z \rangle).$$

Let Γ be a finite index subgroup of $\mathrm{Sp}(2, \mathbb{Z}) := G \cap \mathrm{M}_4(\mathbb{Z})$. We say F is a Siegel modular form of weight k if F is holomorphic on $\mathbb{H}_2$ and satisfies the condition $F = F|_k M$ for any $M \in \Gamma$. We denote the space of all Siegel modular forms of weight k by $\mathbb{M}_k(\Gamma)$.

Let $F \in \mathbb{M}_k(\Gamma)$. For each $M \in \mathrm{Sp}(2, \mathbb{Z})$, F has a Fourier expansion

$$(F|_k M)(Z) = \sum_{n,l,m} a_M(n, l, m)\, q^n \zeta^l p^m,$$

where $q^n := \mathbf{e}(n\tau) := \exp(2\pi\sqrt{-1}n\tau)$, $\zeta^l := \mathbf{e}(lz)$ and $p^m := \mathbf{e}(m\omega)$. It is well known as Koecher principle that $a_M(n, l, m) = 0$ if $4nm - l^2 < 0$ or $m < 0$. we say $F \in \mathbb{M}_k(\Gamma)$ is a cusp form if F satisfies the condition $a_M(n, l, m) = 0$ unless $4nm - l^2 > 0$. We denote the space of all cusp forms of weight k by $\mathbb{M}_k^c(\Gamma)$.

1.2 Jacobi Forms

Jacobi forms were first studied by Eichler and Zagier in their book [6]. In their book, mainly they treated Jacobi forms with respect to the full modular group. In this section, we review Jacobi forms for arbitrary finite index subgroup based on their book.

First, we name some elements of G:

$$C(g) := \begin{pmatrix} a & 0 & b & 0 \\ 0 & 1 & 0 & 0 \\ c & 0 & d & 0 \\ 0 & 0 & 0 & 1 \end{pmatrix} \quad \text{and} \quad U(s, t, u) := \begin{pmatrix} 1 & 0 & 0 & t \\ s & 1 & t & u \\ 0 & 0 & 1 & -s \\ 0 & 0 & 0 & 1 \end{pmatrix}$$

for $g = \begin{pmatrix} a & b \\ c & d \end{pmatrix} \in \mathrm{SL}(2, \mathbb{R})$ and $s, t, u \in \mathbb{R}$. Let

$$G^J := \{M \in G \mid M^{-1}U(0, 0, 1)M = U(0, 0, 1)\}.$$

It is easy to show that G^J is generated by above $C(g), U(s, t, u)$ and $-E_4$. For a function $\varphi(\tau, z)$ on $\mathbb{H} \times \mathbb{C}$, $k \in \mathbb{Z}$ and $m \in \mathbb{Q}$, define

$$(\varphi|_{k,m} M)(\tau, z) := ((\varphi(\tau, z)p^m)|_k M)\, p^{-m}.$$

It is easy to show that the right-hand side of the above equation is independent of ω. Hence this definition makes sense and G^J acts on the set of all functions on $\mathbb{H} \times \mathbb{C}$.

Let Γ^J be a finite index subgroup of $G^J \cap \mathrm{Sp}(2, \mathbb{Z})$. Suppose φ be a holomorphic function on $\mathbb{H} \times \mathbb{C}$ satisfying $\varphi|_{k,m} M = \varphi$ for any $M \in \Gamma^J$. Define $u_0 > 0$ by

$$u_0\mathbb{Z} = \left\{u \in \mathbb{Z} \mid U(0,0,u) \in \Gamma^J\right\}.$$

It is easy to show that φ should be zero if $mu_0 \notin \mathbb{Z}$. According to the book of Eichler and Zagier [6, Theorem 1.2], we can show the following two properties:

- If $m < 0$, φ should be zero.
- If $m = 0$, φ only depends on τ.

For $M \in G^J \cap \mathrm{Sp}(2, \mathbb{Z})$, φ has a Fourier expansion

$$(\varphi|_{k,m} M)(\tau, z) = \sum_{n,l} c_M(n,l)\, q^n \zeta^l.$$

We say φ is a Jacobi form of weight k and index m if $c_M(n,l) = 0$ when $4nm - l^2 < 0$. We say φ is a Jacobi cusp form, weak Jacobi form, or weakly holomorphic Jacobi form if $c_M(n,l) = 0$ when $4nm - l^2 \leq 0$, $n < 0$, or $n << 0$, respectively. We denote the space of all Jacobi forms, Jacobi cusp forms, weak Jacobi forms, or weakly holomorphic Jacobi forms of weight k and index m by $\mathbb{J}_{k,m}(\Gamma^J)$, $\mathbb{J}^{\mathrm{c}}_{k,m}(\Gamma^J)$, $\mathbb{J}^{\mathrm{w}}_{k,m}(\Gamma^J)$, or $\mathbb{J}^{\mathrm{wh}}_{k,m}(\Gamma^J)$, respectively. We remark that $\mathbb{J}^{\mathrm{c}}_{k,m}(\Gamma^J) \subset \mathbb{J}_{k,m}(\Gamma^J) \subset \mathbb{J}^{\mathrm{w}}_{k,m}(\Gamma^J) \subset \mathbb{J}^{\mathrm{wh}}_{k,m}(\Gamma^J)$.

1.3 Fourier-Jacobi Expansion

From now on, we set $\Gamma^J := \Gamma \cap G^J$. The Fourier-Jacobi expansion is a p-expansion of a Siegel modular form. For $F \in \mathbb{M}_k(\Gamma)$, we have a family of Jacobi forms (φ_m) by

$$F(Z) = \sum_{m \geq 0,\ mu_0 \in \mathbb{Z}} \varphi_m(\tau, z) p^m.$$

Namely, the Fourier-Jacobi expansion induces an injection

$$\mathbb{M}_k(\Gamma) \ni F \longmapsto (\varphi_m) \in \prod_{m \geq 0,\ mu_0 \in \mathbb{Z}} \mathbb{J}_{k,m}(\Gamma^J).$$

When $F \in \mathbb{M}^{\mathrm{c}}_k(\Gamma)$, the Fourier-Jacobi expansion induces an injection

$$\mathbb{M}^{\mathrm{c}}_k(\Gamma) \ni F \longmapsto (\varphi_m) \in \prod_{m > 0,\ mu_0 \in \mathbb{Z}} \mathbb{J}^{\mathrm{c}}_{k,m}(\Gamma^J).$$

Now we assume

$$S := \begin{pmatrix} 0&1&0&0 \\ 1&0&0&0 \\ 0&0&0&1 \\ 0&0&1&0 \end{pmatrix} \in \Gamma.$$

We remark that G is generated by S and G^J. On the Fourier expansion

$$\varphi_m(\tau, z) = \sum_{n,l} c_m(n,l)\, q^n \zeta^l,$$

our assumption $S \in \Gamma$ induces a "symmetry" $c_m(n,l) = (-1)^k c_n(m,l)$. Let

$$\left(\prod_{m \geq 0,\ mu_0 \in \mathbb{Z}} \mathbb{J}_{k,m}(\Gamma^J) \right)^{\mathrm{sym}} := \left\{ (\varphi_m) \in \prod_{m \geq 0,\ mu_0 \in \mathbb{Z}} \mathbb{J}_{k,m}(\Gamma^J) \;\middle|\; c_m(n,l) = (-1)^k c_n(m,l) \right\}$$

and

$$\left(\prod_{m > 0,\ mu_0 \in \mathbb{Z}} \mathbb{J}^{\mathrm{c}}_{k,m}(\Gamma^J) \right)^{\mathrm{sym}} := \left\{ (\varphi_m) \in \prod_{m > 0,\ mu_0 \in \mathbb{Z}} \mathbb{J}^{\mathrm{c}}_{k,m}(\Gamma^J) \;\middle|\; c_m(n,l) = (-1)^k c_n(m,l) \right\}.$$

Under the assumption $S \in \Gamma$, the Fourier-Jacobi expansion induces two injections

$$\mathrm{FJ} : \mathbb{M}_k(\Gamma) \ni F \longmapsto (\varphi_m) \in \left(\prod_{m \geq 0,\ mu_0 \in \mathbb{Z}} \mathbb{J}_{k,m}(\Gamma^J) \right)^{\mathrm{sym}}$$

and

$$\mathrm{FJ}^{\mathrm{c}} : \mathbb{M}^{\mathrm{c}}_k(\Gamma) \ni F \longmapsto (\varphi_m) \in \left(\prod_{m > 0,\ mu_0 \in \mathbb{Z}} \mathbb{J}^{\mathrm{c}}_{k,m}(\Gamma^J) \right)^{\mathrm{sym}}.$$

1.4 Our Conjecture

Here we give one conjecture about Fourier-Jacobi expansion.

Conjecture 1. We assume that Γ is generated by Γ^J and S. Then two maps FJ and FJ^{c} are bijective.

A reason to believe this conjecture is the paper [1] by the author in 2000, where we determined the ring structure of $\mathbb{M}_k(\mathrm{Sp}(2,\mathbb{Z}))$ by using the map FJ cleverly.

In this paper, our interest is the case

$$\Gamma = \Gamma_0^{(2)}(N) := \left\{ M = \begin{pmatrix} A & B \\ C & D \end{pmatrix} \in \mathrm{Sp}(2,\mathbb{Z}) \;\middle|\; C \equiv O_2 \pmod{N} \right\},$$

where $N \in \mathbb{N}$. In this case, $S \in \Gamma$ and $u_0 = 1$. By [2, Lemma 6.2], this Γ satisfies our assumption, namely, Γ is generated by Γ^J and S. As we have already known the ring structure of $\mathbb{M}_k(\Gamma_0^{(2)}(N))$ for $N = 1, 2, 3, 4$ (cf. [1, 2]), we have the following theorem.

Theorem 2. *For* $\Gamma = \Gamma_0^{(2)}(N)$ $(N = 1, 2, 3, 4)$, *Conjecture 1 is true.*

This theorem is useful to show the convergence of formal series of Jacobi forms. For example, we can construct many modular forms by Maass lifts or Borcherds products. The explicit formula of Maass lifts and Borcherds products is given by power series of Jacobi forms. However, especially on Borcherds products, to show the convergence of the power series is not easy. Our conjecture is aimed at constructing the general theory of the convergence of these lifts. We remark that our conjecture is also true for the Siegel paramodular group $K(N)$ $(N = 2, 3, 4)$, which was recently shown by Ibukiyama, Poor and Yuen [10].

2 Borcherds Products

2.1 *Maass Lifts*

For the sake of simplicity, here we assume $N = 1$ for a while. In this case $\Gamma = \mathrm{Sp}(2,\mathbb{Z})$. The following proposition is helpful.

Proposition 3. *Let* $(\varphi_m)_{m=0}^{\infty}$ *be a family of weakly holomorphic Jacobi forms. On the Fourier expansion*

$$\varphi_m(\tau, z) = \sum_{n,l \in \mathbb{Z}} c_m(n, l)\, q^n \zeta^l \in \mathbb{J}_{k,m}^{wh}(\Gamma^J).$$

If this family has a symmetry

$$c_m(n, l) = (-1)^k c_n(m, l),$$

and if φ_0 *is a Jacobi form (hence an elliptic modular form), these weakly holomorphic Jacobi forms are Jacobi forms, namely*

$$(\varphi_m)_{m=0}^{\infty} \in \left(\prod_{m=0}^{\infty} \mathbb{J}_{k,m}(\Gamma^J) \right)^{\mathrm{sym}}.$$

Especially, if $\varphi_0 = 0$*, these weakly holomorphic Jacobi forms are Jacobi cusp forms, namely*

$$(\varphi_m)_{m=1}^{\infty} \in \left(\prod_{m=1}^{\infty} \mathbb{J}_{k,m}^{c}(\Gamma^J) \right)^{\mathrm{sym}}.$$

Proof. Because each φ is a weakly holomorphic Jacobi form, we have

$$c_m(n,l) = c_m(n + 2lx + mx^2, l + mx)$$

for any $x \in \mathbb{Z}$. By using this equation and the symmetry $c_m(n,l) = (-1)^k c_n(m,l)$, for each $(m_0, l_0, n_0) \in \mathbb{Z}^3$, we can find $(m_1, l_1, n_1) \in \mathbb{Z}^3$ such that

$$\begin{aligned} &|c_{m_0}(n_0, l_0)| = |c_{m_1}(n_1, l_1)|, \\ &4m_0n_0 - l_0^2 = 4m_1n_1 - l_1^2, \\ &|l_1| \le |m_1| \text{ or } m_1 = 0, \\ &|l_1| \le |n_1| \text{ or } n_1 = 0. \end{aligned}$$

Now we assume $c_{m_0}(n_0, l_0) \neq 0$ for $m_0 > 0, n_0 > 0$ and $4m_0n_0 - l_0^2 < 0$ (resp. $4m_0n_0 - l_0^2 \le 0$). Then above $(m_1, l_1, n_1) \in \mathbb{Z}^3$ satisfies the condition $4m_1n_1 - l_1^2 < 0$ (resp. $4m_1n_1 - l_1^2 \le 0$). Hence m_1 or n_1 should be negative (resp. non-positive). However, from the assumption of our proposition and the symmetry, we have $c_m(n,l) = 0$ when $m < 0$ or $n < 0$ (resp. $m \le 0$ or $n \le 0$). □

Maass lift is a lift from Jacobi cusp forms of index 1 to Siegel modular forms of degree 2. Let $k \in 2\mathbb{Z}$ and

$$\varphi(\tau, z) = \sum_{n \in \mathbb{Z}} \sum_{\substack{l \in \mathbb{Z}, \\ 4n - l^2 > 0}} c(n,l) q^n \zeta^l \in \mathbb{J}_{k,1}^{c}(\Gamma^J).$$

For $m \in \mathbb{N}$, the Hecke operator V_m is defined by

$$(\varphi | V_m)(\tau, z) := \sum_{ad = m} \sum_{b=0}^{d-1} a^k \varphi\left(\frac{a\tau + b}{d}, az \right).$$

This V_m induces a map from $\mathbb{J}_{k,1}^{c}(\Gamma^J)$ to $\mathbb{J}_{k,m}^{c}(\Gamma^J)$. Maass lift of φ is given by the series

$$
\begin{aligned}
(\mathrm{ML}(\varphi))(Z) &:= \sum_{m=1}^{\infty} \frac{1}{m} (\varphi | V_m)(\tau, z) p^m \\
&= \sum_{m=1}^{\infty} \sum_{n=1}^{\infty} \sum_{l \in \mathbb{Z}} \left(\sum_{a|(n,l,m)} a^{k-1} c\left(\frac{mn}{a^2}, \frac{l}{a} \right) \right) q^n \zeta^l p^m.
\end{aligned}
$$

It is clear that this $\mathrm{ML}(\varphi)$ is S-invariant; hence the problem is about the convergence. It can be shown directly from the properties of Jacobi forms. However, if we admit our conjecture, the convergence of Maass lifts is clear from it.

When $\varphi \in \mathbb{J}_{k,1}(\Gamma^J)$, we need slight modification on above definition. By adding the suitable constant term $\varphi | V_0$, that is given by Eisenstein series, we have a Maass lift

$$
(\mathrm{ML}(\varphi))(Z) := (\varphi | V_0)(\tau) + \sum_{m=1}^{\infty} \frac{1}{m} (\varphi | V_m)(\tau, z) p^m \in \mathbb{M}_k(\Gamma).
$$

If we admit our conjecture, the convergence of this series is clear.

Moreover, our conjecture can be applied to Maass lifts of weak Jacobi forms. Borcherds constructed Maass lifts of weak Jacobi forms. For $\varphi \in \mathbb{J}^{\mathrm{w}}_{k,1}(\Gamma^J)$, he constructed $(\varphi | V_0)(\tau, z)$ by using the derivatives of Weierstrass $\wp$ function with its poles at $z \in \mathbb{Z} + \mathbb{Z}\tau$. In this case, the Fourier coefficients of $\mathrm{ML}(\varphi)$ have our symmetry, but $\mathrm{ML}(\varphi)$ does not converge. Actually, Borcherds showed that $\mathrm{ML}(\varphi)$ represents a meromorphic Siegel modular form by analytic continuation, by a bit complicated calculation (cf. [3]).

However, if we admit our conjecture, we can obtain his conclusion easily. Let $\varphi_{10,1} \in \mathbb{J}^{\mathrm{c}}_{10,1}(\mathrm{Sp}(2, \mathbb{Z})^J)$ be a unique Jacobi cusp form of weight 10 and index 1 up to constant (cf. [6]) and let $\Delta_{10} := \mathrm{ML}(\varphi_{10,1})$ be a unique Siegel modular forms of weight 10 up to constant. Since $\varphi_{10,1}$ has its zeros at $z \in \mathbb{Z} + \mathbb{Z}\tau$, by taking sufficiently large $t \in \mathbb{N}$, $\varphi_{10,1}(\tau, z)^t \, ((\varphi | V_0)(\tau, z))$ is holomorphic. Therefore, by Proposition 3, the formal p-expansion of $\Delta_{10}^t \, \mathrm{ML}(\varphi)$ is in the image of our conjecture. Hence we know $\mathrm{ML}(\varphi)$ is a meromorphic Siegel modular form, whose denominator is the power of Δ_{10}.

2.2 *Borcherds Products*

In this subsection, we assume $\Gamma = \mathrm{Sp}(2, \mathbb{Z})$ continuously.

Now let

$$
\varphi(\tau, z) = \sum_{n,l \in \mathbb{Z}} c(n, l) \, q^n \zeta^l \in \mathbb{J}^{\mathrm{wh}}_{0,1}(\Gamma^J).
$$

Calculate the Maass lift of φ, although it does not converge:

$$\begin{aligned}(\mathrm{ML}(\varphi))(Z) &= \sum_{m=1}^{\infty}\sum_{n,l\in\mathbb{Z}}\left(\sum_{a|(n,l,m)} a^{-1} c\left(\frac{mn}{a^2},\frac{l}{a}\right)\right) q^n\zeta^l p^m \\ &= \sum_{m=1}^{\infty}\sum_{n,l\in\mathbb{Z}} c(mn,l)\left(\sum_{a=1}^{\infty} a^{-1}\left(q^n\zeta^l p^m\right)^a\right) \\ &= -\sum_{m=1}^{\infty}\sum_{n,l\in\mathbb{Z}} c(mn,l)\left(\log\left(1-q^n\zeta^l p^m\right)\right).\end{aligned}$$

Hence, formally,

$$\exp(-\mathrm{ML}(\varphi)) = \prod_{m=1}^{\infty}\prod_{n,l\in\mathbb{Z}} \left(1-q^n\zeta^l p^m\right)^{c(mn,l)}$$

is a Γ^J-invariant function of weight 0. By slight modification, Borcherds constructed a Γ-invariant function of weight $c(0)/2$:

$$(\mathrm{BP}(\varphi))(Z) := q^a\zeta^b p^c \prod_{(n,l,m)>0} \left(1-q^n\zeta^l p^m\right)^{c(mn,l)},$$

where

$$a = \frac{1}{24}\sum_{l\in\mathbb{Z}} c(0,-l), \quad b = -\frac{1}{2}\sum_{l>0} c(0,-l)\, l, \quad c = \frac{1}{2}\sum_{l>0} c(0,-l)\, l^2$$

and $(n,l,m)>0$ means $m>0$ or $m=0, n>0$ or $m=n=0, l>0$.

Roughly, Borcherds product is an exponential of Maass lift:

$$\mathrm{BP}(\varphi) = \mathrm{JP}(\varphi)\exp(-\mathrm{ML}(\varphi)),$$

where $\mathrm{JP}(\varphi)$ is a weakly holomorphic Jacobi form of weight $c(0)/2$ and index c. Precisely, we can show

$$(\mathrm{JP}(\varphi))(Z) := q^a\zeta^b p^c \prod_{(n,l)>0} \left(1-q^n\zeta^l\right)^{c(0,l)},$$

where $(n,l)>0$ means $n>0$ or $n=0, l>0$, is a weakly holomorphic Jacobi form by Jacobi triple product formula. We remark that, strictly, we must assume $a,b,c\in\mathbb{Z}$ and $c(0,0)\in 2\mathbb{Z}$, if we do not admit modular forms with characters.

This idea, an exponential of Maass lift gives a new modular form, was first discovered by Borcherds [3, 4] for modular forms on the symmetric domain of type four with respect to unimodular lattices. And then this idea was applied to Siegel modular forms of degree 2 by Gritsenko and Nikulin [7, 8]. Originally, Borcherds proved his theorem with very hard calculation. To show the convergence (and analytic continuation) of Borcherds products is not easy. However, if we admit our conjecture, we can show the convergence easily, under the assumption of the positivity of some Fourier coefficients. (This assumption corresponds to the property that $\mathrm{BP}(\varphi)$ has no pole.)

Now we assume c is sufficiently large such that all coefficients of formal p-expansion of $\mathrm{BP}(\varphi)$ are with positive power of p. On

$$\mathrm{BP}(\varphi) = \mathrm{JP}(\varphi)\exp(-\mathrm{ML}(\varphi)),$$

$\mathrm{JP}(\varphi)$ is a weak Jacobi form. On the formal p-expansion of $\exp(-\mathrm{ML}(\varphi))$, each term is a finite sum of φ with Hecke operators. Hence it is a weakly holomorphic Jacobi form. Therefore, on the formal p-expansion of $\mathrm{BP}(\varphi)$ is a series of weakly holomorphic Jacobi forms. Then by Proposition 3, $\mathrm{BP}(\varphi)$ is in the image of FJ^c, under our assumption $c >> 0$. Hence by Theorem 2, $\mathrm{BP}(\varphi)$ is a holomorphic Siegel modular form of degree 2.

2.3 *Borcherds Products with Levels*

At the end of this paper, in this subsection, we give some remarks on Borcherds products with levels. Constructing Siegel modular forms of degree 2 with levels from Jacobi forms of index 1, most part of the procedure of Maass lifts and Borcherds products is similar to the case of $\Gamma = \mathrm{Sp}(2,\mathbb{Z})$. For arbitrary level, Maass lifts, or Saito-Kurokawa lifts, were investigated in [9, 11] and some other papers. Then, according to the way in previous subsection, or more precisely in [2], the exponential of Maass lifts can be modified to formal series of Jacobi forms with symmetry. Hence, the only difficulty to construct Borcherds products with levels is to show the convergence. As $\Gamma_0^{(2)}(N)$ $(N \neq 1)$ cannot be recognized as a lattice automorphism of signature $(2,3)$, this difficulty is not cleared by the general theory of Borcherds products with theta integrals in [5]. However, when $N \leq 4$, we can clear this difficulty by our Theorem 2. Actually, we constructed some good Siegel modular forms with levels in [2].

Acknowledgements This paper is based on the talk at International Workshop on Mathematics 2012 at Muscat in Oman, organized by Sultan Qaboos University and the German University of Technology. The author would like to thank the organizers Professor Bernherd Heim, Professor Mehiddin Al-Baali and Professor Tomoyoshi Ibukiyama for giving me a chance to talk about my researches in such a nice conference. Also the author would like to thank the referee for carefully reading our manuscript and for giving some constructive comments.

References

1. H. Aoki, Estimating Siegel modular forms of genus 2 using Jacobi forms. J. Math. Kyoto Univ. **40-3**, 581–588 (2000)
2. H. Aoki, T. Ibukiyama, Simple graded rings of Siegel modular forms of small levels, differential operators and Borcherds products. Int. J. Math. **16**(3), 249–279 (2005)
3. R.E. Borcherds, Automorphic forms on $O_{s+2,2}(R)$ and infinite products. Invent. Math. **120**(1), 161–213 (1995)
4. R.E. Borcherds, Automorphic forms with singularities on Grassmannians. Invent. Math. **132-3**, 491–562 (1998)
5. J. Bruinier, *Borcherds Products on $O(2,l)$ and Chern Classes of Heegner Divisors*. Lecture Notes in Mathematics, vol. 1780 (Springer, Berlin, 2002)
6. M. Eichler, D. Zagier, *The Theory of Jacobi Forms* (Birkhäuser, Boston, 1985)
7. V.A. Gritsenko, V.V. Nikulin, Siegel automorphic form corrections of some Lorentzian Kac-Moody Lie algebras. Am. J. Math. **119**, 181–224 (1997)
8. V.A. Gritsenko, V.V. Nikulin, Automorphic forms and Lorentzian Kac-Moody algebras. I, II. Int. J. Math. **9-2**, 153–199; 201–275 (1998)
9. T. Ibukiyama, Saito-Kurokawa liftings of level N and practical construction of Jacobi forms. Kyoto J. Math. **52-1**, 141–178 (2012)
10. T. Ibukiyama, C. Poor, D. Yuen, Jacobi forms that characterize paramodular forms. Abh. Math. Semin. Univ. Hambg. **83-1**, 111–128 (2013)
11. M. Manickam, B. Ramakrishnan, On Saito-Kurokawa correspondence of degree two for arbitrary level. J. Ramanujan Math. Soc. **17-3**, 149–160 (2002)

Finite or Infinite Number of Solutions of Polynomial Congruences in Two Positive Integer Variables

Thomas Bier

1 Basic Definitions, Formulation of the Problem, and Statement of Result

Assume that x, y are two variables taking positive integral values.

Also assume that $f(x), g(y)$ are polynomials with nonnegative and integral coefficients and thus taking nonnegative integer function values for positive integer values of the variables.

Now let $h = \deg(f(x))$ and $k = \deg(g(y))$ be the degrees of these polynomials. The *polynomial system of congruences* is the system

$$g(y) \equiv 0 \mod x \tag{1}$$
$$f(x) \equiv 0 \mod y\,.$$

Immediately the question arises whether there can be an infinity of (positive integral distinct) solutions (x, y) for this system. To make this question meaningful we want to assume that $f(0) > 0$ and $g(0) > 0$ hold. We can almost completely answer this question under this assumption.

Although this problem may be difficult to decide (or at least difficult to prove) for certain specific systems, we can provide an easy answer in terms of the pair of degrees (h, k) of the polynomials as given above. In the following we only consider the case of nonconstant polynomials as it is clear that in case one of the polynomials above, $f(x)$ say, is a positive integer constant, then there are only finitely many

T. Bier (✉)
Department of Mathematics University of Botswana, Gaborone, Botswana

Twiskenweg 43 B D 26129 Oldenburg Germany
e-mail: thomas.bier@gmail.com; 044153881@t-online.de

B. Heim et al. (eds.), *Automorphic Forms*, Springer Proceedings in Mathematics & Statistics 115, DOI 10.1007/978-3-319-11352-4_2

positive y dividing that constant and then there are only finitely values $g(y)$ and only finitely many values x. Let $\mathcal{L}_{fin}$ be the set of pairs $(h,k) \in \mathbb{N} \times \mathbb{N}$ such that each pair of positive polynomials $f(x), g(y)$ with the said degrees and with $f(0), g(0) > 0$ has only finitely many positive integer solutions. Thus we have

$$\mathcal{L}_{inf} = \{(h,k) : \exists f(x), g(y);\ \ h = \deg f(x),\ k = \deg g(y) \text{ and} \qquad (2)$$

$$\text{the system (1) has infinitely many positive solutions.}\}$$

and its complement is

$$\mathcal{L}_{fin} = \mathbb{N} \times \mathbb{N} \backslash \mathcal{L}_{inf}.$$

We shall prove that

Theorem 1.

$$\mathcal{L}_{fin} = \{(1,1), (1,2), (1,3), (2,1), (3,1)\}\ .$$

2 The Examples

Example 1. Let f_n be the sequence of Fibonacci numbers. Then it is well known that for $n \geq 3$

$$f_{n+2} f_{n-2} - f_n^2 = (-1)^{n-1}.$$

By shifting indices up by two we see that for odd n any pair $(x,y) = (f_n, f_{n+2})$ satisfies the system

$$y^2 + 1 \equiv 0 \mod x \quad \text{and} \quad x^2 + 1 \mod y \equiv 0 \mod y. \qquad (3)$$

Thus (3) has infinitely many positive solutions.

This example shows that $(2,2) \in \mathcal{L}_{inf}$.

We can now generalize this example. Assume that $k, h > 0$ are fixed integers and that a_0, a_1, b_0, b_1 are four positive integers satisfying the two equations

$$a_0 a_1 = b_1^h + 1\ ,\ \ b_0 b_1 = a_0^k + 1\ . \qquad (4)$$

We easily see that b_1 divides $a_1^k + 1$. Indeed

$$(a_0 a_1)^k = (b_1^h + 1)^k = b_1^h \cdot X_0 + 1$$

for some positive integer X_0 coming out of the binomial expansion of $(b_1^h + 1)^k$. Thus

$$a_0^k a_1^k + a_0^k = b_1^h \cdot X_0 + a_0^k + 1 = b_1^h \cdot X_0 + b_1 b_0 = b_1(b_1^{h-1} X_0 + b_0).$$

Assume now that (a priori rational) numbers a_2, b_2 are defined by

$$b_1 b_2 = a_1^k + 1 \,, a_1 a_2 = b_2^h + 1 \,. \tag{5}$$

First note that from the above divisibility $b_1 \mid (a_1^k + 1)$ we get that b_2 is an integer. Then we have from the second equation in the system (4) that

$$(b_1 b_2)^h = (a_1^k + 1)^h = a_1^k \cdot X_1 + 1$$

for some positive integer X_1 that comes out of the binomial expansion of $(a_1^k + 1)^h$. Thus from the first equation in the system (4)

$$b_1^h b_2^h + b_1^h = a_1^k X_1 + b_1^h + 1 = a_1^k X_1 + a_0 a_1 = a_1 \cdot Y_1$$

for some positive integer Y_1.

This implies that a_1 divides the product $b_1^h (b_2^h + 1)$. But from the first part of (4) we get $\gcd(a_1, b_1) = 1$. Thus a_1 divides $b_2^h + 1$. Therefore also a_2 is an integer.

We also may verify the divisibility property b_2 divides $a_2^k + 1$ for this step. From the second equation of (5) we get

$$(a_1 a_2)^k = (b_2^h + 1)^k = b_2^h \cdot Z_1 + 1$$

for some positive integer Z_1 coming from the corresponding binomial expansion. Now

$$a_1^k a_2^k + a_1^k = b_2^h Z_1 + a_1^k + 1 = b_2 \cdot (b_2^{h-1} Z_1 + b_1)$$

and hence b_2 divides $a_1^k (a_2^k + 1)$. As again $\gcd(a_1, b_2) = 1$ this shows that b_2 divides $a_2^k + 1$.

We now assume in general that we have given integers $a_0, a_1, \ldots, a_n$ and $b_0, b_1, \ldots, b_n$ satisfying the equations

$$a_i a_{i-1} = b_i^h + 1 \,, \quad b_j b_{j+1} = a_j^k + 1 \,, \tag{6}$$

for $i = 0, 1, \ldots, n$ and for $j = 0, 1, \ldots, n-1$ and the divisibility relation b_n divides $a_n^k + 1$.

Then let b_{n+1}, a_{n+1} be defined for $n \in \mathbb{Z}$ by

$$b_n b_{n+1} = a_n^k + 1 \,, \quad a_n a_{n+1} = b_{n+1}^h + 1 \,. \tag{7}$$

Then we can show that also a_{n+1}, b_{n+1} are actually integers.

First note that from the above divisibility that b_{n+1} is an integer.

Then we have from the second equation in the system (7) that

$$(b_n b_{n+1})^h = (a_n^k + 1)^h = a_n^k \cdot X_n + 1$$

for some positive integer X_n that comes out of the binomial expansion of $(a_n^k + 1)^h$. Thus from the first equation in the system (7)

$$b_n^h b_{n+1}^h + b_n^h = a_n^k X_n + b_n^h + 1 = a_n^k X_n + a_{n-1}a_n = a_n \cdot Y_n$$

for some positive integer Y_n. Thus a_n divides the product $b_n^h(b_{n+1}^h + 1)$. Now as $\gcd(a_n, b_n) = 1$ by the first equation in (7) we obtain that a_n divides $b_{n+1}^h + 1$. This then shows via the second equation of (7) that also a_{n+1} is an integer.

We also may verify the divisibility property b_{n+1} divides $a_{n+1}^k + 1$ for this step. From the second equation of (7) we get

$$(a_n a_{n+1})^k = (b_{n+1}^h + 1)^k = b_{n+1}^h \cdot Z_n + 1$$

for some positive integer Z_n coming from the corresponding binomial expansion. Now

$$a_n^k a_{n+1}^k + a_n^k = b_{n+1}^h Z_n + a_n^k + 1 = b_{n+1} \cdot (b_{n+1}^{h-1} Z_n + b_n)$$

and hence b_{n+1} divides $a_n^k(a_{n+1}^k + 1)$. As again $\gcd(a_n, b_{n+1}) = 1$ this shows that b_{n+1} divides $a_{n+1}^k + 1$.

This completes the verification of the induction step.

We have proved that all the numbers a_n, b_n are positive integers for $n \geq 2$ once the four numbers a_0, a_1, b_0, b_1 are positive integers satisfying (4).

It remains to show that for all $h, k > 0$ there exists at least one such example. But this is trivial; we may take $a_0 = 1, a_1 = 2$ and $b_0 = 2, b_1 = 1$.

We may also continue the process downward obtaining $a_{-1}, b_{-1}, a_{-2}, b_{-2}$, etc., thus obtaining two 2–way sequences a_n, b_n with $n \in \mathbb{Z}$.

This starts from the condition that a_0 divides $b_0^h + 1$. Details are exactly the same as above and may be omitted.

Pompe [3] observed the special case $h = k$ of this example. He introduced the sequences A003818, A003819, and A 003820 in the online encyclopedia of integer sequences which are the same as ours for $h = k$.

Floor v Lamoen [3] has obtained (and he characterized) in A002310 the sequence a_n of the special case $k = 4, h = 1$.

3 Existence Results

3.1 Existence Part of Theorem 1

We note that the construction in the previous section provides positive polynomials with strictly positive constant coefficients. Thus if we take a pair $(f(x) = x^h + 1, g(y) = y^k + 1)$ of such polynomials with degree pair (h, k), we need only to inquire whether the existing sequences a_n, b_n for the initial values $a_0 = 1, b_0 = 2$, $a_1 = 2, b_1 = 1$ are infinite or not.

For the existence result of infinitely many solutions it is enough to show that for all pairs $(h, k) \neq (1, 1), (1, 2), (2, 1), (1, 3), (3, 1)$ the sequence a_n is eventually strictly increasing.

We first consider the case $h, k \geq 2$. We first check that $a_1 = 2, b_2 = 2^k + 1, a_2 = \frac{(2^k+1)^h+1}{2}$. Thus it holds that $a_1 < b_2 < a_2$. Now make the inductive assumption $a_{n-1} < b_n < a_n$. Then we compute

$$\frac{b_{n+1}}{a_n} = \frac{a_n^k + 1}{a_n b_n} > \frac{a_n^k}{a_n b_n} = \frac{a_n^{k-1}}{b_n} \geq \frac{a_n}{b_n} > 1 . \tag{8}$$

Similarly by using this inequality just proved we get

$$\frac{a_{n+1}}{b_{n+1}} = \frac{b_{n+1}^h + 1}{a_n b_{n+1}} > \frac{b_{n+1}^h}{a_n b_{n+1}} = \frac{b_{n+1}^{h-1}}{a_n} \geq \frac{b_{n+1}}{a_n} > 1 . \tag{9}$$

Of course in the first case we have used $k \geq 2$ and in the second case $h \geq 2$.

Now consider the case when one of the numbers h, k is 1. Without loss of generality we may assume that $k \geq h = 1$.

In this case we use as inductive assumption $a_{n-1}^2 < b_n < a_n^2$. This is true for $n = 2$ and $k \geq 4$ as $a_1 = 2, b_2 = 2^k + 1, a_2 = 2^{k-1} + 1$ holds. For the induction step

$$\frac{b_{n+1}}{a_n^2} = \frac{a_n^k + 1}{a_n^2 b_n} > \frac{a_n^k}{a_n^2 b_n} = \frac{a_n^{k-2}}{b_n} \geq \frac{a_n^2}{b_n} > 1 . \tag{10}$$

Using the inequality just proved we also get

$$\frac{a_{n+1}^2}{b_{n+1}} = \frac{(b_{n+1} + 1)^2}{a_n^2 b_{n+1}} > \frac{b_{n+1}^2}{a_n^2 b_{n+1}} = \frac{b_{n+1}}{a_n^2} > 1 . \tag{11}$$

Of course in the first case we have used $k \geq 4$.

In either case we have that a_n is a strictly increasing and thus infinite sequence of positive integers. Thus in all cases except for
$(h, k) = (1, 1), (1, 2), (1, 3), (2, 1), (3, 1)$ there exist polynomials with an infinity of solutions for the system (1).

We note that the above sequences for the cases $(h,k) = (1,1), (1,2), (2,1), (1,3), (3,1)$ become periodic and thus provide only a finite number of solutions.

3.2 *A More General Existence Result in the Biquadratic Case*

We describe now another more general existence result (without proof) for the biquadratic case. Let $s, t \geq 1$ be integers with $s + t \geq 2$ to avoid trivialities.

Then the system

$$x^2 + st - 1 \equiv 0 \mod y \tag{12}$$

$$y^2 + t \equiv 0 \mod x \tag{13}$$

has infinitely many solutions in the form of a pair of two-way double recursion sequences with initial values

$$(x_0, y_0) = (t+1, t);\ (x_1, y_1) = (s^2t + 1, st)$$

and with recursion

$$x_{n+1} = \theta \cdot x_n - x_{n-1},\ y_{n+1} = \theta \cdot y_n - y_{n-1},$$

where $\theta = s^2t - 2st + t + 2 = t(s-1)^2 + 2$.

4 Finiteness Results

4.1 *The Linear Case*

In this section we use matrix notation.

It was shown by the author in his University of Malaya (KL) lecture notes [1] that for positive integers a_{12}, a_{21}, b_1, b_2 the system

$$a_{12}x_2 + b_1 = d_1x_1 \tag{14}$$

$$a_{21}x_1 + b_2 = d_2x_2 \tag{15}$$

for positive integers d_1, d_2 has only finitely many positive solutions. Some estimates for the number of solutions were also given.

Here we sketch the argument of [1]. We first can cross multiply with d_2, d_1, respectively. This leads to the system

$$(d_1d_2 - a_{12}a_{21}) \cdot x_2 = a_{21}b_1 + d_1b_2 \tag{16}$$
$$(d_1d_2 - a_{12}a_{21}) \cdot x_1 = a_{12}b_2 + d_2b_1 \,. \tag{17}$$

Put $\Delta = d_1d_2 - a_{12}a_{21}$. If $\Delta \leq 0$, then there are no positive solutions, so we may assume that Δ is a positive integer. We shorten the notation to

$$\Delta \cdot x_2 = a_{21}b_1 + d_1b_2 \tag{18}$$
$$\Delta \cdot x_1 = a_{12}b_2 + d_2b_1 \tag{19}$$

We now derive a tight bound on the size of x_1 and x_2.

Proposition 1. *For any positive solution of (14) and (15) we have*

$$x_1 \leq a_{12}b_2 + b_1(1 + a_{12}a_{21}) \,. \tag{20}$$

$$x_2 \leq a_{21}b_1 + b_2(1 + a_{12}a_{21}) \,. \tag{21}$$

By symmetry we only need to prove (20). Consider the quantity $x_1 - a_{12}b_2$. If it is not positive, then $x_1 \leq a_{12}b_2$ and (20) holds with strict inequality.
Hence we may assume that $x_1 - a_{12}b_2 > 0$ is a positive integer. Then we get

$$x_1 - a_{12}b_2 \leq d_1(x_1 - a_{12}b_2) \text{ and using (14)} \tag{22}$$
$$= a_{12}x_2 + b_1 - a_{12}d_1b_2 \tag{23}$$
$$\leq a_{12}\Delta x_2 + b_1 - a_{12}d_1b_2 \text{ and using (18)} \tag{24}$$
$$= a_{12}a_{21}b_1 + a_{12}b_2d_1 + b_1 - a_{12}d_1b_2 \tag{25}$$
$$= b_1(1 + a_{12}a_{21}). \tag{26}$$

This shows (20). The inequality (21) follows by symmetry.

We note that the case of equality occurs in (20) iff both $d_1 = 1$ and $\Delta = 1$. Hence if $x_1 = a_{12}b_2 + b_1(1 + a_{12}a_{21})$, then by (18) we get $x_2 = a_{21}b_1 + b_2$. This then is the solution of (14) and (15) with maximal x_1. Similarly for $d_2 = 1$ and $\Delta = 1$ there is the solution $x_2 = a_{21}b_1 + b_2(1 + a_{12}a_{21})$ and $x_1 = a_{12}b_2 + b_1$ with maximal x_2. We call these two solutions the *two maximal* solutions.

We now assume more generally that only $\Delta = 1$ holds. Then d_1, d_2 are a pair of complementary divisors of $1 + a_{12}a_{21}$. For any such pair of divisors by (18) and (19) we obtain the solutions

$$x_1 = a_{12}b_2 + d_2b_1, \; x_2 = a_{21}b_1 + d_1b_2 \,. \tag{27}$$

For an integer m let $\tau(m)$ be the number of positive divisors of m. We see that there are $\tau(1 + a_{12}a_{21})$ distinct solutions to (14) and (15). We refer to these solutions as the *large* $\Delta = 1$ *solutions.*

The statement given is also valid if one of b_1, b_2 (but not both!) is zero.

From that a certain classification of the solutions can be obtained. Details are omitted.

4.2 The Semiquadratic Case

Here we show that for nonnegative integers a, b, c, d, e with $c > 0, e > 0$ and $a > 0, d > 0$ the system

$$ax^2 + bx + c \equiv 0 \mod y \quad \text{and} \quad dy + e \equiv 0 \mod x$$

has only finitely many (positive integral) solutions.

Clearly there exist positive integers h, k such that

$$ky = ax^2 + bx + c \tag{28}$$

$$hx = dy + e \tag{29}$$

Since x divides $k(dy + e) = dky + ke = adx^2 + bdx + cd + ke$, there exists a positive integer m with $mx = cd + ke$.

Now consider the positive integral expression

$$(mx - cd)hxy = e(ax^2 + bx + c)(dy + e) =$$

$$ehkxy = adex^2y + ae^2x^2 + bdexy + be^2x + cdey + ce^2. \tag{30}$$

Note first that from

$$(mh - ade)x^2y = cdhxy + bdexy + ae^2x^2 + be^2x + cdey + ce^2 > 0$$

it follows that $mh > ade$.

The above expression (30) also gives

$$(mh - ade)x^2y - (cdh + bde)xy - ae^2x^2 - be^2x - cdey = ce^2 .$$

Since $x \geq 1$ we get

$$(mh - ade)x^2y - d(ch + be)xy - e^2(a + b)x^2 - cdexy \leq ce^2 ,$$

and hence

$$x \cdot ((mh - ade)xy - d(ch + be)y - e^2(a + b)x - cdey) \leq ce^2 . \tag{31}$$

If there exists an unbounded (in x strictly increasing) sequence of solutions $x_i \mapsto +\infty, y_i \mapsto +\infty$, then clearly y_i also becomes unbounded and we must have that for such solutions $(x, y) = (x_i, y_i)$ eventually (for all but finitely many values of i)

$$(mh - ade)xy - d(ch + be + ce)y - e^2(a + b)x \leq 0\,. \tag{32}$$

But then introducing the nonnegative rational bounded constants

$$A = \frac{d(ch + be + ce)}{mh - ade} \tag{33}$$

$$B = \frac{e^2(a + b)}{mh - ade} > 0 \tag{34}$$

it follows that

$$xy - Ax - By + AB = (x - B)(y - A) \leq AB.$$

This contradicts the fact that (x_i, y_i) are an unbounded sequence.

Thus the set of positive integer solutions is bounded and hence it is finite. We thus have established the following result.

Proposition 2. *For nonnegative integers $a, b, c, d, e \geq 0$ with $a > 0, d > 0,\ c > 0, e > 0$ the system*

$$ax^2 + bx + c \equiv 0 \mod y \quad \textit{and} \quad dy + e \equiv 0 \mod x$$

has only finitely many positive integral solutions (x, y).

4.3 The Semicubic Case

In this part we show that for positive integers $a, b, c, d > 0$ and for integers $e, f \geq 0$ and positive integral $x,\ y$ the system

$$ay + b \equiv 0 \mod x \tag{35}$$

$$cx^3 + ex^2 + fx + d \equiv 0 \mod y \tag{36}$$

has only finitely many positive integer solution pairs (x, y).

We first need some notation. Whenever (x, y) is a (positive) solution of (35) and (36), then we may define positive integers h, k (depending on x, y) by

$$hx = ay + b \tag{37}$$

$$ky = cx^3 + ex^2 + fx + d\ . \tag{38}$$

There are four further positive integral quantities (all depending on x, y) called *divisants* and generically denoted by Δ which are associated to this situation. In the following lemma we derive some of these quantities and analyze their respective relationships.

Lemma 1. *There exist positive integers Δ_m, Δ_n such that*

$$\Delta_m \cdot x = ad + bk\,; \tag{39}$$

$$\Delta_n \cdot y = b^3c + b^2eh + bfh^2 + dh^3\,.\,. \tag{40}$$

For the proof of (39) multiply (37) by k and use (38) to get

$$hkx = (ay + b)k = acx^3 + aex^2 + afx + ad + bk,$$

and after subtracting we see that

$$x(hk - acx^2 - aex - af) = ad + bk,$$

so that we may choose

$$\Delta_m = hk - acx^2 - aex - af. \tag{41}$$

For the proof of (40) multiply (38) by h^3 and use (37) to get

$$\begin{aligned} kh^3y &= h^3(cx^3 + ex^2 + fx + d) \\ &= c(ay + b)^3 + eh(ay + b)^2 + fh^2(ay + b) + dh^3 \\ &= y(a^3cy^2 + 3a^2bcy + 3ab^2c + a^2ehy + 2abeh + afh^2) + b^3c + b^2eh + bfh^2 + dh^3. \end{aligned}$$

After subtracting the bracket we get

$$(kh^3 - a^3cy^2 - 3a^2bcy - 3ab^2c - a^2ehy - 2abeh - afh^2)y = b^3c + b^2eh + bfh^2 + dh^3,$$

so that we may choose

$$\Delta_n = h^3k - a^3yc^2 - 3a^2bcy - 3ab^2c - a^2ehy - 2abeh - afh^2\,. \tag{42}$$

This completes the proof of Lemma 1.

Lemma 2. *There exists a positive integer Δ_s such that*

$$\Delta_s \cdot h = \Delta_n + ab^2c. \tag{43}$$

Use the number (42) and add ab^2c and then simplify with (38):

$$\Delta_n + ab^2c = h(h^2k - a^2ey - 2abe - afh) - ac(a^2y^2 - b^2) - 3abc(ay + b)$$

$$= h(h^2k - a^2ey - 2abe - afh) - achx(ay - b) - 3abchx$$
$$= h(h^2k - a^2ey - 2abe - afh - a^2cxy - 2abcx) \; ;$$

so that we may choose

$$\Delta_s = h^2k - a^2ey - 2abe - afh - a^2cxy - 2abcx \; . \tag{44}$$

This proves Lemma 2.

Lemma 3. *The two product equations involving* Δ_s *give*

$$\Delta_s x = b\Delta_m + abf + adh \; , \tag{45}$$
$$\Delta_s y = b^2e + bfh + dh^2 + b^2cx \; . \tag{46}$$

First we prove (46) by multiplying (43) by h and using (40)

$$\begin{aligned}\Delta_s hy &= \Delta_n y + ab^2cy \\ &= b^3c + b^2eh + bfh^2 + dh^3 + ab^2cy \\ &= +b^2eh + bfh^2 + dh^3 + b^2c(b + ay) \\ &= +b^2eh + bfh^2 + dh^3 + b^2chx \\ &= h(b^2e + bfh + dh^2 + b^2cx)\end{aligned}$$

so by cancelling h we get (46).

We now use (46) and multiply it by x to obtain (45):

$$\begin{aligned}\Delta_s xy &= b^2ex + bfhx + dh^2x + b^2cx^2 \\ &= b^2ex + b^2cx^2 + bd(ay + b) + ch(ay + b) \\ &= b\Delta_m y + abfy + adhy \; .\end{aligned}$$

Cancelling y we get (45). This proves Lemma 3.

Lemma 4. *The divisants* Δ_m, Δ_n *also satisfy the equations*

$$\Delta_m y = bcx^2 + bex + dh + bf \; , \tag{47}$$
$$\Delta_n x = b\Delta_s + ab^2e + abfh + adh^2 \; . \tag{48}$$

Multiply (39) by y and simplify using (37)

$$\begin{aligned}\Delta_m xy &= ady + bky \\ &= ady + bcx^3 + bex^2 + bfx + bd\end{aligned}$$

$$= bcx^3 + bex^2 + bfx + d(ay + b)$$
$$= x(bcx^2 + bex + bf + dh)\ .$$

Now cancelling x gives (47).

Similarly multiply (40) by x and simplify using (37) and (46) to get

$$\begin{aligned}\Delta_n xy &= b^3cx + b^2ehx + bfh^2x + dh^3x\\ &= b^3cx + b^3e + b^2fh + bdh^2 + y(ab^2e + abfh + adh^2)\\ &= b\Delta_s y + y(ab^2e + abfh + adh^2)\end{aligned}$$

and cancel y to get (48). This proves Lemma 4.

There is a formula for the product $h \cdot \Delta_m$:

$$h \cdot \Delta_m = \Delta_s + abe + abcx\ . \tag{49}$$

For the proof we simplify $h\Delta_m - \Delta_s$ as follows. First we use the formula for Δ_m in (41) and then multiply it by h and then in the third line substituting (44) we get

$$\begin{aligned}h\Delta_m - \Delta_s &= h^2k - acx(hx) - ae(hx) - afh - \Delta_s\\ &= h^2k - acx(ay + b) - ae(ay + b) - afh - \Delta_s\\ &= abcx + abe\end{aligned}$$

From the above we obtain an equation involving several of the divisants.

Proposition 3. *The following equation is valid:*

$$(\Delta_s - abc)(x - 1) + (h - b)(\Delta_m - ad) = ab(c + d + e + f)\ . \tag{50}$$

The proof can be obtained from the above relations (49) and (45):

$$\begin{aligned}(\Delta_s - abc)(x - 1) + (h - b)(\Delta_m - ad) &= \Delta_s x - abcx - \Delta_s + abc + h\Delta_m - b\Delta_m - adh + abd\\ &= b\Delta_m + abf + adh - abcx - \Delta_s + abc\\ &\quad + \Delta_s + abe + abcx - b\Delta_m - adh + abd\\ &= abf + abc + abe + abd\ .\end{aligned}$$

In order to apply the proposition we need to introduce one more parameter.

Lemma 5. *There exists a positive integer t such that*

$$k \cdot t = a^3cd^3 + a^2d^2e\Delta_m + adf\Delta_m^2 + d\Delta_m^3 \,. \tag{51}$$

For the proof we use

$$\begin{aligned} ky\Delta_m^3 &= \Delta_m^3(cx^3 + ex^2 + fx + d) \\ &= c(\Delta_m x)^3 + e\Delta_m(\Delta_m x)^2 + f\Delta_m^2(\Delta_m x) + d\Delta_m^3 \\ &= c(ad + bk)^3 + e\Delta_m(ad + bk)^2 + f\Delta_m^2(ad + bk) + d\Delta_m^3 \\ &= k(b^3ck^2 + 3ab^2cdk + 3a^2bcd^2 + b^2ek\Delta_m + 2abde\Delta_m + bf\Delta_m^2) \\ &\quad + a^3cd^3 + a^2d^2e\Delta_m + adf\Delta_m^2 + d\Delta_m^3 \,. \end{aligned}$$

Now bringing the terms with k to the left-hand side we get

$$\begin{aligned} &k(y\Delta_m^3 - b^3ck^2 - 3ab^2cdk - 3a^2bcd^2 - b^2ek\Delta_m - 2abde\Delta_m - bf\Delta_m^2) \\ &\quad = a^3cd^3 + a^2d^2e\Delta_m + adf\Delta_m^2 + d\Delta_m^3 \,. \end{aligned}$$

Thus we may take t to be

$$t = y\Delta_m^3 - b^3ck^2 - 3ab^2cdk - 3a^2bcd^2 - b^2ek\Delta_m - 2abde\Delta_m - bf\Delta_m^2 \,. \tag{52}$$

First we show that

Lemma 6. *The integer d divides the integer t. Thus we may introduce a further divisant as $\Delta_t = \dfrac{t}{d}$.*

Into the expression (52) we substitute the formula (47) to get

$$\begin{aligned} t = {} &\Delta_m^2(bcx^2 + bex + dh + bf) - b^3ck^2 - b^2ek\Delta_m - bf\Delta_m^2 \\ &- 2abde\Delta_m - 3ab^2cdk - 3a^2bcd^2 \end{aligned}$$

where the third term in the bracket and the last three terms are already divisible by d. Thus there exists an integer z such that using (39) in the third line below

$$\begin{aligned} t &= d \cdot z + \Delta_m^2(bcx^2 + bex + bf) - b^3ck^2 - b^2ek\Delta_m - bf\Delta_m^2 \\ &= d \cdot z + bc(\Delta_m x)^2 + be\Delta_m(\Delta_m x) + bf\Delta_m^2 - b^3ck^2 - b^2ek\Delta_m - bf\Delta_m^2 \\ &= d \cdot z + bc(ad + bk)^2 + be\Delta_m(ad + bk) - b^3ck^2 - b^2ek\Delta_m \\ &= d \cdot z + d(a^2bcd + 2ab^2k \cdot c + abe\Delta_m) \end{aligned}$$

which proves the lemma.

Since z in the above proof is $z = h\Delta_m^2 - 2abe\Delta_m - 3ab^2ck - 3a^2bcd$ we obtain another expression for t as

$$t = d\cdot(h\Delta_m^2 - 2abe\Delta_m - 3abcdk - 3a^2bcd + a^2bcd + 2ab^2kc + abe\Delta_m) . \quad (53)$$

$$\Delta_t = h\Delta_m^2 - 2abe\Delta_m - 3abcdk - 3a^2bcd + a^2bcd + 2ab^2kc + abe\Delta_m . \quad (54)$$

Lemma 7. *For the positive integer t and the divisants Δ_m, Δ_s, the following relation holds:*

$$d \cdot \Delta_m \cdot \Delta_s = t + a^2bcd^2. \quad (55)$$

For the proof of Lemma 7 we use the expression (52) and substitute (47) where we can cancel the term $bf\Delta_m^2$; and then in the third line we cancel further expressions. This gives

$$\begin{aligned} t &= \Delta_m^2 x^2 bc + be\Delta_m^2 x + d\Delta_m^2 h - b^3ck^2 - 3ab^2cdk - 3a^2bcd^2 - b^2ek\Delta_m - 2abde\Delta_m \\ &= bc(\Delta_m x)^2 + be\Delta_m(\Delta_m x) + d\Delta_m(\Delta_s + abe + abcx) - b^2ek\Delta_m - 2abde\Delta_m \\ &\quad -b^3ck^2 - 3ab^2cdk - 3a^2bcd^2 \\ &= bc(ad+bk)^2 + be\Delta_m(ad+bk) + d\Delta_m\Delta_s + abcd\Delta_m x - b^2ek\Delta_m - abde\Delta_m \\ &\quad -b^3ck^2 - 3ab^2cdk - 3a^2bcd^2 \\ &= a^2bcd^2 + 2ab^2cdk + d\Delta_m\Delta_s + abcd(ad+bk) - 3ab^2cdk - 3a^2bcd^2 \\ &= +d\Delta_m\Delta_s - a^2bcd^2 \ . \ \text{This proves (55)} . \end{aligned}$$

Next we give the product formula for $h \cdot t$ in terms of Δ_s and Δ_m :

$$h \cdot t = d\Delta_s^2 + abde\Delta_s + ab^2cd\Delta_m + a^2b^2cdf \ . \quad (56)$$

For the proof of (56) we combine (55) with (49) and (45):

$$\begin{aligned} ht &= dh\Delta_m\Delta_s - a^2bcd^2h \\ &= d\Delta_s(\Delta_s + abe + abcx) - a^2bcd^2h \\ &= d\Delta_s^2 + abde\Delta_s + abcd\Delta_s x - a^2bcd^2h \\ &= d\Delta_s^2 + abde\Delta_s + ab^2cd\Delta_m + a^2b^2cdf \ . \ \text{This proves (56).} \end{aligned}$$

Now we come to the crucial formula which shows that the product $\Delta_n \cdot t$ is a cubic polynomial function of Δ_s :

$$\Delta_n \cdot t = d \cdot \Delta_s^3 + abde \cdot \Delta_s^2 + a^2b^2cdf \cdot \Delta_s + a^3b^3c^2d^2 \ . \quad (57)$$

For the proof multiply (56) by Δ_s and use (55) again:

$$\begin{aligned} ht\Delta_s &= d\Delta_s^3 + abde\Delta_s^2 + ab^2cd\Delta_s\Delta_m + a^2b^2cdf\Delta_s \\ &= d\Delta_s^3 + abde\Delta_s^2 + a^2b^2cdf\Delta_s + ab^2ct + ab^2c(a^2bcd^2) . \end{aligned}$$

Hence by (43)

$$\Delta_n \cdot t = h \cdot \Delta_s \cdot t - ab^2ct = d\Delta_s^3 + abde\Delta_s^2 + a^2b^2cdf\Delta_s + a^3b^3c^2d^2 , \quad (58)$$

which proves (57). Thus the product of the two divisants Δ_n, Δ_t is

$$\Delta_n\Delta_t = \Delta_s^3 + abe\Delta_s^2 + a^2b^2cf\Delta_s + a^3b^3c^2d \ . \quad (59)$$

We remark that in a different interpretation of these results we can make the following integral substitution:

$$x_1 = \Delta_s,\ y_1 = \Delta_n,\ h_1 = h,\ k_1 = \frac{t}{d} = \Delta_t, \quad (60)$$

$$s_1 = a^2b^3c^2x,\ n_1 = a^3b^3c^2y,\ m_1 = ab^2c\Delta_m,\ t_1 = a^3b^9c^3dk \ . \quad (61)$$

Then with the constants

$$e_1 = abe, f_1 = a^2b^2cf, d_1 = a^3b^3c^2d, b_1 = ab^2c, c_1 = a_1 = 1 \quad (62)$$

and with $\Delta_{t,1} = b^6ck$ the system $(x_1, y_1, h_1, k_1, m_1, n_1, s_1, \Delta_{t,1})$ also satisfies all the above equations.

In particular the *new* equation corresponding to (38) then is

$$k_1y_1 = x_1^3 + e_1x_1^2 + f_1x_1 + d_1$$

which after multiplication with the constant factor d gives us exactly (57).

From the formulas given above we can now derive the following main theorem.

Theorem 2. *For any integers $a, b, c, d > 0$ and for integers $e, f \geq 0$ the number of solutions of the system (35) and (36) is finite.*

We prove this by contradiction. Assume that (37) and (38) have infinitely many solutions. For each such solution (x_i, y_i) we may derive all the other quantities $h_i, k_i, \Delta_{m,i}, \Delta_{n,i}, \Delta_{s,i}, t_i$ as given above.

Among these solutions we will find a sequence of pairs (x_i, y_i) such that either the first or the second coordinates of this solution sequence is an unbounded sequence of integers $x_i \to \infty$ or $y_i \to \infty$. The equations (37) and (38) then show that both sequences x_i, y_i must be unbounded. Then the associated sequence k_i is unbounded because of (39), and the associated sequence h_i is unbounded because of (40). The associated sequence $\Delta_{m,i}$ is unbounded because of (51). The associated sequence $\Delta_{n,i}$ is unbounded because of (43). Finally the associated sequence $\Delta_{s,i}$ is unbounded because of (57). But then there exists a positive integer i_0 such that for all $i > i_0$ all bracket terms in (50), that is, all of $\Delta_{s,i} - def, x_i - 1, h_i - d, \Delta_{m,i} - ef$, are strictly positive. Thus the left-hand side of (50) becomes positive and as large as we may like for sufficiently large i. But this implies a contradiction as the right-hand side of (50) is a constant. Thus an infinite sequence of solutions cannot exist. This proves the theorem.

We remark that in [2] for a large class of positive polynomials $(p(x), q(x))$ of degrees $(\deg(p), \deg(q)) \neq (1,1), (1,2), (2,1), (1,3), (3,1)$ the system

$$x \mid p(y), \; y \mid q(x)$$

was shown to have infinitely many positive integral solutions.

5 Open Problem and Comment

We have recently shown [4] that for any positive integers $b, d > 0$ and for $a = c = 1$ the biquadratic system

$$ax^2 + ex + b \equiv 0 \mod y \text{ and } cy^2 + fy + d \equiv 0 \mod x$$

for $e, f \geq 0$ and for $b = d$ always has infinitely many positive solutions. The general biquadratic case is still open.

We may ask whether there may exist any such polynomial congruence with $(\deg(p(x)), \deg(q(x))) \neq (1,1), (1,2), (1,3), (2,1), (3,1)$ which has only finitely many solutions.

Acknowledgements I should like to thank the referee for useful comments on a previous version of this text. My particular gratitude goes to Prof Bernhard Heim for the opportunity to contribute to the wonderful Oman conference of GUTech in 2012, for his untiring commitment before, during and after this conference, and for his patience with my efforts during the preparation of this paper.

References

1. T. Bier, *Non Standard Topics in Number Theory: Division Systems*. Lecture Notes (University of Malaysia, KL, Malaysia, 2004)
2. T. Bier, O. Dira, Construction of integer sequences (submitted)
3. Online Encyclopedia of Integer Sequences. Sequences A 002310, A 002320 (contributed by C. Kimberling) and A003818 ff (contributed by W Pompe) in the website www.oeis.org
4. T. Bier, Classifications of solutions of certain biquadratic division systems (submitted)

On the Genus Version of the Basis Problem II: The Case of Oldforms

Siegfried Böcherer

A famous result of Waldspurger [22] asserts that for a given square-free level $N > 1$ we can choose a genus $\mathfrak{G}$ of positive definite even integral quadratic forms of rank $m = 2k$ with determinant a perfect square such that all newforms of level N can be written as linear combinations of theta series attached to lattices from that genus ($m > 4$). Later on it was shown [4] that the statement above is true for *all* genera of lattices of precise level N provided that $p^m \nmid det(\mathfrak{G})$ for all prime divisors of N. Like Waldspurger most other authors working on the basis problem only treat the case of newforms. This seemed to be sufficient, because the oldforms can then be obtained from theta series attached to lattices of lower levels or scaled versions of them (provided that they exist!). Already for prime level and weight congruent 2 mod 4 there is a problem because even unimodular lattices do not exist. The purpose of the present paper is to complete the result of Waldspurger and others by a precise description, which cusp forms beyond the newforms are linear combinations of theta series attached to lattices in a given genus. Due to Siegel's theorem it is reasonable to restrict attention to cusp forms. The answer is simply "ALL" as long as the genus in question is not maximal or adjoint to maximal at any prime divisor of the level N. It seems that only in these cases the arithmetic of these lattices is "strong enough" to have some influence on the kind of modular forms presented. Locally this is quite natural because (in the case adjoint to the maximal case) the p-modular Jordan-component is anisotropic mod p.

In the delicate cases each local maximality (or adjoint to maximality) condition puts a condition on the oldforms represented. In the simplest case of prime level this means that for a level one Hecke eigenform f only the multiples of a certain explicitly given linear combination of f and $f(p \cdot z)$ are represented by a genus of maximal lattices. We refer to the text for a precise formulation. A special case

S. Böcherer (✉)
Kunzenhof 4B, 79117 Freiburg, Germany
e-mail: boech@rumms.uni-mannheim.de

B. Heim et al. (eds.), *Automorphic Forms*, Springer Proceedings
in Mathematics & Statistics 115, DOI 10.1007/978-3-319-11352-4_3

was already sketched in our joint work with Nebe on extremality [6], which was a starting point for the present paper.

The basic method (a precise explicit version of the "doubling method for a genus Eisenstein series") was established in [4] and used there for the case of newforms. To apply it to oldforms, we must study the bad Euler factors for this case in more detail.

Remarks.

(A) Our methods are basically local and could as well be applied to a situation of non-primitive real nebentypus. This will be done in subsequent work.

(B) The situation becomes considerably simpler, if we allow several genera at the same time: Then we can even treat Siegel modular forms; see [10]. For elliptic modular forms, we prove in Sect. 5 a much more subtle version: It is shown there that we just have to consider a given genus $\mathfrak{G}$ together with all those genera, which are locally adjoint to $\mathfrak{G}$ at those places, where $\mathfrak{G}$ is locally maximal or adjoint to maximal.

(C) The results of this paper have some surprising application to the surjectivity of the Witt operator in case of weights ≥ 4 for square-free levels; see [5, 15]

1 Preliminaries

1.1 Modular Forms

For most standard notations concerning modular forms we refer to the literature [1, 12, 16, 20].

We use $\mathbf{e}(z)$ for $e^{2\pi i z}$; let $\mathbb{H}_n$ be Siegel's upper half space with the usual action of the proper symplectic similitude group $GSp^+(n, \mathbb{R})$; this group also acts on functions f on $\mathbb{H}_n$ by the slash operator:

$$(f \mid_k g)(Z) = \det(g)^{\frac{k}{2}} j(g, Z)^{-k} f(g \cdot Z)$$

with $j(g, Z) = \det(CZ + D)$ for $g = \begin{pmatrix} A & B \\ C & D \end{pmatrix}$. We denote by $M_k^n(N)$ and $S_k^n(N)$, respectively, the space of Siegel modular forms of degree n and weight k for the usual congruence subgroup $\Gamma_0^n(N)$. We omit the superscript n if $n = 1$.

For any subgroup G of $Sp(n, \mathbb{R})$ we define G_∞ to consist of those elements $\begin{pmatrix} A & B \\ C & D \end{pmatrix}$ in G with $C = 0$.

We define the Hecke operators $T(n)$ (for $(n, N) = 1$) and $U(d)$ (for $d \mid N^\infty$ in the usual way (as operators on $M_k(N)$); if $f \in S_k(N)$ is an eigenform of all $T(p)$ with $p \nmid N$, $f \mid T(p) = \lambda_p \cdot f$, we attach to it the symmetric square L-function

$$L^{(N)}(f,s) := \prod_{p \nmid N} \frac{1}{(1-\alpha_p^2 p^{-s})(1-\beta_p^2 p^{-s})(1-p^{k-1-s})},$$

where α_p, β_p are defined by

$$\alpha_p + \beta_p = \lambda_p, \quad \alpha_p \cdot \beta_p = p^{k-1}.$$

We also use the Atkin-Lehner involutions $W_R = W_R^N$, defined for $R \mid N$ by any integral matrix

$$W_R^N = \begin{pmatrix} Rx & y \\ N & R \end{pmatrix}, \qquad R^2 x - yN = R.$$

Following [4] we define for $M \mid N$ and $\Re(s) >> 0$ an endomorphism of $S_k(N)$ by

$$\begin{aligned} f \longrightarrow f \mid Sym_M(s) &:= \sum_{d \mid M^\infty} f \mid \Gamma_0(N) \cdot \begin{pmatrix} d^{-1} & 0 \\ 0 & d \end{pmatrix} \Gamma_0(N) d^{-s} \\ &= \sum_{d \mid M^\infty} f \mid U(d^2) d^{-k+2-s}. \end{aligned}$$

1.2 Quadratic Forms

Throughout the paper, we fix a square-free integer $N > 1$ and a genus $\mathfrak{G}$ of even lattices of rank $m = 2k$ (with m divisible by 4) in a positive definite quadratic space (V, Q) over $\mathbb{Q}$; we denote by B the bilinear form associated to Q; the determinant of L should be a perfect square D and L should be of level N; moreover we request that for all primes p dividing N we have

$$p \mid D, \qquad p^m \nmid D,$$

i.e., N is the exact level of L and L does not arise from a lattice of lower level by scaling. We use the notion of maximality of lattices and of dual lattice (locally and globally) as in [19]. We call a lattice $L \in \mathfrak{G}$ *comaximal at* p for $p \mid D$ if $(L \otimes \mathbb{Z}_p)^\sharp$ is maximal as a lattice over $\mathbb{Z}_p$ for the quadratic space $(V, p \cdot Q)$.

For a quadratic space (V, Q) over $\mathbb{Q}$ we define the local Witt invariants $s_p(V) = s_p(V \otimes \mathbb{Q}_p, Q \otimes \mathbb{Q}_p)$ as in [21, p. 80]. This normalization is very convenient for our purposes; in particular we will use the following lemma from [9].

Lemma. *Let L be an even lattice of level $N \cdot p$ with $p \nmid N$ in the quadratic space (V, Q). Then the following statements are equivalent:*

(i) $s_p(V) = 1$.

(ii) *V carries (even) lattices of level N.*
(iii) *If $L_p = L_p^{(0)} \perp L_p^{(1)}$ denotes the Jordan splitting of L at the prime p, then $L_p^{(1)}$ is an orthogonal sum of hyperbolic planes.*

Remark. We tacitly use that under the assumptions above, we have

$$s_p(V, Q) = s_p(V, pQ)$$

We define the degree n theta series attached to the lattice L in the usual way by

$$\Theta^n(L)(Z) = \sum_{\mathbf{x}=(x_1,\dots x_n)\in L^n} \mathbf{e}(tr(Q(\mathbf{x}) \cdot Z)), \quad (Z \in \mathbb{H}_n)$$

with the $n \times n$ matrix $Q(\mathbf{x}) = \frac{1}{2}B(x_i, x_j)$.

2 Review of the Doubling Method for the Genus Eisenstein Series

For a given genus $\mathfrak{G}$ of positive definite quadratic forms in $m = 2k$ variables of (exact) level N with a perfect square as determinant we want to study the map

$$\Lambda_{\mathfrak{G}} : \begin{cases} S_k(N) \longrightarrow & S_k(N) \\ \quad f \quad \longmapsto & \frac{1}{m(\mathfrak{G})} \sum_i \frac{1}{\epsilon(L_i)} < f, \theta(L_i) > \theta(L_i) \end{cases}$$

Here, $m(\mathfrak{G})$ denotes the mass of the genus $\mathfrak{G}$, the L_i run over representatives of the isometry classes of lattices in $\mathfrak{G}$, $\epsilon(L)$ denotes the number of units of L, and $<,>$ is the Petersson inner product. Note that $\Lambda_{\mathfrak{G}}(f)$ is indeed cuspidal, as can be seen by looking at the values in all cusps.

Evidently, the forms in the image of this map are linear combinations of theta series and the forms f in the kernel of this map are orthogonal to all the theta series $\theta(L_i)$.

We can describe the map $\Lambda_{\mathfrak{G}}$ also by an integration of f against the restriction of the degree two genus theta series $\mathcal{E}^2(\mathfrak{G})$ of $\mathfrak{G}$, defined by

$$\mathcal{E}^2(\mathfrak{G}) = \frac{1}{m(\mathfrak{G})} \sum_i \frac{1}{\epsilon(L_i)} \theta^2(L_i).$$

By Siegel's theorem, we may write $\mathcal{E}^2(\mathfrak{G})$ as a linear combination of the degree two Eisenstein series of level N attached to the "cusps" $R \in Sp(2, \mathbb{Z})$, defined by the series

$$E_k^2(Z,R,s) := \sum_{\gamma \in (R^{-1}\Gamma_0^2(N)R)_\infty \backslash R\cdot\Gamma_0^2(N)} j(\gamma,Z)^{-k} \det(\Im(\gamma\cdot Z)^s.$$

This series is known to converge absolutely for $2\Re(s)+k>3$; it only depends on the double cosets

$$\Gamma_0^2(N)_\infty\backslash Sp(2,\mathbb{Z})/\Gamma_0^2(N).$$

Explicit representatives for these double cosets may be chosen as follows (see [4] for details): We fix a decomposition

$$N = N_0\cdot N_1\cdot N_2$$

and choose (any) $R = R_{N_0,N;1,N_2} = \begin{pmatrix} A & B \\ C & D \end{pmatrix} \in Sp(2,\mathbb{Z})$ such that

$$rank_{\mathbb{F}_p}(C) = \begin{cases} 0 \text{ for } p \mid N_0 \\ 1 \text{ for } p \mid N_1 \\ 2 \text{ for } p \mid N_2 \end{cases}$$

If $N = N_0\cdot N_1\cdot N_2$ runs over all such decompositions, then a complete set of representatives for the double cosets above (a set of representatives for the $\Gamma_0^2(N)$-inequivalent cusps) will be given by matrices $R = R_{N0,N_1,N_2}$. We then write $E_k^2(Z,N_0,N_1,N_2,s) := E_k^2(Z,R_{N_0,N_1,N_2},s)$. We can express the genus theta series explicitly in terms of these Eisenstein series; the coefficients can be determined by inspecting the values of the theta series $\theta^2(L)$ in the cusps; using the values as given, e.g., in [7, Lemma 8.2] we get

Proposition.

$$\mathcal{E}^2(\mathfrak{G}) = \sum_{N_0,N_1,N_2} c(\mathfrak{G},N_0,N_1,N_2)\cdot E_k^2(Z,N_0,N_1,N_2,0)$$

with

$$C(\mathfrak{G},N_0,N_1,N_2) = s(V,N_1)\cdot \det(\mathfrak{G})_{N_1}^{-\frac{1}{2}}\cdot \det(\mathfrak{G})_{N_2}^{-1}.$$

Here

$$s(V,N_1) := \prod_{p\mid N_1} s_p(V)$$

denotes the product over the local Witt invariants for $p \mid N_1$ and we put

$$\det(\mathfrak{G})_{N_1} := \prod_{p \mid N_1} \det(S)_p$$

for any quadratic form $S \in \mathfrak{G}$.

With a considerable amount of computation, we determined the exact image of the map $\Lambda_{N_0,N_1,N_2}(s) : S_k(N) \longrightarrow S_k(N)$, defined by

$$f \longmapsto \Lambda_{N_0,N_1,N_2}(s)(f)(w) := \int_{\Gamma_0(N)\backslash\mathbb{H}} f(z) \cdot \overline{E_k^2(\begin{pmatrix} z & 0 \\ 0 & -\bar{w} \end{pmatrix}, N_0, N_1, N_2, \bar{s})}\, y^{k-2} dx dy$$

Theorem. *Let $f \in S^k(N)$ be a eigenform for all Hecke operators away from N. Then*

$$\Lambda_{N_0,N_1,N_2}(f)(s) = W_k(s) \cdot L^{(N)}(f, 2s + 2k - 2) \cdot I_{N_0,N_1,N_2}(f)$$

with

$$W_k(s) = (-1)^{\frac{k}{2}} 2^{3-2s-k} \cdot \frac{\pi}{k + s - 1} \cdot \frac{1}{\zeta^{(N)}(2s + k)\zeta^{(N)}(4s + 2k - 2)}$$

and

$$\begin{aligned} I_{N_0,N_1,N_2}(s)(f) &= N_0^{1-2s-\frac{3k}{2}} N_2^{1-\frac{k}{2}} \\ &\times f \Big| \Big(U(N_0) Sym_{N_0}(2s + k) W_{N_0} W_{N_2} U(N_2) Sym_{N_2}(2s + k) \\ &\times \prod_{p \mid N_1} (Sym_p(2s + k) + p^{2-2k-2s} W_p U(p^2) Sym_p(2s + k) W_p \Big). \end{aligned}$$

Taking into account that the Hecke operators involved in the "bad part" decompose into their p-primary parts ($p \mid N$) and that the quantity $c(\mathfrak{G}, N_0, N_1, N_2)$ is also multiplicative, we obtain as a reformulation of the above (again for an eigenform of the Hecke operators away from N)

$$\begin{aligned} &\Lambda_{\mathfrak{G}}(s)(f) = \\ &= \sum_{N_0 \cdot N_1 \cdot N_2} c(\mathfrak{G}, N_0, N_1, N_2) \Lambda_{N_0,N_1,N_2}(s)(f) \\ &= W_k(s) L_N(f, 2s + k) \times f \Big| \Big(\prod_{p \mid N} \mathcal{T}_p(s) \Big). \end{aligned}$$

Here $\mathcal{T}_p = \mathcal{T}_p(s)$ denotes the (infinite sum of) Hecke operator(s) consisting of 4 basic contributions

$$\mathcal{T}_p = \mathcal{T}_p^0 + \mathcal{T}_p^{1,a} + \mathcal{T}_p^{1,b} + \mathcal{T}_p^2,$$

where the upper indices indicate whether the contributions come from N_0, N_1, or N_2 being divisible by p . Note that the contributions of N_2 always carry 2 summands for each prime dividing N_2.

$$\mathcal{T}_p^0(s) = p^{1-2s-3\frac{k}{2}} U(p) Sym_p(2s+k) W_p$$

$$\mathcal{T}_p^{1,a}(s) = s_p(V_p) det(\mathfrak{G})_p^{-\frac{1}{2}} Sym_p(2s+k)$$

$$\mathcal{T}_p^{1,b}(s) = s_p(V_p) \det(\mathfrak{G})_p^{-\frac{1}{2}} p^{2-2k-2s} W_p U(p^2) Sym_p(2s+k) W_p$$

$$\mathcal{T}_p^2(s) = \det(\mathfrak{G})_p^{-1} p^{1-\frac{k}{2}} W_p U(p) Sym_p(2s+k)$$

3 The Contribution of the $\mathcal{T}_p^i$ for Oldforms

The calculus described in Sect. 2 was sufficient to solve the genus version of the basis problem for newforms. We refer the reader to [4] for details. For newforms the nature of the factors $\mathcal{T}_p(s)$ is rather simple; note that in [4] the subtle point was the case of primitive quadratic nebentypus. If there is anything new in the present paper, then it is the somewhat tedious calculation of the Euler factors arising from the $\mathcal{T}_p^i$ in the case of oldforms:

Throughout this section, let $f = \sum a(n)\mathbf{e}(nz) \in S_k(N)$ be a normalized newform of level R, eigenform of all Hecke operators, and let p be a prime dividing N, which is coprime to R. We write

$$a(p) = \alpha + \beta, \qquad \alpha \cdot \beta = p^{k-1}$$

and we put

$$f_\alpha := f - \alpha \cdot f(p\tau), \quad f_\beta := f - \beta \cdot f(p\tau)$$

Then we will frequently use the following formulas

$$f_\alpha \mid U(p) = \beta \cdot f_\alpha, \quad f_\beta \mid U(p) = \alpha \cdot f_\beta$$

$$f(p\tau) = \frac{f_\alpha - f_\beta}{\beta - \alpha}, \quad f(\tau) = \frac{\beta \cdot f_\alpha - \alpha \cdot f_\beta}{\beta - \alpha}$$

$$f \mid W_p = p^{\frac{k}{2}} \cdot f(p\tau), \quad f(p\tau) \mid W_p = p^{-\frac{k}{2}} \cdot f$$

$$f_\alpha \mid W_p = p^{\frac{k}{2}} f(p\tau) - \alpha p^{-\frac{k}{2}} f, \quad f_\beta \mid W_p = p^{\frac{k}{2}} f(p\tau) - \beta p^{-\frac{k}{2}} f$$

Whenever necessary, we tacitly assume $\alpha \neq \beta$; this assumption will not harm our final formulas, which will hold without this assumption (by considering that degenerate case separately). Also, we use the formal power series

$$\mathsf{S}_p(X) := \sum_{j=0}^{\infty} U(p^{2j}) X^j$$

We note the following identities

$$Sym_M(s)(f) = f \mid \left(\prod_{p|M} S_p(p^{-s-k+2}) \right)$$

and for an eigenform $f \in S_k(N)$ for the Hecke operator $U(p)$ with $f \mid U(p) = \lambda_p \cdot f$

$$f \mid Sym_p(s) = \frac{1}{1 - \lambda_p^2 p^{-s-k+2}} \cdot f$$

3.1 The Contributions for $i = 1$

The series relevant for us are $Sym_p(X)$ and $W_p U(p^2) Sym_p W_p$, evaluated at f and f_p, defined by $f_p(\tau) := f(p \cdot \tau)$:

$$\begin{aligned}
&(\beta - \alpha) f \mid Sym_p \\
&= (\beta f_\alpha - \alpha f_\beta) \mid Sym_p \\
&= \left(\frac{\beta}{1 - \beta^2 X} f_\alpha - \frac{\alpha}{1 - \alpha^2 X} f_\beta \right) \\
&= \left(\frac{\beta}{1 - \beta^2 X} (f - \alpha f(p\tau)) - \frac{\alpha}{1 - \alpha^2 X} (f - \beta f(p\tau)) \right) \\
&= \left(\left(\frac{\beta}{1 - \beta^2 X} - \frac{\alpha}{1 - \alpha^2 X} \right) f - p^{k-1} \left(\frac{1}{1 - \beta^2 X} - \frac{1}{1 - \alpha^2 X} f(p\tau) \right) \right)
\end{aligned}$$

which gives then

$$f \mid Sym_p(X) = \frac{1}{(1-\alpha^2X)(1-\beta^2X)}\left((1+p^{k-1}X)f - p^{k-1}a(p)Xf_p\right).$$

This is the prototype of all the other series to be considered. We just write down the results for the other contributions; they all follow in the same way, using the formulas collected at the beginning of this section.

To compute $f_p \mid Sym_p(X)$ we use the identity

$$f_p \mid Sym_p(X) = f_p + X\cdot(f \mid Sym_p(X) \mid U(p))$$

We obtain

$$\begin{aligned} &f_p \mid Sym_p(X) \\ &= \frac{1}{(1-\alpha^2X)(1-\beta^2X)}\cdot\left\{Xa(p)f + (1-a(p^2)X)f(p\tau)\right\}. \end{aligned}$$

We find it convenient to describe the action of $\mathcal{T}_p^{1,a}(s)$ in the obvious way by a square matrix $\mathcal{A}_p^{1,a}(s)$ of size two:

$$\mathcal{A}_p^{1,a}(s) := \frac{s_p(V)\det(\mathfrak{S})^{-\frac{1}{2}}}{(1-\alpha^2X)(1-\beta^2X)}\begin{pmatrix} 1+p^{k-1}X & -p^{k-1}a(p)X \\ a(p)X & 1-a(p^2)X \end{pmatrix}|_{X=p^{-2s-2k+2}}.$$

Similarly, we get

$$\begin{aligned} &f \mid W_pU(p^2)Sym_pW_p \\ &= \frac{1}{(1-\alpha^2X)(1-\beta^2X)}\times\left(p^ka(p)f(p\tau) + (-p^{k-1}-p^{2k-2}X)f\right) \end{aligned}$$

and

$$\begin{aligned} &f_p \mid W_pU(p^2)Sym_pW_p \\ &= \frac{1}{(1-\alpha^2X)(1-\beta^2X)}\times(a(p^2)-p^{2k-2}X)f_p - p^{-1}a(p)f \end{aligned}$$

and hence

$$\begin{aligned} \mathcal{A}_p^{1,b}(s) = {} & \frac{s_p(V)\det(\mathfrak{S})_p^{-\frac{1}{2}}}{(1-\alpha^2X)(1-\beta^2X)}\cdot p^{2-2k-2s}\times \\ & \begin{pmatrix} -p^{k-1}-p^{2k-2}X & p^ka(p) \\ -\frac{1}{p}a(p) & a(p^2)-p^{2k-2}X \end{pmatrix}_{|X=p^{-2s-2k+2}} \end{aligned}$$

3.2 The Contribution for i = 0

Here we must apply the operator $U(p)Sym_pW_p$ to f and f_p:

$$f \mid U(p) \mid Sym_p \mid W_p$$
$$= \frac{1}{(1-\alpha^2X)(1-\beta^2X)} \times \left(-p^{\frac{k}{2}-1}(1+X)f + a(p)p^{\frac{k}{2}} f_p\right)$$

and

$$f_p \mid U(p) \mid Sym_p \mid W_p$$
$$= \frac{1}{(1-\alpha^2X)(1-\beta^2X)} \times \left(-a(p)p^{\frac{k}{2}-1}Xf + (1+p^{k-1}X)p^{\frac{k}{2}} f_p\right)$$

and the matrix for $\mathcal{T}_p^0(\mathrm{s})$ is then

$$\mathcal{A}_p^0(s) = \frac{p^{-2s-k}}{(1-\alpha^2X)(1-\beta^2X)} \times$$
$$\begin{pmatrix} -1-p^{k-1}X & a(p)p \\ -a(p)X & p+p^kX \end{pmatrix}_{|X=p^{-2s-2k+2}}$$

3.3 The Contribution for i = 2

The relevant operator is $W_pU(p)Sym_p$:

We get

$$f \mid W_pU(p)Sym_p(X) = p^{\frac{k}{2}} f \mid Sym_p$$
$$= \frac{1}{(1-\alpha^2X)(1-\beta^2X)} \cdot p^{\frac{k}{2}} \times$$
$$((1+p^{k-1}X)f - p^{k-1}a(p)Xf_p)$$

and

$$f_p \mid W_pU_pSym_p$$
$$= \frac{p^{-\frac{k}{2}}}{(1-\alpha^2X)(1-\beta^2X)} \times \left(a(p)f + (-p^{k-1} - p^{2k-2}X)f_p\right)$$

The matrix describing $\mathcal{T}_p^2(s)$ is then

$$\mathcal{A}_p^2(s) = \frac{\det(\mathfrak{G})_p^{-1} \cdot p^{1-\frac{k}{2}}}{(1-\alpha^2 X)(1-\beta^2 X)} \cdot \begin{pmatrix} p^{\frac{k}{2}}(1+p^{k-1}X) & -p^{3\frac{k}{2}-1}a(p)X \\ p^{-\frac{k}{2}}a(p) & -p^{\frac{k}{2}-1} - p^{3\frac{k}{2}-2}X \end{pmatrix}_{|X=p^{-2s-2k+2}}$$

4 The Bad Euler Factors in General

For a decomposition $N = R \cdot S$ we consider a normalized newform $f = \sum_n a(n)\mathbf{e}(n\tau)$ of level R together with all oldforms generated by it, i.e., for $T \mid S$ we put $f_T(z) := f(T \cdot z)$ and we consider the space $M(f)$ spanned by the f_T, for $T \mid S$. The doubling method provides (via the map $\Lambda_{\mathfrak{G}}(s)$) a linear endomorphism of $M(f)$, since all the Hecke operators involved in the $\mathcal{T}_p$ (with $p \mid N$) map $M(f)$ into itself.

4.1 The Contribution of p | R

The contribution of the primes $p \mid R$ is easy (it was analyzed in [4]):

$$f_T \mid \prod_{p|R} \mathcal{T}_p(0) = \left(\prod_{p|R} \frac{B_p}{1 - a(p)^2 p^{-2k+2}} \right) \cdot f_T$$

with

$$B_p = -p^{-k} - \det(\mathfrak{G})_p^{-1} + s_p(V)\det(\mathfrak{G})_p^{-\frac{1}{2}}(1 + p^{-k}),$$

which is nonzero unless $\det(\mathfrak{G})_p$ is 1 or p^m. For details we refer to [4, Sect. 8.1].

4.2 The Contribution of p | S

We first analyze the action of $\mathcal{T}_p$ on f and f_p: We get

$$f \mid \mathcal{T}_p = \frac{1}{(1-\alpha_p^2 p^{-2s-2k+2})(1-\beta_p^2 p^{-2s-2k+2})}(A_p f + B_p f_p) \qquad (1)$$

$$f_p \mid \mathcal{T}_p = \frac{1}{(1-\alpha_p^2 p^{-2s-2k+2})(1-\beta_p^2 p^{-2s-2k+2})}(C_p f + D_p f_p) \qquad (2)$$

This defines a matrix

$$\mathcal{C}_p(s) = \mathcal{C}_p(\mathfrak{G})(s) = \begin{pmatrix} A_p & B_p \\ C_p & D_p \end{pmatrix}$$

4.3 *The Complete Analysis of the Contribution of Bad Primes*

The action of $\mathcal{T}_p$ commutes with the $|_k \begin{pmatrix} Q & 0 \\ 0 & 1 \end{pmatrix}$-action for $Q \in \mathbb{N}$, if Q is coprime to p; therefore the same equation as above in (1) and (2) holds with the same matrix $\mathcal{C}_p$ when we consider the pair $(f_Q, f_{p \cdot Q})$ with Q coprime to p. This implies that (for $f \in S_k(R)$ as at the beginning of this section) the complete operation of $|_{p|S}\, \mathcal{T}_p$ on the space $M(f)$ is given by

$$\prod_{p|S} \frac{1}{(1-\alpha_p^2 p^{-2s-2k+2})(1-\beta_p^2 p^{-2s-2k+2})} \cdot \otimes_{p|S} \mathcal{C}_p(s)$$

5 The Basis Problem

As before, starting from a genus $\mathfrak{G}$ of exact level N, we fix a normalized newform $f \in S_k(R)$ and we fix a prime $p \mid N$, $p \nmid R$. It is enough to study the matrices $\mathcal{C}_p(\mathfrak{G})(s)$ at $s = 0$, and the factor $\frac{1}{(1-\alpha_p^2 p^{-2s-2k+2})(1-\beta_p^2 p^{-2s-2k+2})}$ does not vanish at $s = 0$.

We get

$$\begin{aligned}
\mathcal{C}_p(\mathfrak{G}))(0) := {} & p^{-k} \begin{pmatrix} -1-p^{-k+1} & a(p)p \\ -a(p)p^{-2k+2} & p+p^{-k+2} \end{pmatrix} + \\
& \det(\mathfrak{G})_p^{-1} p^{1-\frac{k}{2}} \cdot \begin{pmatrix} p^{\frac{k}{2}} + p^{-\frac{k}{2}+1} & -a(p)p^{-\frac{k}{2}+1} \\ p^{-\frac{k}{2}} a(p) & -p^{\frac{k}{2}-1} - p^{-\frac{k}{2}} \end{pmatrix} + \\
& s_p(V) \det(\mathfrak{G})_p^{-\frac{1}{2}} \cdot \begin{pmatrix} 1+p^{-k+1} & -p^{-k+1}a(p) \\ a(p)p^{-2k+2} & 1 \end{pmatrix} + \\
& s_p(V) \det(\mathfrak{G})_p^{-\frac{1}{2}} p^{2-2k} \cdot \begin{pmatrix} -p^{k-1} - 1 & p^k a(p) \\ -\frac{1}{p} a(p) & -1 \end{pmatrix}.
\end{aligned}$$

Note that in the last two sums there were cancellations on the diagonal of contributions from p^{-k+1} and of $a(p^2)$.

Remark. The matrix $\mathcal{C}_p(\mathfrak{G})(0) = \begin{pmatrix} A_p(0) & B_p(0) \\ C_p(0) & D_p(0) \end{pmatrix}$ satisfies a *functional equation*:

$$\mathcal{C}_p(p \cdot \mathfrak{G}^\sharp)(0) = p^{2t-k} \begin{pmatrix} D_p(0) & C_p(0) \\ B_p(0) & A_p(0) \end{pmatrix} \tag{3}$$

as can be seen by inspection of the entries of the matrix. Another possibility is to apply Atkin-Lehner-involutions W_p^N to the map $\Lambda_{\mathfrak{G}}$.

Now we examine this matrix separately for the cases of maximal, adjoint to maximal, and the remaining case (always assuming that the genus is of exact level N with $p \mid N$).

5.1 $\mathfrak{G}$ is Maximal at p

In this case, $\det(\mathfrak{G})_p = p^2$ and $s_p(V) = -1$ (see [9, Lemma 4.5] and

$$\mathcal{C}_p(\mathfrak{G})(0) =$$

$$\begin{pmatrix} 0 & 0 \\ a(p)\{-p^{-3k+2} + p^{-1-k} - p^{-2k+1} + p^{-2k}\} & p^{-k+1} + p^{-2k+2} - p^{-2} - p^{-1-k} - p^{-1} + p^{1-2k} \end{pmatrix}$$

This means in particular that the matrix $\mathcal{C}_p(\mathfrak{G})$ is not of maximal rank; we only get the linear combination

$$(-p^{-3k+2} + p^{-1-k} - p^{-2k+1} + p^{-2k})\{a(p)f - (p+1)p^{k-1} f_p\}. \tag{4}$$

We remark that there is another description in terms of traces of modular forms:

$$\{g := \lambda f + \mu f_p \mid \lambda, \mu \in \mathbb{C}, \quad tr^N_{\frac{N}{p}}(g) = 0\} = \mathbb{C} \cdot (a(p)f - (p+1)p^{k-1} f_p).$$

This fits well with our characterization of lattices L maximal at p [6]: they are characterized by the property $tr^N_{\frac{N}{p}}(\theta_L) = 0$; we recall that the trace map $tr^N_{\frac{N}{p}}$: $M_k(N) \longrightarrow M_k(\frac{N}{p})$ is defined by

$$tr^N_{\frac{N}{p}}(f) := \sum_{\gamma} f \mid_k \gamma,$$

where γ runs over representatives of $\Gamma_0(N) \backslash \Gamma_0(\frac{N}{p})$.

5.2 $\mathfrak{G}$ Is Comaximal at p

In this case $\det(\mathfrak{G})_p = p^{2k-2}$ and $s_p(V) = -1$.
and

$$\mathcal{C}_p(\mathfrak{G})(0) =$$

$$\begin{pmatrix} -p^{-k} - p^{-2k+1} + p^{3-2k} + p^{4-3k} - p^{1-k} + p^{3-3k} & a(p)\{p^{1-k} - p^{4-3k} + p^{2-2k} - p^{3-2k}\} \\ 0 & 0 \end{pmatrix}$$

Again the matrix is not of maximal rank; we only get linear combinations

$$\left(p^{-k} - p^{3-3k} + p^{1-2k} - p^{2-2k}\right)\{(p+1)f - a(p)pf_p\} \tag{5}$$

and we also have a description in terms of traces:

$$\{g = \lambda f + \mu f_p \mid \lambda, \mu \in \mathbb{C},\ tr^N_{\frac{N}{p}}(g \mid W_p) = 0\} = \mathbb{C} \cdot \left((p+1)f - a(p)pf_p\right)$$

5.3 The Remaining Cases

We assume now that $s_p(V) = 1$ and $det(\mathfrak{G})_p = p^{2t}$ with $1 \leq t \leq k-1$ or $s_p(V) = -1$ and $2 \leq t \leq k-2$. By the functional equation we may relate t and $k-t$ to each other. We have therefore only to consider the cases

$$\text{(A)}\ \ 2 \leq t \leq \frac{k}{2}, \qquad (s_p = s_p(V) \in \{\pm\})$$

$$\text{(B)}\ \ t = 1, s_p(V) = 1$$

In both cases, we write $B_p(0) = a(p)b_p$ and $C_p(0) = a(p) \cdot c_p$.
Then $\det(\mathcal{C}(\mathfrak{G}_p) = 0$ gives

$$a(p)^2 = \frac{A_p(0)D_p(0)}{b_p \cdot c_p} \tag{6}$$

and we will show that this equation implies that $|a(p)|^2$ is too big.

The case (A):

$$\begin{aligned} A_p(0) &= -p^{-k} - p^{-2k+1} + p^{-2t+1} + p^{-2t+2-k} + s_p(V)p^{-t} - s_p(V)p^{-t+2-2k} \\ &= s_p(V)p^{-t}\{1 - s_p(V)p^{-k+t} - s_p(V)p^{-2k+1-t} + s_p(V)p^{-t+1} \\ &\qquad + s_p(V)p^{-t+2-k} - p^{2-2k}\} \end{aligned}$$

$$
\begin{aligned}
b_p &= p^{-k+1} - p^{-2t+2-k} - s_p(V)p^{-t+1-k} + s_p(V)p^{-t+2-k} \\
&= p^{-k+1}\{1 - p^{1-2t} - s_p(V)p^{-t} + s_p(V)p^{-t+1}\}
\end{aligned}
$$

$$
\begin{aligned}
c_p &= -p^{-3k+2} + p^{-2t+1-k} + s_p(V)p^{-t-2k+2} - s_p(V)p^{-t+1-2k} \\
&= p^{-2t+1-k}\{1 - p^{-2k+1+2t} + s_p(V)p^{t+1-k} - s_p(V)p^{t-k}\}
\end{aligned}
$$

$$
\begin{aligned}
D_p(0) &= p^{-k+1} + p^{-2k+2} - p^{-2t} - p^{-2t+1-k} + s_p(V)p^{-t} - s_p(V)p^{-t+2-2k} \\
&= s_p(V)p^{-t}\{1 + s_p(V)p^{-k+1+t} + s_p(V)p^{-2k+2+t} - s_p(V)p^{-t} \\
&\quad - s_p(V)p^{-t+1-k} - p^{2-2k}\}
\end{aligned}
$$

This gives estimates (keeping in mind that $k \geq 4, 2 \leq t \leq \frac{k}{2}$)

$$|A_p(0)| \geq p^{-t}(1 - p^{-1} - p^{-2} - p^{-4} - p^{-6} - p^{-9})$$

$$|D_p(0)| \geq p^{-t}(1 - p^{-1} - p^{-2} - p^{-4} - p^{-5} - p^{-6})$$

$$|b_p| \leq p^{-k+1}(1 + p^{-2} + p^{-1} - p^{1-2t})$$

$$|c_p| \leq p^{-t}(1 + p^{-1} + p^{-2} + p^{-4} + p^{-5} + p^{-6})$$

Using these estimates, we see that (6) implies

$$
|a(p)^2| \geq 4p^{k-1}\cdot \underbrace{\frac{1}{4}p^{k-1}\frac{(1 - p^{-1} - p^{-2} - p^{-4} - p^{-6} - p^{-9})(1 - p^{-1} - p^{-2} - p^{-4} - p^{-5} - p^{-6})}{(1 + p^{-2} + p^{-1} - p^{1-2t})(1 + p^{-1} + p^{-2} + p^{-4} + p^{-5} + p^{-6})}}_{(*)}
$$

A simple calculation shows that (*) is larger than 1 if

$$p \geq 7, k \geq 4, p \geq 3, k \geq 6, \qquad p = 2, k \geq 8.$$

We get therefore a contradiction to the conjecture of Ramanujan-Petersson for $a(p)$ (which assures that $|a(p)|^2 \geq 4p^{k-1}$). The remaining cases

$$k = 4, p = 3, \qquad k = 4, p = 5,, \qquad p = 2, k = 4, 6$$

can be handled using the precise formulas above.

The case (B): $t = 1, s_p(V) = 1$.

Then

$$A_p(0) = 2p^{-1} - p^{-k+1} - p^{1-2k}$$
$$b_p = 2p^{-k+1} - 2p^{-k}$$
$$c_p = p^{-1-k} + p^{1-2k} - p^{-2k} - p^{-3k+2}$$
$$D_p(0) = p^{-1} - p^{-2} - p^{-1-k} + p^{2-2k} - p^{1-2k} + p^{1-k}$$

Keeping in mind that $k \geq 4$ we get the simple estimates

$$|A_p(0)| \geq 2p^{-1}(1 - \frac{1}{2p^2} - \frac{1}{2p^6})$$

$$|D_p(0)| \geq p^{-1}(1 - p^{-1})$$

$$|b_p| \leq 2p^{-k+1}$$

$$|c_p| \leq p^{-1-k}(1 + p^{2-k})$$

and then

$$|a(p)^2| \geq 4p^{k-1} \cdot \underbrace{\frac{1}{4}p^{k-1}\frac{(1 - \frac{1}{2p^2} - \frac{1}{2p^6})(1 - p^{-1})}{1 + p^{2-k}}}_{(*)}$$

This factor (*) is larger than one if $p > 2$ or $k \geq 6$. We get a contradiction to the Ramanujan-Petersson conjecture for $a(p)$. The case $p = 2$ and $k = 4$ can be treated directly.

5.4 The Main Results

We may glue all the (essentially local) computations above together:

Theorem. *Let $\mathfrak{G}$ be a genus of quadratic forms of level N in $m = 2k > 4$ variables. We put*

$$P = P(\mathfrak{G}) := \prod_{\substack{p|N \\ \mathfrak{G}_p \text{ maximal}}} p, \qquad Q = Q(\mathfrak{G}) := \prod_{\substack{p|N \\ \mathfrak{G}_p \text{ comaximal}}} p;$$

then

$$\Theta(\mathfrak{G})^{cusp} = \{f \in S_k(N) \mid \begin{array}{l} \forall p \mid P: \quad tr^N_{\frac{N}{p}}(f) = 0 \\ \forall p \mid Q: tr^N_{\frac{N}{p}}(f \mid W^N_p) = 0 \end{array}\}$$

To reformulate the theorem in terms of spaces generated by newforms we need more notation:

Let $f = \sum_n a(n)\mathbf{e}(n\tau) \in S_k(N)$ be a normalized newform of level $R \mid N$ (eigenform of all Hecke operators). For two arbitrary (coprime) divisors P and Q of N we put $P' := \frac{P}{(R,P)}$ and $Q' := \frac{Q}{(Q,R)}$ and

$$\mathcal{L}(f)(P,Q) := \mathbb{C}\{ \sum_{\substack{d_1 \cdot d_2 = P' \\ e_1 e_2 = Q'}} \sigma_1(d_1)\sigma_1(e_2)\mu(d_2)\mu(e_2) d_2 e_2^{k-1} a(d_2) a(e_1) f_{t d_2 e_2} \Big| t \mid \frac{N}{RP'Q'}\}$$

Theorem'. *Let $\mathfrak{G}$ be a s above, then we have for $P = P(\mathfrak{G})$ and $Q = Q(\mathfrak{G})$*

$$\Theta(\mathfrak{G})^{cusp} = \oplus_f \mathcal{L}(f)(P,Q),$$

where f runs over all normalized newforms of level R, $R \mid N$.

This statement can be viewed as the global version of (4) and (5).

The theorem cannot be extended to the spaces of non-cusp forms (unless $PQ = N$).

We may remove some of the conditions in the theorem above by taking genera together with their adjoints:

Proposition. *Let S be any subset of $\{p \mid \mathfrak{G}_p$ is maximal or comaximal$\}$. Then*

$$\sum_{\mathfrak{G}'} \Theta(\mathfrak{G}')^{cusp} = \{f \in S^k(N) \mid \begin{array}{l} \forall p \mid P, p \notin S: \quad tr^N_{\frac{N}{p}}(f) = 0 \\ \forall p \mid Q, p \notin S: tr^N_{\frac{N}{p}}(f \mid W^N_p) = 0 \end{array}\}$$

Here $\mathfrak{G}'$ runs over all genera such that

$$\mathfrak{G}'_p = \begin{cases} \mathfrak{G}_p & \text{if } p \notin S \\ \mathfrak{G}_p \text{ or } p \cdot \mathfrak{G}_p^\sharp & \text{if } p \in S \end{cases}$$

An equivalent statement in terms of newforms is

$$\sum_{\mathfrak{G}'} \Theta(\mathfrak{G}')^{cusp} = \oplus_f \mathcal{L}(f)(P_S, Q_S);$$

here $P_S := \prod_{p \mid P, p \notin S} p$ and $Q_S := \prod_{p \mid Q, p \notin S} p$ and f runs again over normalized newforms of level $R \mid N$.

Proof. The equality of the right-hand sides is clear. The inclusion "$\subseteq$" is also clear. To prove "$\supseteq$" we proceed by induction over the number of primes contained in S. We increase the set S by a prime q, assuming that q divides P; we put $\tilde{S} := S \cup \{q\}$. By induction, we may apply the proposition for the cases (P, Q) and $P' := \frac{P}{q}$ and $Q' = q \cdot Q$ and the given set S. Let $\mathfrak{G}''$ run over all genera of level N such that

$$\mathfrak{G}''{}_p = \begin{cases} \mathfrak{G}_p & \text{if } p \notin S' \\ \mathfrak{G}_p \text{ or } p \cdot \mathfrak{G}_p^{\sharp} & \text{if } p \in S' \end{cases}$$

Then

$$\sum_{\mathfrak{G}''} \Theta(\mathfrak{G}'') = \oplus_f \mathcal{L}(f)(P_S, Q_S) + \mathcal{L}(f)(P_{\tilde{S}}, q \cdot Q_S)$$

where again f runs over all normalized newforms of level R with $R \mid N$. It is enough to show that (for $q \nmid R$)

$$\mathcal{L}(f)(P_S, Q_S) + \mathcal{L}(f)(P_{\tilde{S}}, q \cdot Q_S) = \mathcal{L}(f)(P_{\tilde{S}}, Q_{\tilde{S}})$$

Ignoring unessential factors, it is enough to show that for fixed d_1, d_2, e_1, e_2 and t with $d_1 d_2 = \frac{P_S}{(P_S, R)}$, $e_1 e_2 = \frac{Q_S}{(Q_S, R)}$ the space generated by $\sigma_1(q) f_{td_2e_2} - qa(q) f_{tqd_2e_2}$ and $-\sigma_1(q) q^{k-1} f_{tqd_2e_2} + a(q) f_{td_2e_2}$ is two-dimensional. This means that the determinant of the coefficient matrix should be nonzero, i.e.,

$$-\sigma_1(q)^2 q^{k-1} + a(q)^2 q \neq 0$$

This is guaranteed by an (elementary) estimate $|a(q)| < q^{\frac{k}{2}}(1 + q)$ for the eigenvalue $a(q)$; see, e.g., [17]. It is amusing that [17] provides the precise estimate necessary for our purpose. □

6 The Case $m = 4$

Mainly due to reasons of convergence (of the degree two Eisenstein series) we do not treat the case of quaternary forms here (although it would be desirable to consider this case along the lines of the present paper). The case $m = 4$ is also remarkable from a second point view: A quaternary lattice L (with $\det(L)_p = p^2$) is maximal at p iff its adjoint has the same property. Therefore it is not surprising that the phenomena described in the previous sections do occur in a somewhat different way:

We quote from [8, 13]

Proposition. *Let D be a definite quaternion algebra over $\mathbb{Q}$ ramified in all primes dividing N_1, and let $N = N_1 \cdot N_2$ be square-free. Let $\mathfrak{G}$ be the genus of quaternary lattices coming from left ideals in the Eichler order of level N in D. Then*

$$\Theta(\mathfrak{G})^{cusp} = \{f \in S_k(N) \mid f \text{ is new for all } p \mid N_1\}$$

Final Remarks.

(A) The results of this paper extend in an immediate way to the problem of representation by theta series involving harmonic polynomials (using differential operators as in [3, 14]).

(B) Our results illustrate a subtle difference between the basis problem in a classical setting and its reformulation in terms of representation theory (theta correspondence). Let us consider the case $N = p$ and $s_p(V) = -1$. Then the representation π_f generated by a modular form of level one will appear in the image of the theta correspondence; our results however show that in the two-dimensional space $\pi_f^{\Gamma_0(p)}$ of $\Gamma_0(p)$-invariant holomorphic vectors only a certain one-dimensional subspace can be expressed by theta lifts using characteristic functions of lattices of level p. From the point of view of individual theta series, this corresponds to the fact that the traces $tr_1^p(\Theta(L)$ of maximal lattices are known to be zero, whereas the traces of comaximal lattices are not understood (see, e.g., [9]).

Acknowledgement I wish to thank the referee for pointing out some inaccuracies.

References

1. A. Andrianov, *Quadratic Forms and Hecke Operators*. Grundlehren, vol. 286 (Springer, New York, 1987)
2. T. Arakawa, S. Böcherer, Vanishing of certain spaces of modular forms and applications. J. Reine Angew. Math. **559**, 25–51 (2003)
3. S. Böcherer, Über die Fourier-Jacobi- Entwicklung Siegelscher Eisensteinreihen II. Math. Z. **189**, 81–100 (1985)
4. S. Böcherer, The genus version of the basis problem I, in *Automorphic Forms and Zeta Functions*, ed. by S. Böcherer, T. Ibukiyama, M. Kaneko, F. Sato (World Scientific, Singapore, 2006)
5. S. Böcherer, T. Ibukiyma, The surjectivity of the Witt operator for $\Gamma_0(N)$ for square-free N. Ann. Inst. Fourier **62**, 121–144 (2012)
6. S. Böcherer, G. Nebe, On theta series attached to maximal lattices and their adjoints. J. Ramanujan Math. Soc. **25**, 265–284 (2010)
7. S. Böcherer, R. Schulze-Pillot, Siegel modular forms and theta series attached to quaternion algebras. Nagoya Math. J. **121**, 35–96 (1991)
8. S. Böcherer, R. Schulze-Pillot, On the central critical value of the triple product L-function, in *Number Theory 1993–1994* (Cambridge University Press, Cambridge, 1996), pp. 1–46
9. S. Böcherer, J. Funke, R. Schulze-Pillot, Trace operators and theta series. J. Number Theory **78**, 119–139 (1999)

10. S. Böcherer, H. Katsurada, R. Schulze-Pillot, On the basis problem for Siegel modular forms with level, in *Modular Forms on Schiermonnikoog* (Cambridge University Press, Cambridge, 2008)
11. M. Eichler, The basis problem for modular forms and the traces of the Hecke operators, in *Modular Functions of One Variable I*. Lecture Notes in Mathematics, vol. 320 (Springer, New York, 1973)
12. E. Freitag, *Siegelsche Modulfunktionen*. Grundlehren, vol. 254 (Springer, New York, 1983)
13. K.-I. Hashimoto, On Brandt matrices of Eichler orders. Memoirs School Sci. Eng. Waseda Univ. Nr. **59**, 153–165 (1996)
14. T. Ibukiyama, On differential operators on automorphic forms and invariant pluriharmonic polynomials. Comment. Math. Univ. St. Pauli **48**, 103–118 (1999)
15. T. Ibukiyama, S. Wakatsuki, Siegel modular forms of small weight and the Witt operator, in *Quadratic Forms-Algebra, Arithmetic and Geometry*. Contemporary Mathematics, vol. 493 (American Mathematical Society, Providence, 2009), pp. 181–209
16. H. Klingen, *Introductory Lectures on Siegel Modular Forms* (Cambridge University Press, Cambridge, 1990)
17. W. Kohnen, A simple remark on eigenvalues of Hecke operators and Siegel modular forms. Abh. Math. Sem. Univ. Hamburg **57**, 33–35 (1987)
18. W. Li, Newforms and functional equations. Math. Ann. **212**, 285–315 (1975)
19. O.T. O'Meara: *Introduction to Quadratic Forms*. Grundlehren, vol. 117 (Springer, Berlin, 1973)
20. T. Miyake, *Modular Forms* (Springer, New York, 1989)
21. W. Scharlau, *Quadratic and Hermitian Forms*. Grundlehren, vol. 270 (Springer, Berlin, 1985)
22. J.-L. Waldspurger, L'engendrement par des series de theta de certains espaces de formes modulaires. Invent. Math. **50**, 135–168 (1979)

On p-Adic Properties of Siegel Modular Forms

Siegfried Böcherer and Shoyu Nagaoka

Mathematics Subject Classification 2010: Primary 11F33; Secondary 11F55.

1 Introduction

Starting with Swinnerton-Dyer [24] and Serre [18], the mod p properties of elliptic modular forms and also their p-adic properties have been deeply studied. Some aspects of this theory were later generalized to other types of modular forms like Jacobi forms [22] and also Siegel modular forms [14]. In our previous works we constructed Siegel modular form congruent 1 mod p; we did this for level one [4] and also for level p with additional good p-adic behavior in the other cusps [5].

In the present paper we are concerned with generalizing some of Serre's results to the case of Siegel modular forms. In the first part we show that Siegel modular forms for congruence subgroups $\Gamma_0^n(p^m)$ are always p-adic modular forms. For $m = 1$ we cannot follow Serre directly [18] because certain modular forms of level p, congruent 1 mod p, and with divisibility by p in the other cusps are not available (there are $n+1$ cusps to be considered!). The generalization to $m > 1$ then works in the same way as in [19], with some delicate new problem concerning the vector-valued case.

S. Böcherer (✉)
Kunzenhof 4B, Freiburg 79117, Germany
e-mail: boech@rumms.uni-mannheim.de

S. Nagaoka
Department of Mathematics, Kinki University, 577-8502 Osaka, Japan
e-mail: nagaoka@math.kindai.ac.jp

B. Heim et al. (eds.), *Automorphic Forms*, Springer Proceedings
in Mathematics & Statistics 115, DOI 10.1007/978-3-319-11352-4_4

The second part of this paper is concerned with derivatives of modular forms as p-adic modular forms. In [4] we generalized the Θ-operator, defined on elliptic modular forms by $\sum a_n q^n \longmapsto \sum n a_n q^n$ to Siegel modular forms. We showed that the algebra of Siegel modular forms mod p is stable under Θ. We generalize the Θ-operator to a wide class of differential operators appearing in certain Rankin-Cohen brackets and show that they define p-adic modular forms; we also correct a mistake in the proof presented in [4].

Most of our results are also valid for modular forms of real nebentypus. Sometimes we just mention this generalization without going into details. At some points our methods give results which are weaker for vector-valued modular forms than for scalar-valued ones. The reason is that we cannot use the pth power of a modular form in the same way as for the scalar-valued case; to take the pth symmetric power is a good substitute, but it changes the representation space. A more detailed treatment of the vector-valued case will be given in a subsequent work [6].

Finally we mention that our paper is not concerned with the intrinsic theory of Siegel modular forms over $\mathbb{F}_p$ (as created by Katz [16] in degree $n = 1$ and in general by Faltings and Chai [10]). We only deal with mod p reductions of characteristic zero modular forms. For an approach to p-adic Siegel modular forms based on the arithmetic theory of Faltings-Chai we refer to [15].

2 Preliminary

2.1 Siegel Modular Forms

Let $\mathbb{H}_n$ denote the Siegel upper half space of degree n. The real symplectic group $Sp_n(\mathbb{R})$ acts on $\mathbb{H}_n$ in usual manner:

$$Z \longmapsto M\langle Z\rangle := (AZ+B)(CZ+D)^{-1}$$

$(Z \in \mathbb{H}_n, M = \left(\begin{smallmatrix} A & B \\ C & D \end{smallmatrix}\right) \in Sp_n(\mathbb{R}))$.

Let (ρ, V_ρ) be a finite-dimensional polynomial representation of $GL_n(\mathbb{C})$. For any V_ρ-valued function $F(Z)$ on $\mathbb{H}_n$ and any element $M = \left(\begin{smallmatrix} A & B \\ C & D \end{smallmatrix}\right) \in Sp_n(\mathbb{R})$, we write

$$(F\mid_{\rho,k} M)(Z) = \det(CZ+D)^{-k}\rho(CZ+D)^{-1}F(M\langle Z\rangle).$$

Let $\Gamma \subset \Gamma^n := Sp_n(\mathbb{Z})$ be a congruence subgroup and ν a character of Γ.

Definition 2.1. A V_ρ-valued holomorphic function F on $\mathbb{H}_n$ is called a (V_ρ-valued) Siegel modular form of type $\rho \otimes \det^k$ on Γ with character ν if

$$(F\mid_{\rho,k} M)(Z) = \nu(M)F(Z) \quad \text{for all } M \in \Gamma$$

(and $F(Z)$ is bounded at the cusps for $n = 1$).

We denote by $M_n^k(\Gamma, \rho, \nu)$ the space of such modular forms. In the case where ρ is the one-dimensional trivial representation, then we are in the scalar-valued case and we write $F\mid_k = F\mid_{\rho,k}$, and $M_n^k(\Gamma, \nu) = M_n^k(\Gamma, \rho, \nu)$ simply.

We will be mainly concerned with the Siegel modular group Γ^n and congruence subgroup

$$\Gamma_0^n(N) := \left\{ \begin{pmatrix} A & B \\ C & D \end{pmatrix} \in Sp_n(\mathbb{Z}) \mid C \equiv O \pmod{N} \right\}.$$

Moreover, we assume that ν comes from a Dirichlet character χ mod N which as usual acts on the determinant of the right lower block. If χ is trivial, we write $M_n^k(\Gamma, \rho)$ for simplicity.

If a Siegel modular form $F(Z)$ is periodic with respect to the lattice $Sym_n(\mathbb{Z})$, then $F(Z)$ admits a Fourier expansion of the form

$$F(Z) = \sum_{0 \le T \in \Lambda_n} a_F(T) \exp(2\pi\sqrt{-1}\operatorname{tr}(TZ)), \quad a_F(T) \in V_\rho,$$

where

$$\Lambda_n := \{T = (t_{ij}) \in Sym_n(\mathbb{Q}) \mid t_{ii}, 2t_{ij} \in \mathbb{Z}\}.$$

2.2 *p-Adic Modular Forms*

By fixing a basis of V_ρ, we may view ρ as a matrix-valued representation ($V_\rho = \mathbb{C}^M$ for some M). Taking $q_{ij} := \exp(2\pi\sqrt{-1}z_{ij})$ with $Z = (z_{ij}) \in \mathbb{H}_n$, we write

$$\boldsymbol{q}^T := \exp(2\pi\sqrt{-1}\operatorname{tr}(TZ)) = \prod_{i<j} q_{ij}^{2t_{ij}} \prod_{i=1}^{n} q_{ii}^{t_{ii}}.$$

Using this notation, we have the generalized q-expansion:

$$F = \sum_{0 \le T \in \Lambda_n} a_F(T)\boldsymbol{q}^T = \sum \Big(a_F(T) \prod_{i<j} q_{ij}^{2t_{ij}} \Big) \prod_{i=1}^{n} q_{ii}^{t_{ii}}$$
$$\in \mathbb{C}^M[q_{ij}^{-1}, q_{ij}][\![q_{11}, \ldots, q_{nn}]\!],$$

$(a_F(T) = (a_F(T)^{(j)}) \in \mathbb{C}^M)$.

For any subring R of $\mathbb{C}$, we shall denote by $M_n^k(\Gamma,\rho,\chi)(R)$ the R-module consisting of those F in $M_n^k(\Gamma,\rho,\chi)$ for which $a_F(T)$ is in R^M for every $T\in\Lambda_n$. From this, any element F in $M_n^k(\Gamma,\rho,\chi)(R)$ may be regarded as an element of the space of formal power series $R^M[q_{ij}^{-1},q_{ij}]\,[\![q_{11},\dots,q_{nn}]\!]$.

For a prime number p, we denote by ν_p the normalized additive valuation on $\mathbb{Q}$ (i.e., $\nu_p(p)=1$). We tacitly extend ν_p to appropriate field extensions $\boldsymbol{K}$ of $\mathbb{Q}$ if necessary.

For a Siegel modular form $F=\sum a_F(T)\boldsymbol{q}^T\in M_n^k(\Gamma,\rho,\chi)(\boldsymbol{K})$, we define $\nu_p(F)$ by

$$\nu_p(F)=\inf_{T\in\Lambda_n}\nu_p(a_F(T)),$$

where $\nu_p(a_F(T))=\min_{1\le j\le M}(\nu_p(a_F(T)^{(j)})$.

Remark 2.2. In the definition above we do not exclude the possibility that $\nu_p(F)$ becomes $-\infty$ in the case of arbitrary ρ. We do not know a (published) statement about boundedness of denominators for general ρ (see however Remark 3.7 for a possible proof and also the preprint [15]). We say that ν_p*-boundedness* holds for $M_n^k(\Gamma,\rho,\chi)$, if $\nu_p(f\mid_{k,\rho}\omega)$ is finite for all $f\in M_n^k(\Gamma,\rho,\chi)$ and all $\omega\in\Gamma^n$.

For two Siegel modular forms $F=\sum a_F(T)\boldsymbol{q}^T\in M_n^k(\Gamma,\rho,\nu)(\boldsymbol{K})$, $G=\sum a_G(T)\boldsymbol{q}^T\in M_n^l(\Gamma,\rho,\nu)(\boldsymbol{K})$, with $\nu_p(F)>-\infty$ and $\nu_p(G)>-\infty$ we write

$$F\equiv G \pmod{p^m}$$

if $\nu_p(a_F(T)-a_G(T))\ge m+\nu_p(F)$ for every $T\in\Lambda_n$.

Definition 2.3. A formal power series

$$F=\sum a_F(T)\boldsymbol{q}^T\in\mathbb{Q}_p^M[q_{ij}^{-1},q_{ij}][\![q_{11},\dots,q_{nn}]\!]$$

is called a (*vector-valued*) *p-adic Siegel modular form* (in the sense of Serre) if there exists a sequence of modular forms $\{F_m\}$ satisfying

$$F_m=\sum a_{F_m}(T)\boldsymbol{q}^T\in M_n^{k_m}(\Gamma^n,\rho)(\mathbb{Q})\quad\text{and}\quad\lim_{m\to\infty}F_m=F,$$

where $\lim_{m\to\infty}F_m=F$ means that

$$\inf_{T\in\Lambda_n}(\nu_p(a_{F_m}(T)-a_F(T)))\longrightarrow+\infty\quad(m\to\infty).$$

This definition also makes sense if we replace $\mathbb{Q}$ and $\mathbb{Q}_p$ by suitable extension fields.

2.3 The Hecke Operator $U(p)$

Let F be a Siegel modular form in
$M_n^k(\Gamma_0^n(N), \rho, \chi)$ with the Fourier expansion $F = \sum a_F(T)\boldsymbol{q}^T$. The action of $U(p)$ on F is defined by

$$F \mid U(p) = \sum a_F(pT)\boldsymbol{q}^T.$$

It is known that $U(p)$ maps the space $M_n^k(\Gamma_0^n(N), \rho, \chi)$ into itself (if $p \mid N$) and maps it into $M_n^k(\Gamma_0^n(\frac{N}{p}), \rho, \chi)$ if $p^2 \mid N$ and χ is defined modulo $\frac{N}{p}$. We recall the following result from [2]:

Theorem 2.4. *The operator $U(p)$ is bijective for $p \parallel N$.*

2.4 Coset Representatives of $\Gamma_0^n(p)\backslash\Gamma^n$

In this subsection, we describe a system of representatives for $\Gamma_0^n(p)\backslash\Gamma^n$, to be used later on to calculate a trace operator on modular forms. For the finite field $\mathbb{F}_p$ let $P = \{\left(\begin{smallmatrix} A & B \\ 0 & D \end{smallmatrix}\right)\} \subset Sp_n(\mathbb{F}_p)$ be the Siegel parabolic subgroup. For $0 \leq j \leq n$ we define "partial involutions"

$$\omega_j = \omega_j(p) = \begin{pmatrix} 1_{n-j} & 0 & 0_{n-j} & 0 \\ 0 & 0_j & 0 & -1_j \\ 0_{n-j} & 0 & 1_{n-j} & 0 \\ 0 & 1_j & 0 & 0_j \end{pmatrix}.$$

Then we have a Bruhat decomposition

$$Sp_n(\mathbb{F}_p) = \coprod_{j=0}^{n} P\omega_j P,$$

where the double coset $P\omega_j P$ consists of the set of elements $\left(\begin{smallmatrix} A & B \\ C & D \end{smallmatrix}\right) \in Sp_n(\mathbb{F}_p)$ with $\mathrm{rank}(C) = j$. Using the Levi decomposition $P = MN$ with Levi factor

$$M = \left\{ m(A) = \begin{pmatrix} A & 0 \\ 0 & (A^{-1})^t \end{pmatrix} \mid A \in GL_n(\mathbb{F}_p) \right\}$$

and unipotent radical

$$N = \left\{ n(B) = \begin{pmatrix} 1 & B \\ 0 & 1 \end{pmatrix} \mid B \in Sym_n(\mathbb{F}_p) \right\},$$

we easily see that

$$(*)\quad \{\,\omega_j\, n(B_j)\, m(A) \mid B_j \in Sym_j(\mathbb{F}_p), A \in P_{n,j}(\mathbb{F}_p)\backslash GL_n(\mathbb{F}_p)\}$$

is a complete set of right coset representatives for $P\backslash P\omega_j P$. Here $M_j(\mathbb{F}_p)$ is embedded into $M_n(\mathbb{F}_p)$ by $B_j \longmapsto \left(\begin{smallmatrix} 0 & 0 \\ 0 & B_j \end{smallmatrix}\right)$ and $P_{n,j} = \{M \in GL_n \mid M = \left(\begin{smallmatrix} * & * \\ 0_{j,n-j} & * \end{smallmatrix}\right)\}$ is a standard maximal parabolic subgroup of GL_n.

Using strong approximation we obtain

Proposition 2.5. *By lifting the coset representatives* $(*)$ *for* $0 \le j \le n$ *to elements of* Γ^n*, we get a complete set of right coset representatives for* $\Gamma_0^n(p)\backslash\Gamma^n$*. (We identify the lifts with their image modulo p.)*

For $F \in M_n^k(\Gamma_0^n(p),\rho)$, we define the trace of F as

$$\mathrm{tr}(F) := \sum_{M\in\Gamma_0^n(p)\backslash\Gamma^n} F \mid_{\rho,k} M.$$

The trace clearly projects $M_n^k(\Gamma_0^n(p),\rho)$ onto $M_n^k(\Gamma^n,\rho)$. Using the coset representatives above, we give an explicit description of the trace operator.

Noting that the action of the $n(B_j)$ comes down to an exponential sum as a factor (equal either to zero or to $p^{\frac{j(j+1)}{2}}$), we obtain

$$\mathrm{tr}(F) = F + \sum_{j=1}^{n-1} p^{\frac{j(j+1)}{2}} F \mid_{\rho,k} \omega_j \mid \tilde{U}_j(p) + p^{\frac{n(n+1)}{2}} F \mid_{\rho,k} \omega_n \mid \tilde{U}_n(p).$$

The action of $\tilde{U}_j(p)$ is defined as follows.

For general j, if

$$F \mid_{\rho,k} \omega_j = \sum_{T\in\Lambda_n} b_j(T)\exp(2\pi\sqrt{-1}\mathrm{tr}(\tfrac{1}{p}TZ)),$$

then $F \mid_{\rho,k} \omega_j \mid \tilde{U}_j(p)$ is defined as

$$F \mid_{\rho,k} \omega_j \mid \tilde{U}_j(p) = \sum_{T\in\Lambda_n}\Big(\sum_{\tilde{T}} b_j(\tilde{T})\Big)\exp(2\pi\sqrt{-1}\mathrm{tr}(TZ)),$$

where $\tilde{T}$ runs over the finite subset of Λ_n determined by T. We do not need the explicit shape of the $\tilde{T}$ in the sequel, except for the case $j = n$, where the description is much simpler:

$$F \mid_{\rho,k} \omega_n \mid \tilde{U}_n(p) = \sum_{T\in\Lambda_n} b_n(p\cdot T)\exp(2\pi\sqrt{-1}\mathrm{tr}(TZ)).$$

3 Modular Forms of Level p^m

From now on p will always be an odd prime.

3.1 *Modular Forms of Level p Are p-adic*

To generalize Serre's result about modular forms for $\Gamma_0^n(p)$ being p-adic modular forms we cannot follow his strategy directly. The problem is the (non-)existence of a modular form with the necessary properties $F \equiv 1 \pmod{p}$ and $F \mid \omega_j \equiv 0 \pmod{p}$ for all $j > 0$. The best result towards the existence of such F is (to the best of our knowledge) our work [5] where we construct F with $F \equiv 1 \pmod{p}$ and $\nu_p(F \mid \omega_j) \geq -\frac{j(j-1)}{2} + 1$. This is however not sufficient to apply Serre's method directly to Siegel modular forms, because $\nu_p(F \mid_{\rho,k} \omega_j)$ is not necessarily positive for $j \geq 2$. We need a variant of Serre's approach (interchanging the roles of the cusps):

We use a modular form $\mathcal{K}_{p-1}$ on $\Gamma_0^n(p)$ with Fourier coefficients in $\mathbb{Z}$ satisfying

$$\mathcal{K}_{p-1} \mid \omega_j \equiv 0 \pmod{p} \qquad (0 \leq j \leq n-1),$$
$$\mathcal{K}_{p-1} \mid \omega_n \equiv 1 \pmod{p}.$$

The existence of such a modular form is not a problem at all: We may use

$$\mathcal{K}_{p-1} := p^n \cdot \theta_L^n$$

where θ_L^n is the theta series associated with any p-special lattice L of rank $2p-2$ and determinant p^2. Here a $\mathbb{Z}$-lattice L is called p-special if there exists an automorphism σ of L such that σ is of order p and acts freely on $L \backslash \{\mathbf{0}\}$. The existence of such lattices (for odd p) is discussed in [4, 5].

Proposition 3.1. *Let p be an odd prime with ν_p-boundedness for $M_n^k(\Gamma_0^n(p), \rho)$. Let f be an element of $M_n^k(\Gamma_0^n(p), \rho)$. Then for any $\alpha \in \mathbb{N}$ there exists $\beta \in \mathbb{N}$ (depending on α, f) and $F \in M_n^{k+\beta\cdot(p-1)}(\Gamma^n, \rho)$ such that*

$$\nu_p(f - F) \geq \nu_p(f) + \alpha.$$

The dependence of β on α will be clarified below.

Proof. As usual, we assume $\nu_p(f) = 0$. We use an extension of ν_p to the field generated by all Fourier coefficients of all the $f \mid_{k,\rho} \omega_j$. For the moment we consider (for an arbitrary modular form $g \in M_n^k(\Gamma_0^n(p), \rho)$ and arbitrary $\beta = \kappa p^\gamma$)

$$\mathrm{Tr}_\beta(g) := p^{-\frac{n(n+1)}{2}} \cdot \mathrm{tr}(g \cdot \mathcal{K}_{p-1}^\beta).$$

The trace decomposes into $n+1$ pieces Y_j which we consider separately: For $0 \leq j \leq n$ we have to look at

$$Y_j := p^{\frac{j(j+1)}{2}-\frac{n(n+1)}{2}} \left(g \mid_{k,\rho} \omega_j \cdot (\mathcal{K}_{p-1} \mid \omega_j)^{\beta}\right) \mid \tilde{U}_j(p).$$

Then for $j < n$ we have

$$\nu_p(Y_j) \geq \frac{j(j+1)}{2} - \frac{n(n+1)}{2} + \nu_p(g \mid_{k,\rho} \omega_j) + \nu_p(\mathcal{K}_{p-1} \mid \omega_j) \cdot \beta.$$

Clearly this becomes large if β is large (note that $\nu_p(\mathcal{K}_{p-1} \mid \omega_j) > 0$).

The contribution for $j = n$ needs a more detailed study: We write $\left(\mathcal{K}_{p-1} \mid \omega_n\right)^{\beta}$ as $1 + p^{\gamma+1}X$ with a Fourier series X with integral Fourier coefficients. Then

$$\left(g \mid_{k,\rho} \omega_n \cdot (\mathcal{K}_{p-1}^{\beta} \mid \omega_n)\right) \mid \tilde{U}_n(p)$$
$$= g \mid_{k,\rho} \omega_n \mid \tilde{U}_n(p) + p^{\gamma+1}\left(g \mid_{k,\rho} \omega_n \cdot X\right) \mid \tilde{U}_n(p).$$

Now we use that the $U(p)$ operator is invertible as a Hecke operator for $\Gamma_0^n(p)$(cf. Theorem 2.4). Therefore we may choose g such that

$$g \mid_{k,\rho} \omega_n \mid \tilde{U}_n(p) = f.$$

With this choice of g the contribution for $j = n$ to the trace of $g \cdot \mathcal{K}_{p-1}^{\beta}$, which we call Y_n, satisfies

$$\nu_p(Y_n - f) \geq \gamma + 1 + \nu_p(g \mid_{k,\rho} \omega_n).$$

Summarizing this, we see that $F := \mathrm{Tr}_{\beta}(g)$ is congruent to f modulo p^{α} if we choose γ to be large enough. □

Remark 3.2. We wrote $\beta = \kappa \cdot p^{\gamma}$ in the proof in order to emphasize the different roles played by β and γ. We have to choose γ large enough to assure the congruence for Y_n, but to make the other Y_j divisible by a high power of p it is sufficient that β becomes large.

To get from the proposition above a statement about p-adic modular forms, we need some rationality and integrality properties:

Proposition 3.3. *Assume that the polynomial representation $\rho : GL(n, \mathbb{C}) \longrightarrow GL(M, \mathbb{C})$ is $\mathbb{Q}$-rational (i.e., the polynomials defining ρ have rational coefficients). Then the following properties hold:*

(a) $M_n^k(\Gamma_0^n(p), \rho) = M_n^k(\Gamma_0^n(p), \rho)(\mathbb{Q}) \otimes \mathbb{C}$.
(b) The ν_p-boundedness holds for $M_n^k(\Gamma_0^n(p), \rho)$.

(c) *If* $f \in M_n^k(\Gamma_0^n(p), \rho)$ *has rational Fourier coefficients, the same is true for* $f \mid \omega_n$ *and for* $tr(f)$.

Proof. First we remark that it is enough to prove these statements for large weights (by multiplying the modular forms in question by a level one modular form G with integral Fourier coefficients and $G \equiv 1 \bmod p$, if necessary). In the scalar-valued case all these properties can be read off from Sturm [23], relying on earlier work of Shimura [20]. To show the validity of these properties for the vector-valued case one may try to extend Sturm's method to this case. We do not know a good reference for this. A more elementary argument goes as follows: In [8] we proved that theta series with pluriharmonic coefficients generate the full space $M_n^k(\Gamma_0(p), \rho)$. This confirms the ν_p-boundedness, because it holds for such theta series. On the other hand, the space of such pluriharmonic polynomials has a basis consisting of such polynomials with rational coefficients (if ρ is $\mathbb{Q}$-rational); therefore one obtains a solution of the basis problem by modular forms with rational coefficients. For such theta series the action of ω_n can be written down explicitly and has the requested property. Furthermore, in [7] we gave explicit formulas for the trace of such theta series (as rational linear combinations of other theta series, again with pluriharmonic polynomials with rational coefficients). □

Theorem 3.4. *Let* p *be an odd prime and* ρ *a* $\mathbb{Q}$*-rational representation. Then any* $f \in M_n^k(\Gamma_0^n(p), \rho)(\mathbb{Q})$ *is a* p*-adic modular form.*

Proof. We apply Proposition 3.1 to f. From Proposition 3.3(a) we see that the inverse image of f under $U(p)$ has again Fourier coefficients in $\mathbb{Q}^M$. The same is then true for g, $tr(g)$ and finally for F in Proposition 3.1. □

Remark 3.5. If we compare our result with Serre's in the degree one case, our result is slightly weaker: It is possible that the application of $\tilde{U}_n(p)^{-1}$ introduces additional powers of p in the denominator (which weakens our congruences somewhat).

Remark 3.6. Theorem 3.4 also holds for the case of nebentypus $(\Gamma_0^n(p), \chi_p)$ with $\chi_p = \left(\frac{(-1)^{\frac{p-1}{2}} p}{*}\right)$. The proof is almost the same; however, we should use as $\mathcal{K}_{\frac{p-1}{2}}$ a theta series attached to a p-special lattice of rank $p-1$ and determinant p.

Remark 3.7. In our proof of Proposition 3.3 we made essential use of theta series. We mention that this is not really necessary: One can (for arbitrary congruence subgroups $\Gamma_0(N)$ and $\mathbb{Q}$-rational representations ρ) prove that the space $M_n^k(\Gamma_0(N), \rho)$ is spanned by modular forms with rational Fourier coefficients with bounded denominators (and cyclotomic Fourier coefficients with bounded denominators in the other cusps) by combining the line of reasoning of Garrett [12] with the use of suitable differential operators as in [8, 13].

In a recent preprint Ichikawa [15] treats these questions from the point of view of arithmetic algebraic geometry.

3.2 The Case of Higher Level

In the previous argument, we showed that Siegel modular forms for $\Gamma_0^n(p)$ are p-adic modular forms. In this section, we shall generalize this fact to the case of higher level. Namely, we show that *scalar-valued* Siegel modular forms for $\Gamma_0^n(p^m)$ are p-adic modular forms for any $m \in \mathbb{N}$. We modify the arguments used by Serre (cf. Theorem 5.4, [19]) such that they apply in our situation.

Let $F_r \in M_n^r(\Gamma_0^n(p))(\mathbb{Z}_{(p)})$ denote any modular form of weight r and level p such that

$$F_r \equiv 1 \pmod{p}.$$

The existence of such F_r is assured by our previous considerations, provided that r is divisible by $p-1$. Furthermore we put for $i \geq 1$

$$\mathcal{E}_i := \prod_{j=0}^{i-1} F_{k(p-1)p^j}.$$

This is a modular form of weight $k(p^i - 1)$ and level p, i.e.,

$$\mathcal{E}_i \in M_n^{k(p^i-1)}(\Gamma_0^n(p))(\mathbb{Z}_{(p)}).$$

Lemma 3.8. *Assume that $m \geq 2$. For $G \in M_n^k(\Gamma_0^n(p^m))(\mathbb{Q})$ with $\nu_p(G) = 0$, there is $H \in M_n^{k\cdot p}(\Gamma_0^n(p^{m-1}))(\mathbb{Z}_{(p)})$ such that*

$$H \equiv G \pmod{p}.$$

Proof. The modular form

$$H := (G^p) \mid U(p)$$

has the properties requested above. □

Proposition 3.9. *Assume that $m \geq 2$. Then for all $G \in M_n^k(\Gamma_0^n(p^m))(\mathbb{Q})$ and all $i \geq 1$ there exists $H \in M_n^{k\cdot p^i}(\Gamma_0^n(p^{m-1}))(\mathbb{Q})$ such that*

$$G \cdot \mathcal{E}_i \equiv H \pmod{p^i}.$$

Proof. We prove this fact by induction on i. The case $i = 1$ is the lemma above. For arbitrary i we may assume that there is $H_i \in M_n^{k\cdot p^i}(\Gamma_0^n(p^{m-1}))(\mathbb{Q})$ such that

$$H_i \equiv G \cdot \mathcal{E}_i \pmod{p^i}.$$

Then we apply the lemma to the p-integral modular form

$$\tilde{G} := \frac{1}{p^i}(G \cdot \mathcal{E}_i - H_i) \in M_n^{k \cdot p^i}(\Gamma_0^n(p^m))(\mathbb{Z}_{(p)})$$

and we obtain a modular form $\tilde{H} \in M_n^{k \cdot p^{i+1}}(\Gamma_0^n(p^{m-1}))(\mathbb{Q})$ such that $\tilde{G} \cdot F_{kp^i(p-1)} \equiv \tilde{H} \pmod{p}$ and therefore

$$G \cdot \mathcal{E}_i \cdot F_{kp^i(p-1)} \equiv H_i \cdot F_{kp^i(p-1)} + p^i \tilde{H} \pmod{p^{i+1}}.$$

If we put $H := H_i \cdot F_{kp^i(p-1)} + p^i \tilde{H} \in M_n^{k \cdot p^{i+1}}(\Gamma_0^n(p^{m-1}))(\mathbb{Q})$, then we obtain

$$G \cdot \mathcal{E}_{i+1} \equiv H \pmod{p^{i+1}}.$$

This completes the induction. □

In the proposition above, the p^{i-1}th power of $\mathcal{E}_i$ is then congruent one mod p^i and we obtain

Corollary 3.10. *Let G, m, and i as above. Then there exists* $H \in M_n^l(\Gamma_0^n(p^{m-1}))(\mathbb{Q})$ *such that*

$$G \equiv H \pmod{p^i}.$$

As weight l we may choose

$$l = k + k(p^i - 1)p^{i-1}.$$

Now we can state the main theorem of this section.

Theorem 3.11. *Every scalar-valued modular form* $G \in M_n^k(\Gamma_0^n(p^m))(\mathbb{Q})$ *is a* p*-adic modular form.*

Proof. We prove the statement by induction on m. The case $m = 1$ was handled before in the general situation of vector-valued modular forms. For $m \geq 2$, we only need to know that G is congruent modulo an arbitrary given power p^i to a modular form of level p^{m-1}. This is guaranteed by the proposition above. □

Remark 3.12. The result of the theorem above also holds for modular forms of quadratic nebentypus χ_p (with obvious modifications of proof).

Remark 3.13. As mentioned in the introduction, the vector-valued case is more difficult, because we cannot use the pth power of a vector-valued modular form. A good substitute for this is pth symmetric power; this however makes things more complicated, because we change the representation space by this procedure (we refer to [6] for details). In the next sections we avoid this problem by showing only congruences involving modular forms of high level. These results will be used in [6] to construct vector-valued p-adic Siegel modular forms (in an appropriate sense).

4 A Large Class of Theta Operators

In the paper [4], we introduced a theta operator Θ (cf. [4], p. 428) and studied the arithmetic properties. For example, we showed that the algebra of Siegel modular forms mod p is stable under the action of Θ ([4], Corollary 3) and proved that $\Theta(F)$ becomes a p-adic Siegel modular form if F is an ordinary Siegel modular form for Γ^n ([4], Theorem 5). However, the proof of Theorem 5 of that paper contains a defective argument. (In fact, the congruence relation of line 7 of p. 432 is not true in general.) Here we give a complete proof of this theorem in a more general version including vector-valued generalizations of the theta operator. Our proof is based on Rankin-Cohen brackets, our Theorem 3.11 on arbitrary levels p^m, and the existence of modular forms congruent 1 mod p . We point out that the method of proof is new even for elliptic modular forms of level one!

4.1 Rankin-Cohen Brackets and General Theta Operators

Rankin-Cohen operators for Siegel modular forms were investigated by Ibukiyama [13], Eholzer/Ibukiyama [9] and many others. Beyond proving the existence of such operators [13], also explicit formulas were considered. We try to avoid such explicit formulas as much as possible. We fix a polynomial representation $\rho : GL_n(\mathbb{C}) \longrightarrow GL(V_\rho)$ and a weight k. We also assume as before that ρ comes up with a *fixed* matrix realization ($V_\rho = \mathbb{C}^M$).

We assume that we are given certain Rankin-Cohen bilinear operators

$$[,]_{k,l} : M_n^k(\Gamma) \times M_n^l(\Gamma) \longrightarrow M_n^{k+l}(\Gamma, \rho).$$

We consider the case $l = (p-1)p^m$ with $m \geq 0$ varying. We make two assumptions

(R-C 1): $[f, g]_{k,l}$ is a polynomial in the holomorphic derivatives of f and g; more precisely, there is a V_ρ-valued polynomial $\mathcal{P} = \mathcal{P}_{k,l,\rho}(R_1, R_2)$ (with rational coefficients) in the matrix variables $R_1, R_2 \in Sym_n$, homogeneous of a certain degree $\lambda = \lambda(\rho)$ such that

$$[f, g]_{k,l} = (2\pi\sqrt{-1})^{2-\lambda}\mathcal{P}(\partial_1, \partial_2)(f(Z_1) \cdot g(Z_2))_{|Z_1=Z_2=Z}.$$

(R-C 2): If we consider $\mathcal{P}$ as a polynomial in the variables R_2 alone then we can decompose it into homogeneous components of degree j:

$$\mathcal{P} = \sum_{j \geq 0} \mathcal{P}_j.$$

Then $\mathcal{P}_0$ should be independent of $l = (p-1)p^m$; we define a V_ρ-valued differential operator $\Theta_{k,\rho}$ by

$$(2\pi\sqrt{-1})^{-\lambda}\mathcal{P}_0(\partial_1,\partial_2)(f(Z_1)\cdot g(Z_2))_{|Z=Z_1=Z_2} = \Theta_{k,\rho}(f)(Z)\cdot g(Z).$$

Implicit in this assumption is a certain normalization of $[,]_{k,l}$. Note also that $\Theta_{k,\rho}(f)$ has rational Fourier coefficients if f has. This operator is a generalization of the well-known theta operator from Serre [18]. When ρ is the one-dimensional representation $\det^2$, $\Theta = \Theta_{k,\rho}$ is just one considered in the previous paper [4].

4.2 Congruences for $\Theta_{k,\rho}(f)$

Under the conditions above, we show

Theorem 4.1. *For any $f \in M_n^k(\Gamma_0^n(p^r))(\mathbb{Q})$ and any $m \geq 0$ there is m' and a modular form $F \in M_n^{k+(p-1)p^{m-1}}(\Gamma_0^n(p^{m'}),\rho)(\mathbb{Q})$ such that*

$$\Theta_{k,\rho}(f) \equiv F \pmod{p^m}.$$

Proof. We may assume that $\nu_p(f) = 0$. We choose a modular form $F_{p-1} \in M_n^{p-1}(\Gamma_0^n(p))(\mathbb{Z})$ such that

$$F_{p-1} \equiv 1 \pmod{p}.$$

We choose arbitrary nonnegative integers m, m' (to be specified later) and we consider the Rankin-Cohen bracket

$$[f, F_{p-1}^{p^{m-1}} \mid V_{m'-1}]_{k,l}$$

where $(F_{p-1}^{p^{m-1}} \mid V_{m'-1})(Z) := F_{p-1}^{p^{m-1}}(p^{m'-1}Z) \in M_n^{p^{m-1}(p-1)}(\Gamma_0^n(p^{m'}))$.

We investigate the $\mathcal{P}_j(\partial_1,\partial_2)(f(Z_1)\cdot F_{p-1}^{p^{m-1}} \mid V_{m'-1}(Z_2))_{|Z=Z_1=Z_2}$ separately: Clearly

$$(2\pi\sqrt{-1})^{-\lambda}\mathcal{P}_0(\partial_1,\partial_2)(f(Z_1)\cdot F_{p-1}^{p^{m-1}} \mid V_{m'-1}(Z_2))_{|Z=Z_1=Z_2} = \Theta_{k,\rho}(f)(1+p^m G_0)$$

where $G_0 \in \mathbb{Z}_{(p)}[q_{ij}^{-1}, q_{ij}][\![q_{11},\dots,q_{nn}]\!]$.

To study the contributions for $j \geq 1$, we write $\mathcal{P}_j$ as finite sum of certain monomials (when considered as polynomials in the matrix variable R_2):

$$\mathcal{P}_j = \sum_{\alpha} P_{j,\alpha}\cdot Q_{j,\alpha}$$

where the $P_{j,\alpha}$ denote polynomials in the variable R_1 and $Q_{j,\alpha}$ denotes a normalized monomial of degree j in the variable R_2.

Returning to the calculation, we write as

$$f(Z) = \sum_{T} a(T)\boldsymbol{q}^{T}, \quad (F_{p-1}^{p^{m-1}} \mid V_{m'-1})(Z) = 1 + \sum_{T\neq 0} p^{m} b'(T)\boldsymbol{q}^{p^{m'-1}T}.$$

Then

$$(2\pi\sqrt{-1})^{-\lambda} P_{j,\alpha}(\partial_1) \cdot Q_{j,\alpha}(\partial_2)(f(Z_1) \cdot F_{p-1}^{p^{m-1}} \mid V_{m'-1}(Z_2))_{|Z=Z_1=Z_2}$$

$$= \left(\sum_{T} a(T) P_{j,\alpha}(T)\boldsymbol{q}^{T}\right)\left(\sum_{T\neq 0} p^{m} b'(T) p^{j(m'-1)} Q_{j,\alpha}(T)\boldsymbol{q}^{p^{m'-1}T}\right)$$

$$= p^{m+j(m'-1)} \cdot G_{j,\alpha}$$

where the possible p-denominators in the Fourier coefficients of $G_{j,\alpha}$ depend only on the polynomial $P_{j,\alpha}$; note that this polynomial may depend on m, but not on m'. Now we choose m' sufficiently large to guarantee that

$$p^{j(m'-1)} G_{j,\alpha} \in \mathbb{Z}_{(p)}[q_{ij}^{-1}, q_{ij}][\![q_{11}, \ldots, q_{nn}]\!]^{V_\rho}.$$

The upper index V_ρ indicates that we deal with polynomials with values in the vector space V_ρ.

We have achieved in this way that $\Theta_{k,\rho}(f)$ is congruent mod p^m to a vector-valued modular form of level $\Gamma_0^n(p^{m'})$. □

Remark 4.2. The same is true in the case of real nebentypus χ_p (by a suitable modification of the proof above).

Remark 4.3. With a little bit more efforts it is also possible to formulate the theorem above for the case where f is already vector-valued.

Remark 4.4. The simplest possible example for the Rankin-Cohen bracket (for $n = 1, \rho = \det^2$) shows that we cannot avoid using the operator $V_{m'}$ in our proof (this is also a good example to illustrate our normalization):

$$[f, g]_{k,(p-1)p^m} = (2\pi\sqrt{-1})^{-1}\left(f' \cdot g - \frac{k}{(p-1)p^m} f \cdot g'\right).$$

To compensate the denominator p^m, we have to use the level raising operator V_m.

The Theorem 4.1 is not completely satisfying because of the possibly very high level of the modular form F. If we impose the additional condition

(R-C 3): The coefficients of the polynomial $\mathcal{P}$ depend continuously (in the p-adic sense) on the weight l.

We can improve Theorem 4.1 significantly by a slight modification of the proof:

Theorem 4.5. *Assume that the Rankin-Cohen operators satisfy the conditions (R-C 1)–(R-C 3). Then, for any $f \in M_n^k(\Gamma_0^n(p^r))(\mathbb{Q})$ and any $m \geq 0$, there is a weight k' and a modular form $G \in M_n^{k'}(\Gamma_0^n(p^r), \rho)(\mathbb{Q})$ such that*

$$\Theta_{k,\rho}(f) \equiv G \pmod{p^m}.$$

In particular, $\Theta_{k,\rho}(f)$ defines a p-adic modular form.

Proof. We proceed as follows:

We start by the same procedure as before, investigating

$$[f, F_{p-1}^{p^{m-1}} \mid V_{m'-1}]_{k,l},$$

assuring the congruence

$$\Theta_{k,\rho}(f) \equiv F \bmod p^m$$

as in Theorem 4.1.

Now we observe that $F_{p-1}^{p^{m-1}} \mid V_{m'-1}$ is congruent to a modular form H of level one, weight l'; more precisely, there is a weight $l' = (p-1)p^{m-1} + \alpha(p-1)p^{N'}$ such that

$$F_{p-1}^{p^{m-1}} \mid V_{m'-1} \equiv H \bmod p^N$$

Now we consider the bracket

$$[f, H]_{k,l'}$$

instead of the bracket above. If we choose N, N' sufficiently large, we do not get new p-denominators (because of the continuity condition) and we get that

$$\Theta_{k,\rho}(f)$$

is congruent modulo a power of p to a modular form of level p^r.

To see that $\Theta_{k,\rho}$ is p-adic, we have to modify the procedure above still further: We substitute f modulo an arbitrary power of p by a form $\tilde{f}$ of level $\Gamma_0(p)$. Then the G as above is also of level $\Gamma_0(p)$. We may then apply Theorem 3.4. □

4.3 A Remark on $\mathcal{P}_0$

Here we show that $\mathcal{P}_0$ is uniquely determined (up to a scalar factor) by a certain invariance property if ρ is irreducible.

The bilinear differential operator $[,]_{k,l}$ defines a polynomial function f : $Sym_n(\mathbb{C}) \longrightarrow V_\rho$ by

$$\{[e_S, e_T]_{k,l}\}_{|T=0} = f(S)e_S,$$

where S and T are symmetric complex matrices of size n and e_S denotes the function on $\mathbb{H}_n$ defined by

$$Z \longmapsto \exp(2\pi\sqrt{-1}\mathrm{tr}(S \cdot Z)).$$

The invariance properties of $[,]_{k,l}$ give

$$f(A^t SA) = \rho(A)f(S),$$

first for all $A \in GL_n(\mathbb{R})$ and hence also for $A \in GL_n(\mathbb{C})$ and then also for $A \in M_n(\mathbb{C})$. Every symmetric complex matrix S can be written as $S = A^t \cdot 1_n \cdot A$ with $A \in M_n(\mathbb{C})$. Therefore f is determined by $\mathbf{v} := f(1_n) \in V_\rho$. This vector $\mathbf{v}$ is $O(n, \mathbb{C})$ invariant. The space of such invariants is at most one-dimensional if ρ is irreducible. This follows from branching rules (see, e.g., [17]) or from $GL(n, \mathbb{C})$, $O(n, \mathbb{C})$ being a Gelfand pair [1].

Remark 4.6. The problem of (non-)vanishing of $\mathcal{P}_0$ deserves further investigation.

4.4 Theta Operators and Maaß Differential Operators

Let us now give some "basic" examples of such operators: Note that the existence of certain Rankin-Cohen brackets is not sufficient; we must know the nature of the "constant term" $\mathcal{P}_0$. It is desirable to show quite generally that $\mathcal{P}_0$ is different from zero (say, if $l = (p-1)p^m$ is large). We consider here only a simple type: Our exposition follows [11, 25]; we also use some elements of Shimura's theory of nearly holomorphic functions [21].

We fix a decomposition $n = r + s$. For a complex matrix A of size n we denote by $A^{[r]}$ the matrix of size $\binom{n}{r}$ whose entries are the r-minors of A; here we fix an order among the subsets of $\{1, \dots, n\}$ with r elements.

We get an irreducible representation $\rho^{[r]}$ of $GL_n(\mathbb{C})$ on $Sym_{\binom{n}{r}}(\mathbb{C})$ by

$$(A, X) \longmapsto \rho^{[r]}(A)(X) := A^{[r]} \cdot X \cdot (A^{[r]})^t.$$

We denote by $\partial := (\partial_{ij})$ the $n \times n$-matrix defined by partial derivatives on $\mathbb{H}_n$:

$$\partial_{ij} = \begin{cases} \frac{\partial}{\partial z_{ii}} & \text{if } i = j \\ \frac{1}{2}\frac{\partial}{\partial z_{ij}} & \text{if } i \neq j \end{cases}.$$

We consider the differential operator $\partial^{[r]}$ which maps C^∞ functions on $\mathbb{H}_n$ to $Sym_{\binom{n}{r}}(\mathbb{C})$-valued functions. The transformation properties of this operator are well known ([11], p. 214)

$$\partial^{[r]}(h \mid_{\frac{r-1}{2}} M) = \left(\partial^{[r]}h\right) \mid_{\det^{\frac{r-1}{2}} \otimes \rho^{[r]}} M \qquad (M \in Sp_n(\mathbb{R})).$$

For arbitrary weight k we consider now the Maaß-type differential operator

$$(\partial_k^{[r]})h := \det(Y)^{-k+\frac{r-1}{2}} \cdot \partial^{[r]}h \cdot \det(Y)^{k-\frac{r-1}{2}}.$$

This operator changes the automorphy factor from $\det^k$ to $\det^k \otimes \rho^{[r]}$.

For arbitrary holomorphic functions f, g on $\mathbb{H}_n$ we consider

$$(\partial_k^{[r]} f) \cdot g.$$

This is a nearly holomorphic function in the sense of Shimura [21]. We think of f and g to carry a $|_k$ and $|_l$ action of $Sp_n(\mathbb{R})$ (respectively). Then a structure theorem of Shimura [21] on such nearly holomorphic functions (with $Sp_n(\mathbb{R})$ acting by $|_{\det^{k+l} \otimes \rho^{[r]}}$) says that (provided that $k + l$ is large enough)

$$(\partial_k^{[r]} f) \cdot g = B_{k,l,\rho}(f, g) + \Delta$$

where $B_{k,l,\rho}(f, g)$ is a holomorphic function and Δ is a finite sum of images of certain holomorphic functions under differential operators of Maaß-Shimura type. Analytically $B_{k,l,\rho}(f, g)$ is the "holomorphic projection" of $(\partial_k^{[r]} f) \cdot g$ (at least if f, g are actually modular forms satisfying certain growth conditions). Actually, the decomposition above is of purely algebraic nature and an inspection of Shimura's proof shows that the expression for $B_{k,l\rho}(f, g)$ is a bilinear form in the derivatives of f and g (only depending on k, l, ρ), i.e., it is a bilinear differential operator of Rankin-Cohen type. The coefficients of the derivatives of f and g are rational functions of k and l over $\mathbb{Q}$. The representation $\rho^{[r]}$ is irreducible and using the reasoning of 4.3 we see that the part of $B_{k,l,\rho}$, which is free of derivatives of g, must be a multiple of $(\partial^{[r]} f) \cdot g$.

Now we consider the coefficient $q_\rho(k, l)$ of $(\partial^{[r]} f) \cdot g$ in $B_{k,l,\rho}(f, g)$. We observe that this rational function is not identically zero because clearly

$$(\partial_{\frac{r-1}{2}}^{[r]} f) \cdot g = B_{\frac{r-1}{2},l,\rho}(f, g) \qquad (l \gg 0).$$

This implies: There is a finite set M such that for all $k \in \mathbb{Z}\setminus M$ we have $q_\rho(k,l) \neq 0$ for $l \gg 0$.

We can rephrase this as follows:

Proposition 4.7. *For $k \notin M$ there is for $l \gg 0$ a Rankin-Cohen differential operator*

$$[f,g]_{k,l,\rho^{[r]}} = (\partial^{[r]}(f)) \cdot g + \sum_{j\geq 1} \mathcal{P}_j(\partial_1, \partial_2)(f(Z_1)g(Z_2))_{|Z=Z_1=Z_2}$$

satisfying the conditions (R-C 1)–(R-C 3).

For $1 \leq r \leq n$, we put

$$\Theta_{k,\rho^{[r]}} := (2\pi\sqrt{-1})^{-r}\,\partial^{[r]}.$$

Theorem 4.8. *Assume that $1 \leq r \leq n$. For any $f \in M_n^k(\Gamma_0^n(p^m))(\mathbb{Q})$ the formal power series $\Theta_{k,\rho^{[r]}}(f)$ is a p-adic modular form.*

Proof. Using Proposition 4.7 we may apply Theorem 4.5. We observe that by multiplying f by a power of F_{p-1} we may avoid the finite set M of Proposition 4.7; furthermore, in the same way, the assumption $l >> 0$ can be achieved. □

The scalar-valued case ($r = n$) of the theorem above repairs a gap in the proof of Theorem 5 in [4] concerning the ordinary Θ-operator:

Corollary 4.9. *For any $f \in M_n^k(\Gamma_0(p^m))(\mathbb{Q})$ the derivative $\Theta(f)$ is a p-adic modular form.*

4.5 A Combinatorial Approach (Note Added in Proof)

Recently we gave a purely combinatorial approach to the differential operators $B_{k,l,\rho}(f,g)$ in the context of hermitian modular forms [3]. The same approach (avoiding the nonholomorphic differential operators of Sect. 4.4.) also works for the Siegel case as follows:

For $0 \leq \alpha \leq r \leq n$ we define a polynomial in the matrix variables $R, S \in Sym_n(\mathbb{C})$ with values in $Sym_{\binom{n}{r}}(\mathbb{C})$ by

$$(R + \lambda S)^{[r]} = \sum_{\alpha=0}^{r} P_{\alpha,r}(R,S)\lambda^\alpha$$

In the notation of [11, III.6] we have

$$P_{\alpha,r}(R,S) = \binom{r}{\alpha} R^{[\alpha]} \sqcap S^{[r-\alpha]}.$$

We put

$$C_h(s) = s \cdot (s + \frac{1}{2}) \cdot \dots \cdot (s + \frac{h-1}{2}) \qquad (0 \leq h \leq n)$$

Then one can show along the same lines as in [3] that

$$D(f,g) := \sum_{\alpha=0}^{r} (-1)^{\alpha} C_{\alpha}(l - \frac{r-1}{2}) \cdot C_{r-\alpha}(k - \frac{r-1}{2}) \binom{r}{\alpha} \partial^{[\alpha]}(f) \sqcap \partial^{[r-\alpha]}(g)$$

is an explicit realization of the Rankin-Cohen bracket $[f,g]_{k,l,\rho^{[r]}}$.

Acknowledgements Crucial work on this paper was done during our stay at the Mathematisches Forschungsinstitut Oberwolfach under the programme "Research in Pairs"; we continued our work during research visits at Kinki University and Universität Mannheim (respectively); a final revision was done, when the first author held a guest professorship at the University of Tokyo. We thank these institutions for the support. We also thank Dr.Kikuta for pointing out some gaps in our presentation and Professor T.Ichikawa for discussions about p-adic modular forms.

References

1. A. Aizenbud, D. Gurevitch, Some regular symmetric pairs. Trans. AMS **362**, 3757–3777 (2010)
2. S. Böcherer, On the Hecke operator $U(p)$. J. Math. Kyoto Univ. **45**, 807–829 (2005)
3. S. Böcherer, S. Das, On holomorphic differential operators equivariant for $Sp(n,\mathbb{R}) \hookrightarrow U(n,n)$. Int. Math. Res. Notices. doi:10.1093/imrn/rns116
4. S. Böcherer, S. Nagaoka, On mod p properties of Siegel modular forms. Math. Ann. **338**, 421–433 (2007)
5. S. Böcherer, S. Nagaoka, On Siegel modular forms of level p and their properties mod p. Manuscripta Math. **132**, 501–515 (2010)
6. S. Böcherer, S. Nagaoka, On vector-valued p-adic Siegel modular forms (in preparation)
7. S. Böcherer, J. Funke, R. Schulze-Pillot, Trace operator and theta series. J. Number Theory **78**, 119–139 (1999)
8. S. Böcherer, H. Katsurada, R. Schulze-Pillot, On the basis problem for Siegel modular forms with level, in *Modular Forms on Schiermonnikoog* (Cambridge University Press, Cambridge, 2008)
9. W. Eholzer, T. Ibukiyama, Rankin-Cohen type differential operators for Siegel modular forms. Int. J. Math. **9**, 443–463 (1998)
10. G. Faltings, C.L. Chai, *Degeneration of Abelian Varieties* (Springer, New York, 1990)
11. E. Freitag, *Siegelsche Modulfunktionen* (Springer, New York, 1983)
12. P.B. Garrett, On the arithmetic of Hilbert-Siegel cusp forms: Petersson inner products and Fourier coefficients. Invent. Math. **107**, 453–481 (1992)

13. T. Ibukiyama, On differential operators on automorphic forms and invariant pluri-harmonic polynomials. Commentarii Math. Univ. St. Pauli **48**, 103–118 (1999)
14. T. Ichikawa, Congruences between Siegel modular forms. Math. Ann. **342**, 527–532 (2008)
15. T. Ichikawa, Vector-valued p-adic Siegel modular forms (preprint)
16. N.M. Katz, p-adic properties of modular schemes and modular forms, in *Modular Functions of One Variable III*. Lecture Notes in Mathematics, vol. 350 (Springer, New York, 1973), pp. 69–190
17. K. Koike, I. Terada, Young diagrammatic methods for the representation theory of the classical groups of type B_n, C_n, D_n. J. Algebra **107**, 466–511 (1978)
18. J.-P. Serre, Formes modulaires et fonctions zêta p-adiques, in *Modular Functions of One Variable III*. Lecture Notes in Mathematics, vol. 350 (Springer, New York, 1973), pp. 191–268
19. J.-P. Serre, Divisibilité de certaines fonctions arithmétiques. L'Ens. Math. **22**, 227–260 (1976)
20. G. Shimura, On the Fourier coefficients of modular forms in several variables. Nachr. Akad. Wiss. Göttingen Math.-Phys. Kl. II **17**, 261–268 (1975)
21. G. Shimura, Nearly holomorphic functions on hermitian symmetric spaces. Math. Ann. **278**, 1–28 (1987)
22. A. Sofer, p-adic aspects of Jacobi forms. J. Number Theory **63**, 191–202 (1997)
23. J. Sturm, The critical values of zeta functions associated to the symplectic group. Duke Math. J. **48**, 327–350 (1981)
24. H.P.F. Swinnerton-Dyer, On l-adic representations and congruences for coefficients of modular forms, in *Modular Functions of One Variable III*. Lecture Notes in Mathematics, vol. 350 (Springer, New York, 1973), pp. 1–55
25. R. Weissauer, *Stabile Modulformen und Eisensteinreihen*. Lecture Notes in Mathematics, vol. 1219 (Springer, New York, 1986)

On the Existence of Hermitian Self-Dual Extended Abelian Group Codes

Lilibeth Dicuangco-Valdez, Pieter Moree, and Patrick Solé

Mathematics Subject Classification: 11N64, 94B05, 11N37

1 Introduction

Binary duadic codes were first introduced in 1984 by Leon et al. [7] as a generalization of quadratic residue codes. Smid [18] generalized these results further by defining duadic codes over arbitrary finite fields in terms of a *splitting* of the length of the code. The Q-codes of Pless [12] are then duadic codes over $\mathbb{F}_4$ in this setting.

Quadratic residue codes have also been generalized in a different direction (i.e., see [19]). In this approach, quadratic residue codes are defined as ideals of abelian group algebras, which is a generalization of cyclic codes. Rushanan [15] proceeded to define duadic codes in this setting.

In [2], Ding, Kohel and Ling defined split group codes as ideals of abelian group algebras. Their construction makes use of a *splitting* of the abelian group. Under this definition, split group codes are seen as a generalization of duadic codes.

L. Dicuangco-Valdez (✉)
Institute of Mathematics, University of the Philippines, Diliman, Quezon City, 1101 Philippines
e-mail: ldicuangco@math.upd.edu.ph

P. Moree
Max-Planck-Institut, Vivatsgasse 7, D-53111 Bonn, Germany
e-mail: moree@mpim-bonn.mpg.de

P. Solé
Department of Comelec, CNRS, LTCI, Telecom- Paristech, 46 rue Barrault, 75013 Paris, France
e-mail: sole@enst.fr

B. Heim et al. (eds.), *Automorphic Forms*, Springer Proceedings
in Mathematics & Statistics 115, DOI 10.1007/978-3-319-11352-4_5

In this paper, we consider the finite field $F = \mathbb{F}_{q^2}$ and an abelian group G of order n such that $\gcd(n, q) = 1$. Following the treatment in [2], we work with the dual group G^* of G and consider the group algebra $F[G^*]$. We prove that every ideal code in $F[G^*]$ whose extension by a suitable parity check is Hermitian self-dual is a split group code (Corollary 3.7). We then give sufficient and necessary conditions on the order of the group G for the existence of Hermitian self-dual extended ideal codes (Theorem 3.11). We conclude the paper by deriving asymptotic estimates on $HSD(x)$, the number of non-isomorphic abelian groups of order less than x for which a Hermitian self-dual extended ideal code exists (Theorem 4.3).

2 Preliminaries

Let R be a finite commutative ring with unity. Let G be its underlying finite abelian group written additively. Denote the order and exponent of G by n and m, respectively. Let q be a power of a prime p_1 such that $\gcd(n, q) = 1$ or equivalently, $\gcd(m, q) = 1$. Let $F = \mathbb{F}_{q^2}$. Let K be the smallest extension of F containing all the mth roots of unity.

Let G^* be the set of all characters of G into K. The groups G and G^* are isomorphic. Let $K[G^*]$ be the group algebra of K over G^*. The elements of $K[G^*]$ are the sums $\sum_{\psi \in G^*} a_\psi \psi$, where the a_ψ's are elements of K. An ideal I of $K[G^*]$ is called an *ideal code*.

The dimension of the commutative group algebra $K[G^*]$ over K is n. The group of units of this group algebra $K[G^*]$ contains a subgroup isomorphic to G^*. For any character ψ in G^*, we also denote by ψ the corresponding element in $K[G^*]$.

If $x \in G$ and $f = \sum_{\psi \in G^*} a_\psi \psi \in K[G^*]$, define $f(x) = \sum_{\psi \in G^*} a_\psi \psi(x)$. Thus we can view the elements of $K[G^*]$ as functions from G to K.

An element e of a ring is called an idempotent if $e^2 = e$. An idempotent is called primitive if for every other idempotent f, either $ef = e$ or $ef = 0$. The primitive idempotents of $K[G^*]$ are the elements

$$e_x = \frac{1}{n} \sum_{\psi \in G^*} \psi(x)^{-1} \psi,$$

for each element x of G. The group algebra $K[G^*]$ decomposes as a direct sum $\oplus_{x \in G} K e_x$. If $f \in K[G^*]$, then f has the form

$$f = \sum_{x \in G} f(x) e_x.$$

Every idempotent e in $K[G^*]$ can be uniquely written in the form

$$e = \sum_{x \in X} e_x,$$

for some non-empty subset X of G.

Let X be a non-empty subset of G and define the ideal

$$I_X = \{f \in K\left[G^*\right] \mid f(x) = 0 \text{ for all } x \in X\}.$$

For every ideal I in $K[G^*]$, there is a unique proper subset X of G such that $I = I_X$ and I is generated by the idempotent $e = \sum_{x \notin X} e_x$.

2.1 Split Group Codes

Let s be an element of R. Consider the endomorphism of G given by $\tau_s : x \longrightarrow sx$. This induces a map μ_s on G^* given by $\mu_s(\psi) = \psi \circ \tau_s$ for each element ψ of G^*. This extends to a map on $K[G^*]$, also denoted by μ_s, defined by $\mu_s(f) = f \circ \tau_s$ for all $f \in K[G^*]$. That is, for $f \in K[G^*]$, $\mu_s(f)(x) = f(sx)$ for every $x \in G$.

A *splitting* of G over Z is a triple (Z, X_0, X_1) which gives a partition $G = Z \cup X_0 \cup X_1$ such that there exists an invertible element s of R with $\tau_s(X_0) = X_1$ and $\tau_s(X_1) = X_0$. Under these conditions, s is said to *split* the triple (Z, X_0, X_1). In addition, an invertible element r of R is said to *stabilize* the splitting if $\tau_r(X_0) = X_0$ and $\tau_r(X_1) = X_1$.

Given a splitting (Z, X_0, X_1), let $C_0(K)$ be the ideal I_{X_0} over K and let $C_1(K)$ be the ideal I_{X_1} over K. The ideal $C_0(K)$ is defined as the *split group code* associated to the splitting, and the ideal $C_1(K)$ is called the *conjugate split group code*. The following notations are used to denote some special subcodes: $C_0^Z(K) = I_{Z \cup X_0}$, $C_1^Z(K) = I_{Z \cup X_1}$ and $C_Z(K) = I_{X_0 \cup X_1}$.

Let s be an invertible element of R. The element s is said to *split* the group code $C_0(K)$ if $\mu_s(C_0(K)) = C_1(K)$ and $\mu_s(C_1(K)) = C_0(K)$, while s is said to *stabilize* the code $C_0(K)$ if $\mu_s(C_0(K)) = C_0(K)$ and $\mu_s(C_1(K)) = C_1(K)$.

It was shown in [2] that if s is a unit in R then a split group code $C_0(K)$ is split or stabilized by s if and only if s splits or stabilizes (Z, X_0, X_1), respectively.

Let (Z, X_0, X_1) be a splitting with associated split group code $C_0(K)$. Then the codes $C_0(K)$ and $C_1(K)$ are generated by the idempotents $e = \sum\limits_{x \notin X_0} e_x$ and $f = \sum_{x \notin X_1} e_x$, respectively. Thus $C_0(K)$ and $C_1(K)$ have dimension $(n + |Z|)/2$. Also the codes $C_0^Z(K)$, $C_1^Z(K)$ and $C_Z(K)$ are generated by $\sum\limits_{x \in X_1} e_x$, $\sum\limits_{x \in X_0} e_x$, and $\sum\limits_{z \in Z} e_z$ which imply that the subcodes $C_0^Z(K)$ and $C_1^Z(K)$ each have dimension

$(n - |Z|)/2$ while the subcode $C_Z(K)$ has dimension $|Z|$. If the splitting is given by s, then μ_s induces an equivalence of $C_0(K)$ with its conjugate $C_1(K)$ and of the subcode $C_0^Z(K)$ with $C_1^Z(K)$. Note also that $K[G^*]$ decomposes as a direct sum $C_Z(K) \oplus C_0^Z(K) \oplus C_1^Z(K)$ (see [2]).

2.2 Split Group Codes Over $\mathbb{F}_{q^2}$

In the previous section, the split group codes are defined over the field K which is assumed to contain all the mth roots of unity. In this paper, we want our split group codes to be defined over the subfield $F = \mathbb{F}_{q^2}$ without requiring $\mathbb{F}_{q^2}$ to contain any mth roots of unity. First, let us consider some results regarding split group codes.

Let $V = V(K)$ be a vector subspace of $K^n = K[G^*]$. Define $V(F) = V(K) \cap F^n$. Clearly, $\dim_F(V(F)) \leq \dim_K(V(K))$. If equality holds then we say that V is *defined over the field* F.

If the vector subspace $C_0(K)$ of $K^n = K[G^*]$ is defined over F, we simply write C_0 for the subcode $C_0(F) = C_0(K) \cap F^n$ in $F^n = F[G^*]$. In this case, we call C_0 *the split group code over* F. Similarly, we write C_1, C_0^Z, C_1^Z and C_Z for the other codes defined over F.

Note that $(m, q) = 1$ by assumption, so the integer $-q$ as an element of the finite ring R is invertible and τ_{q^2} is a well-defined automorphism of G. The action of the group generated by τ_{q^2} on the elements of G partitions G into disjoint orbits. These $\langle \tau_{q^2} \rangle$-orbits play the same role as the cyclotomic cosets for the cyclic codes.

It can be shown that the idempotents of $F[G^*]$ are those e in $K[G^*]$ of the form $e = \sum_{x \in Y} e_x$, where Y is a union of $\langle \tau_{q^2} \rangle$-orbits in G. An idempotent e in $F[G^*]$ is primitive if and only if $Y = \langle \tau_{q^2} \rangle x$ for some $x \in G$. Also, if $\{e_1, e_2, \ldots, e_r\}$ is the set of all primitive idempotents of $F[G^*]$ then every nonzero ideal I of $F[G^*]$ is generated by $e = \sum_{i \in T} e_i$ where T is a non-empty subset of $\{1, 2, \ldots, r\}$.

3 Hermitian Duality and Extended Ideal Codes in $F[G^*]$

In this section, we present our results concerning the Hermitian orthogonality of ideal codes in $F[G^*]$. We prove that every Hermitian self-orthogonal ideal code in $F[G^*]$ is a subcode of a split group code for some splitting of G given by $-q$. We then proceed to define an extension of ideal codes and determine conditions for the extended code to be Hermitian self-dual. We also give necessary and sufficient conditions on the order of the abelian group G for the existence of Hermitian self-dual extended ideal codes.

3.1 Hermitian Orthogonality of Ideal Codes in $F[G^*]$

Let $f = \sum_{\psi \in G^*} a_\psi \psi$ and $g = \sum_{\psi \in G^*} b_\psi \psi$ be elements of $F[G^*]$. The *Hermitian inner product* between f and g is defined as $\langle f, g \rangle_H = \sum_{\psi \in G^*} a_\psi {b_\psi}^q$.

Let C be a code in $F[G^*]$. The Hermitian dual of C is the ideal $C^{\perp_H} = \{f \in F[G^*] \mid \langle f, g \rangle_H = 0 \text{ for all } g \in C\}$. The code C is said to be *Hermitian self-orthogonal* if $C \subseteq C^{\perp_H}$ and is *Hermitian self-dual* if $C = C^{\perp_H}$.

Theorem 3.5 states the main result of this section. It is a generalization of Proposition 4.4 in [1] to split group codes. We first prove some basic results concerning the Hermitian duals of split group codes.

Proposition 3.1. *Let $f = \sum_{\psi \in G^*} a_\psi \psi$ and $g = \sum_{\psi \in G^*} b_\psi \psi$ be elements of $F[G^*]$. Then the Hermitian inner product of f and g is*

$$\langle f, g \rangle_H = \frac{1}{n} \sum_{x \in G} f(x) g(-q^{-1}x)^q$$

Proof. This follows from the orthogonality relations. □

Proposition 3.2. *Let C be an ideal in $K[G^*]$ which is defined over F. Suppose $C = I_X$ for some non-empty subset X of G. Then $C^{\perp_H} = I_{X'}$ where $X' = G \setminus \tau_{-q}(X)$.*

Proof. Note that X is a union of $\langle \tau_{q^2} \rangle$-orbits. Let $X' = G \setminus \tau_{-q}(X)$. Let $f \in I_{X'}$ and $g \in C = I_X$. By Proposition 3.1, $\langle f, g \rangle_H = \frac{1}{n} \sum_{x \in G} f(x)(g(-q^{-1}x))^q$. Since X and $\tau_{-q}(X)$ are unions of $\langle \tau_{q^2} \rangle$-orbits and clearly x and $q^2 x$ belong to the same τ_{q^2}-orbit, it follows that $-q^{-1}x \in X$ if and only if $(-q)^2(-q^{-1})x \in X$ if and only if $-qx \in X$ if and only if $(-q)(-q)x \in \tau_{-q}(X)$ if and only if $x \in \tau_{-q}(X)$. Thus $g(-q^{-1}x) = 0$ for all $x \in \tau_{-q}(X)$. Since $f \in I_{X'}$, $f(x) = 0$ for every $x \in X' = G \setminus \tau_{-q}(X)$. Therefore $\sum_{x \in G} f(x)(g(-q^{-1}x))^q = 0$, implying that $\langle f, g \rangle_H = 0$. Thus $I_{X'} \subseteq C^{\perp_H}$. Comparing dimensions, we get $C^{\perp_H} = I_{X'}$. □

Proposition 3.3. *Let (Z, X_0, X_1) be a splitting of G over Z where Z, X_0 and X_1 are unions of $\langle \tau_{q^2} \rangle$-orbits. The ring element $-q$ splits or stabilizes C_0 if and only if $-q^{-1}$ splits or stabilizes C_0, respectively.*

Proposition 3.4. *Let (Z, X_0, X_1) be a splitting of G over Z. Assume that Z, X_0 and X_1 are unions of $\langle \tau_{q^2} \rangle$-orbits. Suppose Z is stabilized by τ_{-q}. Then $C_Z^{\perp_H} = C_0^Z \oplus C_1^Z$. If $-q$ splits C_0, then $C_0^{\perp_H} = C_0^Z$. If $-q$ stabilizes C_0, then $C_0^{\perp_H} = C_1^Z$.*

Proof. Note that $\tau_{-q^{-1}}(Z) = \tau_{(-q)^2}(\tau_{-q^{-1}}(Z)) = \tau_{-q}(Z)$ and by assumption, $\tau_{-q}(Z) = Z$. Let $f \in C_Z = I_{X_0 \cup X_1}$ and let $g \in C_0^Z \oplus C_1^Z$. Then

$$\langle f, g \rangle_H = \frac{1}{n} \sum_{x \in G} f(x)(g(-q^{-1}x))^q = 0,$$

since $f(x) = 0$ for all $x \in X_0 \cup X_1$ and $g(-q^{-1}x) = 0$ for all $x \in Z$. Thus $C_0^Z \oplus C_1^Z \subseteq C_Z^{\perp_H}$. Comparing dimensions, we have $C_Z^{\perp_H} = C_0^Z \oplus C_1^Z$.

Suppose $-q$ splits C_0. Let $f \in C_0$ and $g \in C_1^Z$. Note that $\tau_{-q^{-1}}(Z) = Z$. By assumption, $\tau_{-q}(X_1) = X_0$, or equivalently, $\tau_{-q^{-1}}(X_1) = X_0$. Clearly $f(x) = 0$ for all $x \in X_0$ and $g(-q^{-1}x) = 0$ for all $x \in Z \cup X_1$. Thus

$$\langle f, g\rangle_H = \frac{1}{n}\sum_{x\in G} f(x)(g(-q^{-1}x))^q = 0,$$

implying that $C_0^Z \subseteq C_0^{\perp_H}$. Comparing dimensions, we have $C_0^{\perp_H} = C_0^Z$.

Suppose $-q$ stabilizes C_0. Then $\tau_{-q^{-1}}(X_1) = \tau_{-q}(X_1) = X_1$. Let $f \in C_0$ and $g \in C_1^Z$. Clearly $f(x) = 0$ for all $x \in X_0$ and $g(-q^{-1}x) = 0$ for all $x \in Z \cup X_1$. Thus

$$\langle f, g\rangle_H = \frac{1}{n}\sum_{x\in G} f(x)(g(-q^{-1}x))^q = 0,$$

implying that $C_1^Z \subseteq C_0^{\perp_H}$. Comparing dimensions, we have $C_1^{\perp_H} = C_0^Z$. □

We now prove the main result of this section.

Theorem 3.5. *Let $C = I_X$ be an ideal in $F[G^*]$. Then C is Hermitian self-orthogonal if and only if $C = C_0^Z$ for some splitting (Z, X_0, X_1) of G which is split by $-q$ (that is, C is a subcode of a split group code which is split by $-q$).*

Proof. ($\Longleftarrow$) Suppose (Z, X_0, X_1) is a splitting of G which is split by $-q$. Let $C = C_0^Z$. By Proposition 3.4, $C^{\perp_H} = (C_0^Z)^{\perp_H} = C_0 \supseteq C_0^Z = C$.
($\Longrightarrow$) Let $C^{\perp_H} = I_{X'}$. By Proposition 3.2, $X' = G \setminus \tau_{-q}(X)$. By assumption, $C \subseteq C^{\perp_H}$. Thus $X' \subseteq X$, or $G \setminus \tau_{-q}(X) \subseteq X$. Note that both X and X' are unions of $\langle\tau_{q^2}\rangle$-orbits. Write $X = Z \cup X_0$ with

$$\tau_{-q}(Z) = Z \text{ and } Z \cap X_0 = \emptyset. \tag{1}$$

Since τ_{-q} either fixes a $\langle\tau_{q^2}\rangle$-orbit or it sends it to another $\langle\tau_{q^2}\rangle$-orbit, Equation (1) says that we are choosing Z as the union of $\langle\tau_{q^2}\rangle$-orbits contained in X which are fixed by τ_{-q} and X_0 as the complement of Z in X, so that clearly $\tau_{-q}(X_0) \cap X_0 = \emptyset$.

We first show that neither Z nor X_0 is empty. If $0 \notin X$ then $0 \notin \tau_{-q}(X)$, which implies that $0 \in G \setminus \tau_{-q}(X)$. But $G \setminus \tau_{-q}(X) \subseteq X$, implying that $0 \in X$, a contradiction. Thus $0 \in X$, and so by our choice of partition of X, $0 \in Z$ proving that $Z \neq \emptyset$. If $X_0 = \emptyset$ then $X = Z$ and $\tau_{-q}(X) = X$. Thus $X' = G \setminus \tau_{-q}(X) = G \setminus X$ which implies that $X' \cap X = \emptyset$, a contradiction since $X' \subseteq X$. Thus $X_0 \neq \emptyset$.

Note that $X' = G \setminus \tau_{-q}(X) = G \setminus Z \cup \tau_{-q}(X_0)$. Since $X_0 \cap Z = X_0 \cap \tau_{-q}(X_0) = \emptyset$, we have $X_0 \subseteq X'$. Consider $\tau_{-q}(X') = G \setminus X$. Since $X' \subseteq X$, it follows that $\tau_{-q}(X') \cap X' \subseteq \tau_{-q}(X') \cap X = \emptyset$, or $\tau_{-q}(X') \cap X' = \emptyset$. So τ_{-q} does not fix any $\langle\tau_{q^2}\rangle$-orbit contained in X'. Thus $X' \subseteq X_0$. Hence $X' = X_0$.

Let $X_1 = \tau_{-q}(X')$. Then (Z, X_0, X_1) gives a splitting of G such that $\tau_{-q}(Z) = Z$, $\tau_{-q}(X_0) = X_1$ and $\tau_{-q}(X_1) = X_0$ and $\mathcal{C} = I_X = I_{Z \cup X_0} = C_0^Z$. □

3.2 Extensions of Ideal Codes in $F[G^*]$

Using the terminology of split group codes, duadic codes are easily seen to be split group codes for splittings of G over $Z = \{0\}$ where G is cyclic (see Example III.1 of [2]). In [1], we defined an extension for an odd-like duadic code and gave a sufficient condition for the extended code to be Hermitian self-dual. In this section, we consider split group codes for the abelian group G with splittings over $Z = \{0\}$ and derive analogous results regarding Hermitian self-duality of the extended split group codes.

Let the order n of the abelian group G be odd. Consider the equation

$$\frac{1}{n} + \gamma^{q+1} = 0, \tag{2}$$

which is solvable in $\mathbb{F}_{q^2}$. Let γ be a solution to (2). For each $f \in F[G^*]$, define $\tilde{f} = (f, -\gamma f(0)) \in F[G^*] \times F$. If C is a code in $F[G^*]$, then the extended code $\widetilde{C}$ is defined as the subspace

$$\widetilde{C} = \{\tilde{f} = (f, -\gamma f(0)) \mid f \in C\} \subseteq F\left[G^*\right] \times F.$$

Proposition 3.6. *Let $(Z = \{0\}, X_0, X_1)$ be a splitting of G where Z, X_0 and X_1 are unions of $\langle \tau_{q^2} \rangle$-orbits. Let C_0 be the corresponding split group code defined over $F = \mathbb{F}_{q^2}$.*

1. *The extended codes $\widetilde{C_0}$ and $\widetilde{C_1}$ are equivalent.*
2. *If $-q$ splits C_0, then $\widetilde{C_0}^{\perp_H} = \widetilde{C_0}$ and $\widetilde{C_1}^{\perp_H} = \widetilde{C_1}$.*
3. *If $-q$ stabilizes C_0, then $\widetilde{C_0}^{\perp_H} = \widetilde{C_1}$ and $\widetilde{C_1}^{\perp_H} = \widetilde{C_0}$.*

Proof. The equivalence of $\widetilde{C_0}$ and $\widetilde{C_1}$ is obvious.

Suppose $-q$ splits C_0. Let f, g be elements of C_0. From Proposition 3.3, $-q^{-1}$ also splits C_0. Thus $\mu_{-q^{-1}}(g) \in C_1$. Note that $-q^{-1}x \in X_0$ if and only if $x \in \tau_{-q^{-1}}(X_0) = X_1$. It follows that $f(x) = 0$ for all $x \in X_0$ and $g(-q^{-1}x) = 0$ for all $x \in X_1$. Hence using Proposition 3.1 we get

$$\langle f, g \rangle_H = \frac{1}{n} f(0) g(0)^q = -\gamma^{q+1} f(0) g(0)^q.$$

Thus $\langle \tilde{f}, \tilde{g} \rangle_H = 0$ and $\widetilde{C_0}^{\perp_H} = \widetilde{C_0}$. By a similar argument it can be shown that $\widetilde{C_1}^{\perp_H} = \widetilde{C_1}$.

Suppose $-q$ stabilizes C_0. Let $f \in C_0$ and let $g \in C_1$. The element $-q^{-1}$ also stabilizes C_0 and $\mu_{-q^{-1}}(g) \in C_0$. Again, $-q^{-1}x \in X_1$ if and only if $x \in \tau_{-q^{-1}}(X_1) = X_1$. So $f(x) = 0$ for all $x \in X_0$ and $g(-q^{-1}x) = 0$ for all $x \in X_1$. Thus

$$\langle f, g\rangle_H = \frac{1}{n} f(0)g(0)^q = -\gamma^{q+1} f(0)g(0)^q,$$

and so $\langle \tilde{f}, \tilde{g}\rangle_H = 0$ and $\widetilde{C_0}^{\perp_H} = \widetilde{C_1}$. Similarly, $\widetilde{C_1}^{\perp_H} = \widetilde{C_0}$. □

Corollary 3.7. *Let $C = I_X$ be a group code defined over F. The extended code $\widetilde{C}$ is Hermitian self-dual if and only if C is a split group code for some splitting $(Z = \{0\}, X_0, X_1)$ of G by $-q$.*

Proof. ($\Longleftarrow$) This follows directly from the previous proposition.
($\Longrightarrow$) Since $\widetilde{C}$ is Hermitian self-dual, the dimension of C is $\frac{n+1}{2}$ and so C cannot be Hermitian self-orthogonal. This fact combined with the assumption that $\widetilde{C}$ is Hermitian self-dual implies that $0 \notin X$. Let $C_e = I_{X\cup\{0\}}$. This subcode C_e is Hermitian self-orthogonal and has dimension $\frac{n-1}{2}$. By Theorem 3.5, C_e is a subcode of a split group code which is split by $-q$, that is, $C_e = C_0^Z$ for some splitting (Z, X_0, X_1) of G by $-q$. Since dim $C_e = \frac{n-1}{2}$ and dim $C_0^Z = \frac{n-|Z|}{2}$, it follows that $Z = \{0\}$. Hence $X = X_0$ and C is a split group code of G which is split by $-q$. □

3.3 Existence of Hermitian Self-Dual Extended Ideal Codes

In view of Theorem 3.5 and Corollary 3.7, it is natural to ask under what conditions we obtain splittings over $Z = \{0\}$ of an abelian group G by $-q$. Such conditions would guarantee existence of Hermitian self-orthogonal codes and Hermitian self-dual extended codes in $F[G^*]$.

Define $ord_r(q)$ to be the smallest positive integer t such that $q^t \equiv 1 \pmod r$. (The multiplicative order plays a role in many mathematical settings, for a survey see, e.g., Moree [8].) If l is a positive odd integer relatively prime to q then l is said to be split by $-q$ over $\mathbb{F}_{q^2}$ if and only if the set $X = \{1, 2, \ldots, l\}$ has a partition $X = X_0 \cup X_1$ such that $(-q)X_0 = X_1$ and $(-q)X_1 = X_0$, where the multiplication is read modulo l.

Proposition 3.8 ([1]). *Let l be a positive odd integer which is relatively prime to q. The integer l has a splitting by $-q$ if and only if $ord_r(q) \not\equiv 2 \pmod 4$ for every prime r dividing l.*

Theorem 3.9. *Let G be an abelian group of order n. The group G has a splitting over $Z = \{0\}$ given by $-q$ if and only if $ord_r(q) \not\equiv 2 \pmod 4$ for every prime r dividing n.*

Proof. The abelian group G is isomorphic to a unique product of cyclic groups of the form

$$\mathbb{Z}_{m_1} \times \mathbb{Z}_{m_2} \times \ldots \times \mathbb{Z}_{m_s},$$

where m_i divides m_{i+1} for $i = 1, 2, \ldots, s-1$, and $m_s = m$, where m denotes the exponent of G.

If each summand $\mathbb{Z}_{m_i}$ has a splitting over $Z = \{0\}$ given by $-q$, then G also has a splitting over $Z = \{0\}$ given by $-q$. Indeed, if $(\{0\}, X_0^{(i)}, X_1^{(i)})$ is a splitting by $-q$ of $\mathbb{Z}_{m_i}$ for each $i = 1, 2, \ldots, s$, then letting

$$\begin{aligned} X_t = \; & X_t^{(1)} \times \mathbb{Z}_{m_2} \times \mathbb{Z}_{m_3} \times \cdots \times \mathbb{Z}_{m_s} \\ & \cup \{0\} \times X_t^{(2)} \times \mathbb{Z}_{m_3} \times \cdots \times \mathbb{Z}_{m_s} \\ & \cup \{0\} \times \{0\} \times X_t^{(3)} \times \mathbb{Z}_{m_4} \times \cdots \times \mathbb{Z}_{m_s} \\ & \vdots \\ & \cup \{0\} \times \{0\} \times \{0\} \times \cdots \times \{0\} \times X_t^{(s)} \end{aligned}$$

for $t = 0, 1$, $(\{0\}, X_0, X_1)$ gives a splitting of G by $-q$. Conversely, suppose that G has a splitting over $Z = \{0\}$ given by $-q$. Let $\mathbb{Z}_i = \{0\} \times \{0\} \times \ldots \times \mathbb{Z}_{m_i} \times \{0\} \times \ldots \{0\}$ be the subgroup of G isomorphic to $\mathbb{Z}_{m_i}$. If $(Z = \{0\}, X_0, X_1)$ gives a splitting for G by $-q$, then $(Z = \{0\}, \mathbb{Z}_i \cap X_0, \mathbb{Z}_i \cap X_1)$ gives a splitting for $\mathbb{Z}_i$ given by $-q$. Hence using Proposition 3.8, G has a splitting over $Z = \{0\}$ given by $-q$ if and only if each summand $\mathbb{Z}_{m_i}$ has a splitting over $Z = \{0\}$ given by $-q$ if and only if m_i is split by $-q$ for all $i = 1, 2, \ldots, m$ if and only if $ord_r(q) \not\equiv 2 \pmod 4$ for every prime r dividing m_i for all $i = 1, 2, \ldots, s$ if and only if $ord_r(q) \not\equiv 2 \pmod 4$ for every prime r dividing m. But the primes dividing m are precisely the primes dividing n. Thus G has a splitting over $Z = \{0\}$ given by $-q$ if and only if $ord_r(q) \not\equiv 2 \pmod 4$ for every prime r dividing n. □

Example. Let $G = \mathbb{Z}_3 \times \mathbb{Z}_9$ and $\mathbb{F}_{4^2} = \mathbb{F}_{16}$. Note that $ord_3(4) = 1$ and by Theorem 3.9 the abelian group G has a partition which is split by -4. The cyclic groups $\mathbb{Z}_3$ and $\mathbb{Z}_9$ have splittings by the multiplier μ_{-4} given by $(\{0\}, A_1, A_2)$ and $(\{0\}, B_1 \cup B_3, B_2 \cup B_6)$, respectively, where A_i is the 16-cyclotomic coset modulo 3 containing i and B_j is the 16-cyclotomic coset modulo 9 which contains j. Define $C_{(i,j)}$ as the orbit of τ_{16} in G containing (i, j). Letting $X_0 = C_{(1,0)} \cup C_{(1,1)} \cup C_{(1,2)} \cup C_{(1,3)} \cup C_{(1,6)} \cup C_{(0,1)} \cup C_{(0,3)}$ and $X_1 = C_{(2,0)} \cup C_{(2,1)} \cup C_{(2,2)} \cup C_{(2,3)} \cup C_{(2,6)} \cup C_{(0,2)} \cup C_{(0,6)}$, the set $(\{(0,0)\}, X_0, X_1)$ gives a splitting of G by -4. Notice that this partition can be obtained from the splittings of $\mathbb{Z}_3$ and $\mathbb{Z}_9$ as described in the proof.

We remark that $ord_r(q) \not\equiv 2 \pmod 4$ means that either $ord_r(q)$ is odd or $ord_r(q)$ is doubly even. It can easily be verified that $ord_r(q)$ is doubly even if and only if $ord_r(q^2)$ is even. Thus Theorem 3.9 can be restated as:

Theorem 3.10. *Let G be an abelian group of order n. The group G has a splitting over $Z = \{0\}$ given by $-q$ if and only if for every prime r dividing n, either $ord_r(q)$ is odd or $ord_r(q^2)$ is even.*

Thus we get the following condition for the existence of extended ideal codes of $F[G^*]$ which are Hermitian self-dual.

Theorem 3.11. *Let G be an abelian group of order n. An ideal code of $F[G^*]$ whose extension is Hermitian self-dual exists if and only if for every prime r dividing n, either $ord_r(q)$ is odd or $ord_r(q^2)$ is even.*

Proof. This is a direct consequence of Corollary 3.7 and Theorem 3.10. □

4 Counting Hermitian Self-Dual Extended Abelian Group Codes

Theorem 3.11 raises the question of counting the number of non-isomorphic abelian groups of order $\leq x$, with x an arbitrary positive real number, for which an ideal code of $F[G^*]$ whose extension is Hermitian self-dual exists. An estimate for this quantity, for x tending to infinity, is provided by Theorem 4.3.

Let $q = p_1^t$ be a prime power and let $\mathcal{P}_q$ be the set of primes $r \neq p_1$ for which $\mathrm{ord}_r(q)$ is odd or $\mathrm{ord}_r(q^2)$ is even. Let $\mathcal{P}_q(x) = \#\{p \leq x : p \in \mathcal{P}_q\}$. The primes r not counted, that is, the primes r such that $\mathrm{ord}_r(q) \equiv 2 (\mathrm{mod}\ 4)$ or the prime $r = p_1$ can be shown (see [1]), to have a natural density $\delta(q)$ that is given by the following formula (with λ the exponent of 2 in the factorization of t):

$$\delta(q) = \delta(p_1^t) = \begin{cases} 7/24 & \text{if } p_1 = 2 \text{ and } \lambda = 0; \\ 1/3 & \text{if } p_1 = 2 \text{ and } \lambda = 1; \\ 2^{-\lambda-1}/3 & \text{if } p_1 = 2 \text{ and } \lambda \geq 2; \\ 2^{-\lambda}/3 & \text{if } p_1 \neq 2. \end{cases}$$

It can be proved (see [1, Lemma A.3]) that

$$\mathcal{P}_q(x) = (1 - \delta(q))\mathrm{Li}(x) + O_q\left(\frac{x(\log\log x)^4}{\log^3 x}\right), \qquad (3)$$

where the subscript q indicates that the implied constant may depend on q and $Li(x) = \int_2^x dt/\log t$ denotes the logarithmic integral.

Let $\mathcal{G}_q$ be the subsemigroup of the natural numbers generated by the primes in $\mathcal{P}_q$. Let $HSD(x)$ count the number of non-isomorphic abelian groups of order n with $(n, q) = 1$ and $n \leq x$ for which an ideal code of $F[G^*]$ whose extension is Hermitian self-dual exists. Then by Theorem 3.11 we have that

$$HSD(x) = \sum_{n \le x,\, n \in \mathcal{G}_q} a(n),$$

where $a(n)$ denotes the number of non-isomorphic abelian groups having n elements.

Thus we are naturally led to study the behaviour of $a(n)$ on subsemigroups $\mathcal{G}$ of the natural numbers. For our purposes it is enough to restrict to subsemigroups $\mathcal{G}$ that are generated by a set $\mathcal{P}$ of primes satisfying

$$\mathcal{P}(x) = \tau \mathrm{Li}(x) + E_{\mathcal{P}}(x), \tag{4}$$

where $0 < \tau < 1$ and the error term $E_{\mathcal{P}}(x)$ satisfies $E_{\mathcal{P}}(x) = O(x \log^{-1-\gamma} x)$ for some positive real constant γ. In case $\gamma > 1$, a sharper result can be obtained (cf. Theorem 4.1).

Although the literature on $a(n)$ is quite extensive, the latter problem does not seem to have been studied before. Before delving into it, we recall some relevant facts on the behaviour of $a(n)$.

4.1 Counting Non-isomorphic Abelian Groups

It is easy to see that $a(n)$ is a multiplicative function with the property that $a(p^k) = P(k)$ for every prime p and every integer $k \ge 1$, where $P(k)$ denotes the number of unrestricted partitions of k. Thus $a(p^k)$ does not depend on p but only on k, so that $a(n)$ is a "prime-independent" multiplicative function.

An analytic approach to $a(n)$ is based on the fact that the Dirichlet series associated with this function may be written as products of the Riemann zeta function, which is defined for $\Re(s) > 1$ as $\zeta(s) = \sum_{n=1}^{\infty} n^{-s} = \prod_p (1 - p^{-s})^{-1}$ and otherwise by analytic continuation. Using the well-known identity

$$\sum_{k=0}^{\infty} P(k) x^k = \prod_{m=1}^{\infty} \frac{1}{1 - x^m}, \quad |x| < 1,$$

one finds that, for $\Re(s) > 1$,

$$\sum_{k=0}^{\infty} \frac{a(p^k)}{p^{ks}} = \sum_{k=0}^{\infty} \frac{P(k)}{p^{ks}} = \prod_{m=1}^{\infty} \frac{1}{1 - \frac{1}{p^{ms}}},$$

and thus, using the multiplicativity of $a(n)$,

$$\sum_{n=1}^{\infty} \frac{a(n)}{n^s} = \prod_p \sum_{k=0}^{\infty} \frac{a(p^k)}{p^{ks}} = \prod_p \prod_{m=1}^{\infty} \frac{1}{1 - \frac{1}{p^{ms}}} = \prod_{m=1}^{\infty} \zeta(ms).$$

Using the standard results from tauberian theory, one obtains

$$\sum_{n\le x} a(n) \sim x \prod_{m=2}^{\infty} \zeta(m),\ x \to \infty,$$

from this. By much more refined methods, it can be shown that

$$\sum_{n\le x} a(n) = \sum_{m=1}^{3} c_m x^{1/m} + E(x),\ c_m = \prod_{\substack{k=1\\ k\ne m}}^{\infty} \zeta\left(\frac{k}{m}\right),$$

where the estimates for the error term $E(x)$ have a long history of improvements, with the best result to date being due to Robert and Sargos [16], who proved that $|E(x)| \ll x^{1/4+\epsilon}$. Furthermore one has (see [3, p. 274]) $c_1 = 2.2948565916\cdots$, $c_2 = -14.6475663016\cdots$ and $c_3 = 118.6924619727\cdots$.

Thus on average $a(n)$ is constant (namely about 2.29). Individual values, however, might get large. In this direction Krätzel [6] proved that

$$\lim_{n\to\infty} \sup \log(a(n)) \frac{\log\log n}{\log n} = \frac{\log 5}{4}, \tag{5}$$

which implies that $a(n) \ll n^{\epsilon}$ for every $\epsilon > 0$.

Ivić [5] has pointed out that $C(x)$, the number of *distinct* values assumed by $a(n)$ for $n \le x$, satisfies the bound

$$C(x) \le \exp((1 + o(1))2\pi \sqrt{\log x/3 \log\log x}). \tag{6}$$

The reason for this (see [5, pp. 130–131]) is that there are

$$\exp((1 + o(1))2\pi \sqrt{\log x/3 \log\log x})$$

integers $n \le x$ of the form

$$n = 2^{a_2} 3^{a_3} \cdots p^{a_p},\ a_2 \ge a_3 \ge \cdots \ge a_p \ge 1, \tag{7}$$

which is a classical result of Hardy and Ramanujan [14, pp. 245–261]. Suppose that $a(n)$ is counted by $C(x)$, and let

$$n = p_{1,1}^{b_1} \cdots p_{1,k}^{b_k},\ b_1 \ge b_2 \ge \cdots \ge b_k \ge 1$$

be the canonical decomposition of n. Then if $m = 2^{b_1} 3^{b_2} \cdots p_k^{b_k}$, we have $m \le n$ and $a(m) = P(b_1) \cdots P(b_k) = a(n)$. Therefore $C(x)$ does not exceed the number of $n \le x$ having the form (7) and hence inequality (6) holds.

Note that if f is *any* prime-independent function, then the number of distinct values assumed by it for $n \le x$ satisfies the same upper bound as in (6).

4.2 Summing $a(n)$ Over $\mathcal{G}$

Let $\chi_{\mathcal{G}}$ be the characteristic function of $\mathcal{G}$, i.e.,

$$\chi_{\mathcal{G}}(n) = \begin{cases} 1 & \text{if } n \text{ is in } \mathcal{G}; \\ 0 & \text{otherwise.} \end{cases}$$

We consider

$$\sum_{n \le x,\, n \in \mathcal{G}} a(n) = \sum_{n \le x} \chi_{\mathcal{G}}(n) a(n).$$

Note that $\chi_{\mathcal{G}}(n)a(n)$ is multiplicative in n.

Theorem 4.1. *If* (4) *is satisfied with* $E_{\mathcal{P}}(x) = O(x \log^{-1-\gamma} x)$ *and* $0 < \gamma < 1$, *then*

$$\sum_{n \le x,\, n \in \mathcal{G}} a(n) = x b_0 \log^{\tau-1} x + O_{\mathcal{G}}(x \log^{\tau-1-\gamma/2} x).$$

If (4) *is satisfied with* $E_{\mathcal{P}}(x) = O(x \log^{-2-\gamma} x)$ *and* $\gamma > 0$, *then*

$$\sum_{n \le x,\, n \in \mathcal{G}} a(n) = x \sum_{0 \le \nu < \gamma} b_\nu \log^{\tau-1-\nu} x + O_{\mathcal{G}}(x \log^{\tau-1-\gamma+\epsilon} x), \tag{8}$$

where $b_0, b_1, \ldots$ *are constants possibly depending on* $\mathcal{G}$ *and*

$$b_0 = \frac{1}{\Gamma(\tau)} \lim_{s \downarrow 1} (s-1)^\tau \sum_{n \in \mathcal{G}} \frac{a(n)}{n^s} > 0.$$

The proof uses the following lemma, which except for the formula for b_0 is taken from [9]. The formula for b_0 is well known.

Lemma 4.2 ([9]). *Let* $f : \mathbb{N}_{\ge 0} \to \mathbb{R}_{\ge 0}$ *be a multiplicative function satisfying*

$$0 \le f(p^r) \le c_1 c_2^r,\ c_1 \ge 1,\ 1 \le c_2 < 2, \tag{9}$$

and

$$\sum_{p \le x} f(p) = \tau \mathrm{Li}(x) + O(x \log^{-2-\gamma} x), \tag{10}$$

where $\tau > 0$ and $\gamma > 0$ are fixed, then, for $\epsilon > 0$,

$$\sum_{n \le x} f(n) = x \sum_{0 \le \nu < \gamma} b_\nu \log^{\tau-1-\nu} x + O(x \log^{\tau-1-\gamma+\epsilon} x),$$

where $b_0 = \frac{1}{\Gamma(\tau)} \lim_{s \downarrow 1} (s-1)^\tau \sum_{n=1}^{\infty} f(n) n^{-s}$.

Proof of Theorem 4.1. The first assertion has been proved by Odoni [11] using a tauberian remainder theorem due to Subhankulov.

In order to prove the second assertion we apply Lemma 4.2 with $f(n) = a(n)\chi_{\mathcal{G}}(n)$. The fact that condition (9) is satisfied follows from the classical result of Hardy and Ramanujan (see [14, p. 240]) that

$$a(p^r) = P(r) = (1 + o(1))(4\sqrt{3}r)^{-1} e^{\pi\sqrt{2r/3}}$$

as r tends to infinity. However, the much more easily proved upper bound $P(r) \le 5^{r/4}$ (see [6]) is already sufficient in order to show that (9) is satisfied. The assumption on $E_{\mathcal{P}}(x)$ ensures that condition (10) is satisfied. On invoking Lemma 4.2 the proof is then completed. □

For our problem at hand we find the following estimate:

Theorem 4.3. *Let $HSD(x)$ count the number of non-isomorphic abelian groups of order n with $(n, q) = 1$ and $n \le x$ for which an ideal code of $F[G^*]$ whose extension is Hermitian self-dual exists. Then*

$$HSD(x) = b_0 \frac{x}{\log^{\delta(q)} x} + O_{\epsilon,q}\left(\frac{x}{\log^{\delta(q)+1-\epsilon} x}\right),$$

where

$$b_0 = \frac{1}{\Gamma(1-\delta(q))} \lim_{s \downarrow 1} (s-1)^{1-\delta(q)} \sum_{n \in \mathcal{G}} \frac{a(n)}{n^s}.$$

4.3 The Connection with Free Arithmetical Semigroups

A much weaker form of Theorem 4.3 is obtained as a straightforward consequence of Bredikhin's Theorem, which is a basic result in the theory of *free arithmetical semigroups*.

Let G be a commutative semigroup with identity element 1, relative to a multiplication operation denoted by juxtaposition. Suppose that G has a finite or countably infinite subset P of generators and that G is *free*. This means that every element n in G has a unique factorization of the form $n = \omega_1^{a_1} \cdot \omega_2^{a_2} \cdots \omega_r^{a_r}$, where the ω_r are distinct elements of P, the a_i are possible integers, and uniqueness is up to order of factors. A free semigroup will be called a free arithmetical semigroup if in addition there exists a homomorphism of G into some multiplicative semigroup $\overline{G}$

consisting of real numbers such that for every $x > 0$, G contains only finitely many elements n with $|n| \leq x$, where $|n|$ denotes the image (or norm) of the element n of G under the homomorphism $|.|$. (In the older literature the generators of $\overline{G}$ are called Beurling's generalized primes.) Bredikhin's theorem, for a proof see e.g. [13, pp. 92–99], then reads as follows:

Theorem 4.4 (Bredikhin). *If G is a free arithmetical semigroup such that*

$$\sum_{|\omega|\leq x,\ \omega\in G} 1 = \tau\frac{x}{\log x} + O\left(\frac{x}{\log^{1+\gamma} x}\right), \tag{11}$$

where $\tau > 0$ and $\gamma > 0$ are fixed, then

$$\sum_{\substack{|n|\leq x\\ n\in G}} 1 = C_G x \log^{\tau-1} x + O(x \log^{\tau-1} x (\log\log x)^{-\gamma_1}),$$

where $\gamma_1 = \min(1, \gamma)$ and $C_G = \Gamma(\tau)^{-1} \lim_{s\downarrow 1}(s-1)^{\tau} \sum_{n\in G} |n|^{-s}$.

Now consider the free arithmetical semigroup G of all non-isomorphic finite abelian groups with as composition the usual direct product operation and as norm function $|A| = \mathrm{card}(A)$. By the fundamental theorem on finite abelian groups, G is a free arithmetical semigroup having $C(p),\ C(p^2),\ C(p^3),\ldots$ as generators, where p runs over all the primes and $C(n)$ is the cyclic group of order n. Since the number of cyclic groups of prime power order whose norm is not prime having norm $\leq x$ is $O(\sqrt{x}\log x)$, by the prime number theorem in the form $\pi(x) = x/\log x + O(x/\log^2 x)$, (11) is satisfied with $\tau = 1$ and $\gamma = 1$. It then follows from Bredikhin's theorem that

$$\sum_{\substack{|n|\leq x\\ n\in G}} 1 = \sum_{n\leq x} a(n) = x \prod_{m=2}^{\infty} \zeta(m) + O\left(\frac{x}{\log\log x}\right),$$

where we have used the observations that $\sum_{n\in G} |n|^{-s} = \sum_n a(n)n^{-s} = \prod_{m=1}^{\infty} \zeta(ms)$ and $\lim_{s\downarrow 1}(s-1)\zeta(s) = 1$.

Now let G_q be the free arithmetical semigroup generated by all cyclic groups of the form $C(p), C(p^2), C(p^3),\ldots$, with $\mathrm{ord}_p(q)$ is odd or $\mathrm{ord}_p(q^2)$ is even. Then similarly using Bredikhin's theorem we obtain the result in Theorem 4.3 with the much weaker error term $O_q(x \log^{-\delta(q)} x (\log\log x)^{-1})$.

4.4 The Maximal Order of $a(n)$ on $\mathcal{G}$

In this section we indicate what Krätzel's result (5) looks like when one considers the maximal order of $a(n)$ on the subsemigroup $\mathcal{G}$.

Theorem 4.5. *Let $A = A(n)$ be the smallest integer such that*

$$\sum_{p\in\mathcal{P},\ p\le A} \log p \ge (\log n)/4.$$

Then as n tends to infinity and runs through the elements of $\mathcal{G}$, the estimate

$$\log a(n) \le \mathcal{P}(A)\log 5 + O(\mathcal{P}(A^{\theta})\log A),$$

holds with $\theta = \log(121)/\log(125) < 0.994$, *and there are infinitely many integers n, $n \in G$, for which one has* $\log a(n) = \mathcal{P}(A)\log 5$.

Proof. Completely similar to that of the (only) theorem in Schwarz and Wirsing [17], who proved this result in case $\mathcal{P}$ is the full set of primes. In their proof one merely intersects every range of primes that occurs with $\mathcal{P}$. □

Remark. The implicit constant in the order term may be chosen as

$$\frac{2\pi^2}{3\log 5\cdot\log 2} = 5.898\cdots$$

Theorem 4.6. *If $\mathcal{P}(x) \sim \tau x/\log x$ as x tends to infinity, then*

$$\lim_{n\to\infty}\sup_{n\in\mathcal{G}} \log(a(n))\frac{\log\log n}{\log n} = \frac{\log 5}{4}.$$

Proof. By a standard argument in elementary number theory it follows that if $\mathcal{P}(x) \sim \tau x/\log x$, then $\sum_{p\in\mathcal{P},\ p\le x}\log p \sim \tau x$, $A(n) \sim (\log n)/(4\tau)$ and $\mathcal{P}(A) \sim \log n/(4\log\log n)$. The result now follows on invoking Theorem 4.5. □

Remark. Let $p_1, p_2, \ldots$ denote the consecutive primes in $\mathcal{P}$. Let $n_r = \prod_{i=1}^{r} p_i^4$. Suppose that $\mathcal{P}(x) \sim \tau x/\log x$ as x tends to infinity. We leave it as an exercise to the reader to show that

$$\lim_{r\to\infty} \log(a(n_r))\frac{\log\log n_r}{\log n_r} = \frac{\log 5}{4}.$$

Remark. It is rather surprising that in Theorem 4.6 the estimate does not depend on τ. A similar situation arises if one compares the maximal order of $\log d(n)$ with that of $\log r(n)$, where $d(n)$ denotes the number of divisors of n and $r(n)$ the number of way n can be written as a sum of two squares. Jacobi proved that $r(n) = 4\{d_1(n) - d_3(n)\}$, where $d_1(n)$ and $d_3(n)$ denote the number of divisors of n of the form $4k+1$ and $4k+3$, respectively. Thus $r(n)$ counts (crudely) the divisors of n made up of primes $\equiv 1 \pmod 4$. These primes have density $1/2$ amongst all primes, but nevertheless the maximal orders of $\log d(n)$ and $\log r(n)$ are the same. Namely, we have

$$\lim_{n\to\infty} \sup \log(d(n)) \frac{\log\log n}{\log n} = \log 2, \quad \lim_{n\to\infty} \sup \log(r(n)) \frac{\log\log n}{\log n} = \log 2.$$

For further details see, e.g., Nicolas [10]. The maximal order for $\log d(n)$ was first determined by S. Wigert in 1907. Hardy and Wright [4, Theorem 338] erroneously give $(\log 2)/2$ instead of $\log 2$ in the result for $\log r(n)$.

4.5 Counting Distinct Values Assumed by $a(n)$ on $\mathcal{G}$

Let $C_{\mathcal{G}}(x)$ denote the number of distinct values assumed by $a(n)$ with $n \in \mathcal{G}$ and $n \le x$.

Theorem 4.7. *Let p_0 be the smallest prime in $\mathcal{P}$. Suppose that there are positive constants c_3 and c_4 such that, for $x \ge p_0$,*

$$c_3 x < \sum_{p\in\mathcal{P},\ p\le x} \log p < c_4 x,$$

then

$$\log C_{\mathcal{G}}(x) \sim \log C(x) \sim (1 + o(1)) 2\pi \sqrt{\log x / 3 \log\log x},$$

as x tends to infinity.

Proof. Very similar to that given in [14, pp. 245–261]. Instead of defining l_n to be the product of the first n consecutive primes, we define it to be the product of the first n consecutive primes in $\mathcal{P}$. Then instead of [14, (3.23)] we find $\phi(s) > c_1 \int_{p_0}^{\infty} e^{-c_1 s x} dx / \log x + O(1)$ and instead of [14, (3.24)] we find $\phi(s) < c_2 \int_{p_0}^{\infty} e^{-c_2 s x} dx / \log x + O(1)$. This, through Lemma 3.4 in [14], then leads to the same asymptotic for $\phi(s)$ as in the paper of Hardy and Ramanujan. This then results in the same asymptotic for $C_{\mathcal{G}}(x)$ as that for $C(x)$. □

Acknowledgements The first author gratefully acknowledges financial support from the University of the Philippines and from the Philippine Council for Advanced Science and Technology Research and Development through the Department of Science and Technology.

The second author would like to thank Alexander Ivić for pointing out reference [16] to him.

The second and the third author like to thank Bernhard Heim for the invitation to participate in the Automorphic Forms conference in Oman. Heim proved again that he is a Grandmaster in conference organization.

References

1. L. Dicuangco, P. Moree, P. Solé, The lengths of Hermitian self-dual extended duadic codes. J. Pure Appl. Algebra **1**, 223–237 (2007)
2. C. Ding, D.R. Kohel, S. Ling, Split group codes. IEEE Trans. Inform. Theory **IT-46**, 485–495 (2000)
3. S.R. Finch, *Mathematical Constants, Encyclopedia of Mathematics and Its Applications*, vol. 94 (Cambridge University Press, Cambridge, 2003)
4. G.H. Hardy, E.M. Wright, *An Introduction to the Theory of Numbers*, 5th edn. (The Clarendon Press, Oxford University Press, New York, 1979)
5. A. Ivić, On the number of abelian groups of a given order and on certain related multiplicative functions. J. Number Theory **16**, 119–137 (1983)
6. E. Krätzel, Die maximale Ordnung der Anzahl der wesentlich verschiedenen abelschen Gruppen n-ter Ordnung. Quart. J. Math. Oxford Ser. **21**, 273–275 (1970)
7. J.S. Leon, J.M. Masley, V.S. Pless, Duadic codes. IEEE Trans. Inform. Theory **IT-30**(5), 709–714 (1984)
8. P. Moree, Artin's primitive root conjecture: a survey. Integers **12A**, A13, 100 pp (2012)
9. P. Moree, J. Cazaran, On a claim of Ramanujan in his first letter to Hardy. Expos. Math. **17**, 289–311 (1999)
10. J.-L. Nicolas, On highly composite numbers, in *Ramanujan Revisited (Urbana-Champaign, Ill., 1987)* (Academic, Boston, 1988), pp. 215–244
11. R.W.K. Odoni, A problem of Rankin on sums of powers of cusp-form coefficients. J. Lond. Math. Soc. **44**(2), 203–217 (1991)
12. V. Pless, Q-codes. J. Combin. Theory Ser. **A 43**, 258–276 (1986)
13. A.G. Postnikov, *Introduction to Analytic Number Theory*. Translations of Mathematical Monographs, vol. 68 (AMS, Providence, 1988)
14. S. Ramanujan, *Collected Papers* (Chelsea, New York, 1962)
15. J.J. Rushanan, Duadic codes and difference sets. J. Combin. Theory Ser. **A 57**, 254–261 (1991)
16. O. Robert, P. Sargos, Three-dimensional exponential sums with monomials. J. Reine Angew. Math. **591**, 1–20 (2006)
17. W. Schwarz, E. Wirsing, The maximal number of non-isomorphic abelian groups of order n. Arch. Math. (Basel) **24**, 59–62 (1973)
18. M. Smid, Duadic codes. IEEE Trans. Inform. Theory **IT-33**(3), 432–433 (1987)
19. H.N. Ward, Quadratic residue codes and divisibility, in *Handbook of Coding Theory*, ed. by V.S. Pless, W.C. Huffman (Elsevier Science, Amsterdam, 1998), pp. 827–870

Restrictions of Jacobi Forms of Several Variables

With Special Emphasis on Quaternionic Jacobi Forms

Till Dieckmann

1 Introduction

In this note we briefly repeat the basics on Jacobi forms on $\mathcal{H} \times \mathbb{C}^n$, where we put emphasis on the classical theory, which was described in [1], and quaternionic Jacobi forms, which were treated in [5]. We study the Jacobi forms which are induced by certain isometric embeddings into the lattice of Hurwitz quaternions. We show that in this case the pullback map of Jacobi forms is injective. Furthermore, the pullbacks only depend on the orbits under the orthogonal group on the Hurwitz quaternions. Finally, we give an application to obtain a family of lifts $J_{k,1} \to J_{k,t}$ for arbitrary t, which can be made explicit.

2 Jacobi Forms of Several Variables

Let $(L, Q) \neq 0$ be a positive definite lattice with quadratic form $Q : L \to \mathbb{Z}$. Its associated bilinear form σ is defined by

$$\sigma : L \times L \to \mathbb{Z}, \quad \sigma(x, y) := Q(x + y) - Q(x) - Q(y).$$

Let $V = L \otimes \mathbb{R}$ be the associated quadratic space and

$$L' = \{\mu \in V : \sigma(l, \mu) \in \mathbb{Z} \text{ for all } l \in L\}$$

T. Dieckmann (✉)
Lehrstuhl A für Mathematik, Templergraben 55, 52062 Aachen, Germany
e-mail: till.dieckmann@matha.rwth-aachen.de

B. Heim et al. (eds.), *Automorphic Forms*, Springer Proceedings
in Mathematics & Statistics 115, DOI 10.1007/978-3-319-11352-4_6

be the dual lattice of L. The level of L is denoted by q, i.e.

$$q = \min\left\{d \in \mathbb{N} : dQ(l) \in \mathbb{Z} \text{ for all } l \in L'\right\}.$$

By Aut(L) we denote the isometry group of L, i.e. the group of all automorphisms η of V such that $Q \circ \eta = Q$ and $\eta(L) = L$. Let $\mathcal{H}$ denote the upper half-plane in $\mathbb{C}$. We write $V_{\mathbb{C}} = V \otimes \mathbb{C}$ for the complexification of V and extend the quadratic form to $V_{\mathbb{C}}$ by $\mathbb{C}$-linearity. We follow [2] for the next definition.

Definition 1. Given $k \in \mathbb{Z}$ a holomorphic function $\phi : \mathcal{H} \times V_{\mathbb{C}} \to \mathbb{C}$ is called *Jacobi form of weight k with respect to* (L, Q) if the functional equations

$$\phi(\tau, z) = \phi\big|_k M(\tau, z) := (c\tau + d)^{-k} e^{2\pi i \frac{-cQ(z)}{c\tau+d}} \phi\left(\frac{a\tau + b}{c\tau + d}, \frac{z}{c\tau + d}\right) \tag{J.1}$$

$$\phi(\tau, z) = \phi\big|[l, l'](\tau, z) := e^{2\pi i(Q(l)\tau + \sigma(l,z))} \phi(\tau, z + l\tau + l') \tag{J.2}$$

$$\phi(\tau, z) = \phi\big|\eta(\tau, z) := \phi(\tau, \eta(z)) \tag{J.3}$$

hold for all $(\tau, z) \in \mathcal{H} \times V_{\mathbb{C}}$, $M = \left(\begin{smallmatrix} a & b \\ c & d \end{smallmatrix}\right) \in \mathrm{SL}_2(\mathbb{Z})$, $l, l' \in L$, $\eta \in \mathrm{Aut}(L)$, and ϕ possesses a Fourier expansion of the form

$$\phi(\tau, z) = \sum_{n \geq 0} \sum_{\mu \in L', Q(\mu) \leq n} c(n, \mu) e^{2\pi i(n\tau + \sigma(\mu, z))}. \tag{J.4}$$

The associated space of Jacobi forms of weight k is denoted by $J_k(L, Q)$.

Every $\phi \in J_k(L, Q)$ has a unique *theta decomposition* of the form

$$\phi(\tau, z) = \sum_{\mu : L'/L} f_\mu(\tau)\theta_\mu(\tau, z), \tag{1}$$

where

$$f_\mu(\tau) = \sum_{n \geq 0, n \equiv -qQ(\mu) \bmod q} c\left(\frac{n}{q} + Q(\mu), \mu\right) e^{\frac{2\pi i n\tau}{q}},$$

$$\theta_\mu(\tau, z) = \sum_{l \in \mu + L} e^{2\pi i(Q(l)\tau + \sigma(l,z))}.$$

As a consequence, the invariance property (J.3) yields

$$f_\mu = f_{\eta(\mu)} \quad \text{for all } \mu \in L', \eta \in \mathrm{Aut}(L). \tag{2}$$

Example 1. (a) For $t \in \mathbb{N}$ we consider the lattice (L_t, Q_t), where $L_t = \mathbb{Z}$ and $Q_t(x) = tx^2$. The dual lattice is given by $L_t' = \frac{1}{2t}L_t$. In this case $J_k(L_t, Q_t)$ can be identified with the space $J_{k,t}$ of Jacobi forms of weight k and index t, which are intensively studied in the book [1].

(b) As a second example we consider the root lattice D_4 in its representation

$$D_4 = \left\{(a,b,c,d)^t : a,b,c,d \text{ all in } \mathbb{Z} \text{ or all in } \mathbb{Z} + \frac{1}{2}\right\}.$$

It may be regarded as the lattice of Hurwitz *integral quaternions* $\mathcal{O}$ (cf. [3]), which has the structure of a non-commutative integral domain. In the ladder representation, using quaternionic multiplication, the quadratic form is given by $N(a) := \overline{a}a$, where $\overline{a}$ is the *conjugate quaternion*. The bilinear form can be expressed by the *real part*, i.e. $\sigma(a,b) = 2\,\mathrm{Re}(\overline{a}b)$. One has $\mathcal{O}' = \frac{1}{2}\wp$, where $\wp$ is the two-sided ideal of even quaternions. The elements of $J_k(\mathcal{O})$ are called *quaternionic Jacobi forms* of weight k and index 1.

2.1 *Quaternionic Jacobi Forms*

The results in this subsection are due to [5]. We write

$$\mathcal{O} = \mathbb{Z} + \mathbb{Z}i_1 + \mathbb{Z}i_2 + \mathbb{Z}\omega, \quad \omega = \frac{1 + i_1 + i_2 + i_3}{2}$$

for the ring of Hurwitz integers. The discriminant group $\mathcal{O}'/\mathcal{O} \simeq \mathbb{F}_4$ is represented by

$$h_0 = 0,\ h_1 = \frac{1+i_1}{2},\ h_2 = \frac{1+i_2}{2},\ h_3 = \frac{1+i_3}{2}.$$

Given $\phi \in J_k(\mathcal{O})$ of the form (1) and abbreviating $f_j := f_{h_j}$, (2) implies

$$f_0(\tau) = \sum_{n=0}^{\infty} c(4n)e^{2\pi i n\tau}, \quad \tau \in \mathcal{H}$$

$$f_1(\tau) = f_2(\tau) = f_3(\tau) = \sum_{n=0}^{\infty} c(4n+2)e^{\pi i(2n+1)\tau}, \quad \tau \in \mathcal{H}$$

for a function $c : \mathbb{N}_0 \to \mathbb{C}$ such that $c(N) = 0$ for $N \equiv 1, 3 \bmod 4$. Now (1) becomes

$$\phi(\tau, z) = f_0(\tau)\theta_0(\tau, z) + f_1(\tau)\theta_1(\tau, z), \tag{3}$$

where $\theta_1 := \sum_{j=1}^{3} \theta_{h_j}$. Due to [5, Corollary 1] f_0 and f_1 are related by the identity

$$f_1 = -\frac{1}{3} f_0 - \frac{2}{3} f_0\big|_{k-2} J$$

and the theta transformation formula [5, Lemma 1] implies

$$\theta_1 = -\theta_0 - 2\theta_0\big|_2 J. \tag{4}$$

Furthermore let

$$[\Gamma_0(2), k]_T := \left\{ f \in [\Gamma_0(2), k] : f + f\big|_k J + f\big|_k JT = 0 \right\},$$

where

$$\Gamma_0(2) = \{ M \in \mathrm{SL}_2(\mathbb{Z}) : c \equiv 0 \bmod 2 \}$$

and

$$J = \begin{pmatrix} 0 & 1 \\ -1 & 0 \end{pmatrix}, \quad T = \begin{pmatrix} 1 & 1 \\ 0 & 1 \end{pmatrix}.$$

Theorem 1 (Krieg). *Let $k > 0$ be even. Then the map*

$$J_k(\mathcal{O}) \longrightarrow [\Gamma_0(2), k-2]_T, \; \phi \mapsto f_0$$

is an isomorphism of the vector spaces and one has

$$\dim_{\mathbb{C}} J_k(\mathcal{O}) = \left\lfloor \frac{k+2}{6} \right\rfloor.$$

2.2 Restrictions of Jacobi Forms

Let (L, Q) and (L_0, Q_0) be even, positive definite lattices, such that there is a isometric embedding $\iota : L_0 \to L$. We extend ι to an isometry of the quadratic spaces

$$(V_0)_{\mathbb{C}} = L_0 \otimes \mathbb{C} \longrightarrow L \otimes \mathbb{C} = V_{\mathbb{C}}.$$

Given $\phi \in J_k(L, Q)$ we define the pullback $\phi \circ \iota$ with respect to ι by

$$(\tau, z) \longmapsto \phi(\tau, \iota(z)), \quad (\tau, z) \in \mathcal{H} \times (V_0)_{\mathbb{C}}.$$

One easily checks that $\phi \circ \iota$ satisfies (J.1) and (J.2) with respect to (L_0, Q_0). The Fourier expansion of $\phi \circ \iota$ is given by

$$(\phi \circ \iota)(\tau, z) = \sum_{n \geq 0} \sum_{\mu_0 \in L_0', Q_0(\mu_0) \leq n} \left(\sum_{\mu \in L', \iota^*(\mu) = \mu_0} c(n, \mu) \right) e^{2\pi i (n\tau + \sigma_{L_0}(\mu_0, z))},$$

where $\iota^* : L' \to L_0'$ is the dual map. Assume that

$$\text{for all } \psi \in \text{Aut}(L_0) \text{ there exists } \eta \in \text{Aut}(L) \text{ such that } \psi = \eta|_{L_0} \tag{A}$$

holds. Now $\phi \circ \iota \in J_k(L_0, Q_0)$ follows immediately.

In this context the following questions arise:

(1) Is the restriction map $J_k(L, Q) \to J_k(L_0, Q_0)$ injective?
(2) How does the choice of the embedding $\iota : L_0 \hookrightarrow L$ influence the arising Jacobi forms?

3 Restrictions of Quaternionic Jacobi Forms

Given $t \in \mathbb{N}$ choose $a \in \mathcal{O}$ such that $N(a) = t$. This defines an isometric embedding

$$\iota_a : L_t \longrightarrow \mathcal{O}, \; z \mapsto az.$$

Of course, every isometry $L_t \to \mathcal{O}$ is of this form. Obviously, assumption (A) holds. The pullback with respect to ι_a is given by

$$J_k(\mathcal{O}) \longrightarrow J_{k,t}, \; (\tau, z) \mapsto \phi_a(\tau, z) := \phi(\tau, az).$$

Then (3) yields a representation

$$\phi_a(\tau, z) = f_0(\tau)\theta_{a,0}(\tau, z) + f_1(\tau)\theta_{a,1}(\tau, z), \tag{5}$$

where

$$\theta_{a,0}(\tau, z) = \sum_{g \in \mathcal{O}} e^{2\pi i (N(g)\tau + 2\text{Re}(\overline{a}g)z)},$$

$$\theta_{a,1}(\tau, z) = \sum_{g \in \frac{1}{2}\wp \setminus \mathcal{O}} e^{2\pi i (N(g)\tau + 2\text{Re}(\overline{a}g)z)}.$$

Lemma 1. *Let $0 \neq a \in \mathcal{O}$ and $t = N(a)$. Then the pullback map*

$$J_k(\mathcal{O}) \longrightarrow J_{k,t}, \ \phi \mapsto \phi_a$$

is injective. For $t = 1$ it is an isomorphism.

Proof. One easily checks that the first non-zero Fourier coefficient of $\theta_{a,0}/\theta_{a,1}$ with respect to τ depends on z. Now the equation

$$0 = \phi_a(\tau, z) = f_0(\tau)\theta_{a,0}(\tau, z) + f_1(\tau)\theta_{a,1}(\tau, z)$$

holds if and only if $f_0 = 0$ and thus if and only if $\phi = 0$. Bijectivity follows from the dimension formula for $J_{k,1}$ in [1] and Theorem 1. □

Remark 1. Classical dimension formulas (cf. [1], p. 105) yield

$$\dim J_{4,2} = \dim J_{6,2} = 1 \text{ and } \dim J_{8,2} = 2.$$

From $\dim J_k(\mathcal{O}) = 1$ for $k = 4, 6, 8$, we obtain, that $\phi \to \phi_\pi, \pi = 1 + i_1$ is bijective for $k = 4, 6$ and not surjective for $k = 8$. Comparison of the constant terms yields $e_{k,2} = e^{\mathbb{H}}_{k,1,\pi}$ for $k = 4, 6$, where $e^{\mathbb{H}}_{k,1}$ are the Jacobi-Eisenstein series of index 1 (cf. [5, p. 537ff], [5, p. 677]).

Let $\phi \in J_k(\mathcal{O})$ and $a, b \in \mathcal{O}$ such that $N(a) = N(b)$. If a and b lie in the same orbit under $\mathrm{Aut}(\mathcal{O})$, then one easily checks $\phi_a = \phi_b$.

Conjecture 1. Let $t \in \mathbb{N}$ and $0 \neq \phi \in J_k(\mathcal{O})$. Then the equivalence

$$\sum_{a \in E} \alpha_a \phi_a = 0 \Longleftrightarrow \sum_{a \in E} \alpha_a \theta_{a,0} = 0 \tag{6}$$

holds for all $E \subset \{a \in \mathcal{O} : N(a) = t\}$.

We have checked this numerically with SAGE for the $e^{\mathbb{H}}_{k,1}$ up to $k = 80$ and up to $t = 200$. Under the assumption of Conjecture 1, we obtain

Conjecture 2. Let $0 \neq \phi \in J_k(\mathcal{O})$. Then for $a, b \in \mathcal{O}$ with $N(a) = N(b)$ one has

$$\phi_a = \phi_b \Longleftrightarrow \theta_{a,0} = \theta_{b,0}.$$

Its a classical fact (cf. [3]) that $\mathrm{Aut}(\mathcal{O})$ is generated by the transformations

$$x \to \delta x \varepsilon, \quad x \to (1 + i_1)^{-1} x (1 + i_1), \quad x \to \overline{x},$$

where δ, ε are the units of $\mathcal{O}$. It acts as $z \mapsto \omega z$, sign-changes and permutations. A closer examination of the Jacobi theta series associated to $\mathcal{O}$ yields

Lemma 2. *Given $a, b \in \mathcal{O}$ such that $N(a) = N(b)$ the identity $\theta_{a,0} = \theta_{b,0}$ holds if and only if there is $\eta \in \mathrm{Aut}(\mathcal{O})$ such that $\eta(a) = b$.*

Combining the lemma and Conjecture 2 we obtain

Conjecture 3. Let $a, b \in \mathcal{O}$ such that $N(a) = N(b)$ and $0 \neq \phi \in J_k(\mathcal{O})$. Then $\phi_a = \phi_b$ holds if and only if there is $\eta \in \mathrm{Aut}(\mathcal{O})$ such that $\eta(a) = b$.

Recently after the paper was submitted, the author proved conjectures 1 to 3 in the case, where N(a) has odd 2-exponent.

4 Index-Raising Operators

Due to Lemma 1 the restriction map $J_k(\mathcal{O}) \to J_{k,1}$ is an isomorphism of the vector spaces. For fixed t this yields a family of injective lifts

$$U_a : J_{k,1} \longrightarrow J_{k,t}, \ t = N(a)$$

where a runs through the set of Hurwitz quaternions of norm t. This is visualized in the diagram below:

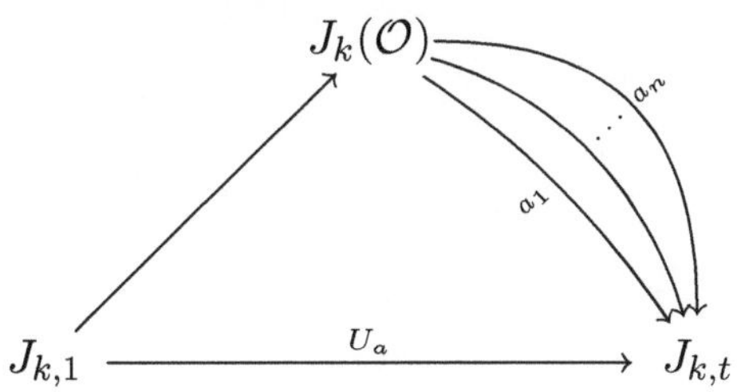

It is clear that for $a \in \mathbb{Z}$ the operator U_a coincides with the one defined in [1, Sect. 4 (1)]. It is also possible to give a more or less explicit representation of $U_a\phi, \phi \in J_{k,1}$. Therefore let $\phi \in J_{k,1}$ with theta decomposition

$$\phi(\tau, z) = h_0(\tau)\vartheta_0(\tau, z) + h_1(\tau)\vartheta_1(\tau, z). \tag{7}$$

Detailed analysis of the Jacobi theta series associated to $\mathcal{O}$ yields

$$\theta_{1,0}(\tau, z) = \vartheta_0(\tau, z)\vartheta_0(\tau)^3 + \vartheta_1(\tau, z)\vartheta_1(\tau)^3,$$
$$\theta_{1,1}(\tau, z) = 3\left(\vartheta_1(\tau, z)\vartheta_0(\tau)^2\vartheta_1(\tau) + \vartheta_0(\tau, z)\vartheta_0(\tau)\vartheta_1(\tau)^2\right).$$

where $\vartheta_j = \vartheta_j(\tau, 0)$. A comparison of the theta coefficients proves

Theorem 2. *Let $\phi \in J_{k,1}$ with theta decomposition* (7). *Then one has*

$$\phi_a(\tau, z) = \frac{h_0\vartheta_0 - h_1\vartheta_1}{\vartheta_0^4 - \vartheta_1^4}\theta_{a,0}(\tau, z) + \frac{1}{3}\frac{h_1\vartheta_0^3 - h_0\vartheta_1^3}{\vartheta_0^5\vartheta_1 - \vartheta_1^5\vartheta_0}\theta_{a,1}(\tau, z). \tag{8}$$

Acknowledgement We would like to thank the organizers for the nice conference and the referee for valuable suggestions.

References

1. M. Eichler, D. Zagier, *The Theory of Jacobi Forms* (Birkhäuser, Boston, 1985)
2. V.A. Gritsenko, Fourier-Jacobi functions of n. J. Sov. Math. **53**, 243–252 (1991)
3. A. Hurwitz, *Vorlesungen über die Zahlentheorie der Quaternionen* (Springer, Berlin, 1919)
4. A. Krieg, in *Modular Forms on the Half-Spaces of Quaternions*, Lecture Notes in Mathematics, vol. 1143 (Springer, Berlin/Heidelberg/New York, 1985)
5. A. Krieg, The Maaß-Space on the Half-Space of Quaternions of Degree 2. Math. Ann. **276**, 675–686 (1987)

Reversing Borcherds Lifts: A Survey

Bernhard Heim and Atsushi Murase

Mathematics Subject Classification (2010): 11F11, 11F50, 11F55, 11G18

1 Introduction

Borcherds lifts [5] are automorphic forms with infinite product expansions analogous to the Dedekind eta function. They have found various applications in geometry, arithmetic, and the theory of Lie algebras. For example, they appear as denominator functions of certain generalized Kac-Moody algebras [4, 27, 28], as new product representations of well-known classical functions [4, 5, 21, 26] or as the partition function of quarter-BPS dyons [14, 15, 17]. The lift is used to obtain automorphic forms on $O(2, n+2)$ with known vanishing properties [8, 10, 19]. We also refer to applications in algebraic geometry [1, 11, 25]. In the context of studying the graded ring of modular form, it is also an interesting question to identify Borcherds and Saito-Kurokawa lifts among the generators (see for example [2, 16, 19]).

B. Heim (✉)
German University of Technology in Oman, Halban Campus
PO Box 1816, Athaibah PC 130, Muscat, Sultanate of Oman
e-mail: bernhard.heim@gutech.edu.om

A. Murase
Department of Mathematical Science, Faculty of Science, Kyoto Sangyo University
Motoyama, Kamigamo, Kita-ku, Kyoto 603-8555, Japan
e-mail: murase@cc.kyoto-su.ac.jp

B. Heim et al. (eds.), *Automorphic Forms*, Springer Proceedings
in Mathematics & Statistics 115, DOI 10.1007/978-3-319-11352-4_7

The main theme of this paper is to identify Borcherds lifts F in the target space, the space of modular forms on the orthogonal group, and to extract from F the input function and the divisor.

In two celebrated papers [5, 7], Borcherds established a correspondence (Borcherds lift) of vector-valued weakly holomorphic elliptic modular forms f of weight $-n/2$ to meromorphic modular forms Ψ_f on the orthogonal group $O(2, n+2)$. The second paper [7] is a systematic generalization of Harvey and Moore's [23] discovery that Borcherds first approach [5] can be understood as a singular theta correspondence. We also refer to the introduction of Bruinier's book [10] and Kontsevich's Bourbaki article [26] for a detailed overview. Famous Borcherds products are given by the denominator formula for the monster Lie algebra [4], the Igusa modular form of weight 10 [21], and the Weyl denominator formula for the fake monster Lie algebra [5].

Let j be the modular invariant, let Δ be the Ramanujan delta function, and let $\chi_{12,1}$ be the Jacobi cusp form of weight 12 and index 1. Let $f_1 = j - 744$, and let $f_2 = 2\chi_{12,1}/\Delta$. The corresponding Borcherds lifts are automorphic forms on $O(2,2)$ and $O(2,3)$. They are given by

$$\Psi_{f_1}(\tau_1, \tau_2) = j(\tau_1) - j(\tau_2) = q_1^{-1} \prod_{m=1,n=-1}^{\infty} (1 - q_1^m q_2^n)^{c(mn)} \tag{1}$$

$$\Psi_{f_2}(\tau_1, z, \tau_2) = \chi_{10}(\tau_1, z, \tau_2) = q_1 \zeta q_2 \prod_{m,r,n}^{\infty} (1 - q_1^m \zeta^r q_2^n)^{a(nm,r)}. \tag{2}$$

Let $\mathfrak{H}$ be the upper half space. We have realized the corresponding Hermitian symmetric domains as $\mathfrak{H} \times \mathfrak{H}$ and $\mathfrak{H}_2$ (Siegel upper half space of degree 2). Here q_1, q_2, ζ are given by $e(w) := \exp(2\pi i w)$ for $w = \tau_1, \tau_2, z$. Moreover $c(n)$ and $a(n,l)$ are the Fourier coefficients of f_1 and f_2, and χ_{10} is the Igusa modular form of weight 10. Let further $f = 1/\Delta$ be a weakly holomorphic modular form of weight -12, then Ψ_f is the Weyl denominator formula for the fake monster Lie algebra (see also [10], Sect. 3.4.1).

From the construction of the correspondence it can be concluded that a large class of Borcherds lifts have the following two significant properties:

Ⓓ The zeros and poles of Borcherds lifts are supported at special divisors $\mathcal{Z}$ (finite sums of Heegner divisors). This has been first observed by Borcherds [5,7].

Ⓢ Borcherds lifts satisfy for all prime numbers p the following symmetry (see also (6, 7)):

$$F|T_{\Pi}^{\uparrow}(p) = \varepsilon_F(p)\; F|T_{\Pi}^{\downarrow}(p), \tag{3}$$

where $\varepsilon_F(p) \in \mathbb{C}^{\times}$ depends on F and p. This property has been discovered by Heim and Murase [24].

It is a delicate task to identify Borcherds products in the target space. For example, in the case of $O(2,3)$ (Siegel modular forms), possible weights of holomorphic Borcherds lifts are $5, 10, 15, 20, 24, 25, 29, 30, 34, 35, \ldots$. But even if we know that a Siegel modular form F has the suitable weight and rational integers as Fourier coefficients, there are no criteria known to read off the Fourier coefficients if F is a Borcherds lift. Borcherds stated in [6], Sect. 16, several open problems. Problem 16.10 is addressed to reverse his lifting (see also [5], Problem 10 in Sect. 17):

> *Can one reverse the correspondence from modular forms to automorphic forms with singularities, and reconstruct modular forms from automorphic forms with singularities? In particular when are there isomorphisms between spaces?*

In the special case of elliptic modular forms of level 1 this problem was completely solved by Borcherds ([5], Theorem 14.1). For $O(2, 2+n)$ with $n \in \mathbb{N}$, Bruinier [10, 13] was the first who has proven converse theorems in the sense of property Ⓓ. He has shown that in a large class of cases every meromorphic modular form on $O(2, n+2)$ whose divisor is supported on a special divisor (linear combination of Heegner divisors) is the Borcherds lift of a weakly holomorphic modular form of weight $-n/2$. See also the work of Bruinier and Freitag [9] and Bruinier and Funke [12]. The case of orthogonal groups of signature $(1,n)$ had been considered by Barnard [3].

We state that the symmetry property Ⓢ identifies Borcherds lifts among the modular forms on $O(2, n+2)$:

- Effective principal divisors $\mathcal{X}$ possessing an additive symmetry (induced by the multiplicative symmetry stated in (3)) are special divisors
- Let $F \in M_k(\Gamma_L)$ be a holomorphic modular form on the orthogonal group with divisor $\mathcal{X}$. Then

$$F \text{ is a Borcherds lift} \Longleftrightarrow \text{Ⓢ} \Longleftrightarrow \text{Ⓓ} \tag{4}$$

- The symmetry in Ⓢ implies symmetry properties for Jacobi forms (the first non-vanishing Fourier-Jacobi coefficient of F). Moreover we have new characterization of modular polynomial among all polynomials with two variables.

The Hecke-like operators $T_{\Pi}^{\uparrow}(n)$ and $T_{\Pi}^{\downarrow}(n)$ can be considered as some kind of multiplicative higher dimensional analogue of the operators V_n introduced in ([18], Sect. 4) and employed recently in a work on mock modular forms ([15], Sect. 4.4). Note that we have twisted the operators V_n in such a way that they do not change the index. Only if n is a pure square the involved matrices are rational. Gritsenko and Nikulin defined several Hecke operators on the space of weakly Jacobi forms $\mathfrak{H} \times \mathbb{C}$ and the space of paramodular forms on $\mathfrak{H}_2$. Their approach (see [20], formula (1.3)) is very general. They introduced additive and multiplicative versions (see for example [20], formula (0.4), Sect. 1, and the Appendix). The operator $T_{\Pi}^{\uparrow}(n)$ in the special case $O(2,3)$ for prime numbers n is similar to the operators defined in [21] (Definition 3.1). Gritsenko and Nikulin prove some interesting functorial properties (see Theorem 3.3, [21]). They are not used in this paper. Similar results had been obtained by Guerzhoy [22] in the split case $O(2,1)$.

Let us now begin to set up the general framework. This makes it possible to relate our results to the work of Bruinier and also give a clear answer to the question proposed by Borcherds. Let (V,q) be a quadratic space over $\mathbb{Q}$ of signature $(2,n+2)$ and let Q be the associated bilinear form. Let $L \subset V$ be an even lattice and let L^* be the dual lattice. Then we denote by $\mathbb{C}[L^*/L]$ the discriminant group and by $\Gamma = \Gamma(L)$ the discriminant kernel (a certain subgroup of the stabilizator of L). We denote by $\mathbb{H}_n \subset \mathbb{C}^{n+2}$ the tube domain realization of the associated Hermitian symmetric space of type IV. Let $g\langle Z\rangle$ be the induced action of the orthogonal group on $\mathbb{H}_n$. We denote by $M_k(\Gamma)$ the space of holomorphic automorphic forms on $\mathbb{H}_n$ with respect to Γ of weight k.

We assume that $L = \mathbb{Z}^{n+4}$ and that

$$Q := \begin{pmatrix} 0 & 0 & J \\ 0 & -S & 0 \\ J & 0 & 0 \end{pmatrix}, \qquad \text{where } J := \begin{pmatrix} 0 & 1 \\ 1 & 0 \end{pmatrix} \text{ and } S > 0 \text{ even.} \tag{5}$$

The tube domain $\mathbb{H}_n$ can be realized as a subset of $\mathfrak{H} \times \mathbb{C}^n \times \mathfrak{H}$. We fix two canonical embeddings $^\uparrow$ and $^\downarrow$ of SL_2 into the orthogonal group $O(2, n+2)$. The images of these embeddings commute with each other. For $h = \left(\begin{smallmatrix} a & b \\ c & d \end{smallmatrix}\right) \in \mathrm{SL}_2(\mathbb{R})$ and $Z = (\tau_1, z, \tau_2) \in \mathbb{H}_n$, we have

$$h^\uparrow\langle Z\rangle = \left(\frac{a\tau_1+b}{c\tau_1+d}, \frac{z}{c\tau_1+d}, \tau_2 - \frac{c}{c\tau_1+d}S'[z]\right),$$
$$h^\downarrow\langle Z\rangle = \left(\tau_1 - \frac{c}{c\tau_2+d}S'[z], \frac{z}{c\tau_2+d}, \frac{a\tau_2+b}{c\tau_2+d}\right).$$

Here $S' := 2^{-1}S$. Let F be any function on $\mathbb{H}_n$, then we can define multiplicative Hecke-like operators:

$$F|T_\Pi^\uparrow(p)\langle Z\rangle := F\left(p\tau_1, \sqrt{p}\,z, \tau_2\right)\prod_{a=1}^{p} F\left(\frac{\tau_1+a}{p}, (\sqrt{p})^{-1}z, \tau_2\right) \tag{6}$$

$$F|T_\Pi^\downarrow(p)\langle Z\rangle := F\left(\tau_1, \sqrt{p}\,z, p\tau_2\right)\prod_{a=1}^{p} F\left(\tau_1, (\sqrt{p})^{-1}z, \frac{\tau_2+a}{p}\right). \tag{7}$$

These operators induce additive Hecke-like operators $T_\Sigma^\uparrow(p)$ and $T_\Sigma^\downarrow(p)$ on divisors $\mathcal{X} \subset \mathbb{H}_n$. Hence we have

$$\operatorname{div}(F)|T_\Sigma^\uparrow(p) = \operatorname{div}\left(F|T_\Pi^\uparrow(p)\right).$$

To state our main result we fix the following notation. Let $F \in M_k(\Gamma)$. Let $F_{m_0}(\tau_1, z)$ be the first non-trivial Fourier-Jacobi coefficient of F, and let $A_{n_0}(z)$ be the first non-trivial Fourier-Jacobi coefficient of $F_{m_0}(\tau_1, z)$.

$$F(\tau_1, z, \tau_2) = F_{m_0}(\tau_1, z) q_2^{m_0} \left(\sum_{m=0}^{\infty} f_m(\tau_1, z) q_2^m \right) \tag{8}$$

$$F_{m_0}(\tau_1, z) = A_{n_0}(z)\, q_1^{n_0} \left(\sum_{n=0}^{\infty} B_n(z) q_1^n \right). \tag{9}$$

Let further $\mathcal{G}_m(X_1, \ldots, X_m) := -X_m - \sum_{j=1}^{m-1} m^{-1} j\, X_{m-j} \mathcal{G}_j(X_1, \ldots, X_j)$ for $m \geq 2$ with $\mathcal{G}_1(X_1) = -X_1$. In this paper we realize the input data of the Borcherds lift by the group $\mathcal{J}^!_{0,S'}(\mathbb{Z})$ of weakly Jacobi forms of weight 0 and index S' with integral principal part instead of $\mathbb{C}[L^*/L]$ valued weakly modular forms. For our purpose this seems to be quite natural since the input function of a Borcherds lift $F \in M_k(\Gamma)$ is given by $-F_{m_0+1}/F_{m_0}$. Let $F \in M_k(\Gamma)$ non-zero with divisor $\mathcal{X}$. To make our arguments transparent we list the following properties, additional to the previous properties Ⓓ and Ⓢ:

Ⓑ F is a Borcherds lift with input function $-f_1$.
Ⓓ $\mathcal{X}$ is a special divisor.
Ⓢ $F|T^{\uparrow}_{\Pi}(p) = \varepsilon_F(p)\; F|T^{\downarrow}_{\Pi}(p)$ (for all prime p).
Ⓡ $\mathcal{G}_n(B_1(z), \cdots, B_n(z)) = \sum_{d|n} d^{-1} B_1(dz)$, and $\mathcal{G}_n(f_1, \cdots, f_n) = -f_1|_{0,S'} V(n)$ (for $n \in \mathbb{N}$).
Ⓐ $\mathcal{X}|T^{\uparrow}_{\Sigma}(p) = \mathcal{X}|T^{\downarrow}_{\Sigma}(p)$ (for all prime p).

The precise definition of the operator $|_{0,S'} V(n)$ is the following: For a positive integer n and a weakly Jacobi form of weight 0 and index S', define

$$(\phi|_{0,S'} V(n))\,(\tau, z) := n^{-1} \sum_{\left(\begin{smallmatrix} a & b \\ c & d \end{smallmatrix}\right)} \mathbf{e}\left(n \frac{-cS'[z]}{c\tau + d} \right) \phi \left(\frac{a\tau + b}{c\tau + d}, \frac{nz}{c\tau + d} \right), \tag{10}$$

where $\left(\begin{smallmatrix} a & b \\ c & d \end{smallmatrix}\right)$ runs over $\mathrm{SL}_2(\mathbb{Z})\backslash \mathrm{M}_2(\mathbb{Z})$ with $ad - bc = n$. We have also

$$(\phi|_{0,S'} V(n))\,(\tau, z) = n^{-1} \sum_{d|n} \sum_{b=0}^{d-1} \phi \left(\frac{d^{-1} n\tau + b}{d}, d^{-1} nz \right).$$

We have the following result:

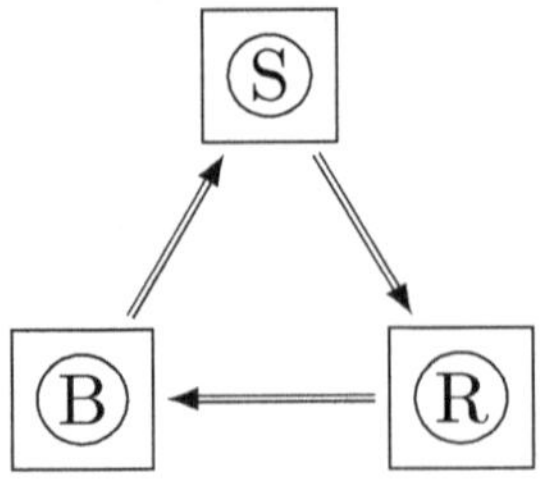

Hence we obtain that F being a Borcherds lift with input function $-f_1$ is equivalent to F having the symmetry property. Here Ⓑ $\Rightarrow$ Ⓢ follows from [24].

Theorem I. *Let $F \in M_k(\Gamma(L))$ be non-trivial. Then we have the following property:*

$$\begin{aligned} F|T_{\Pi}^{\uparrow}(p) &= \varepsilon_F(p)\; F|T_{\Pi}^{\downarrow}(p) \textit{ for all prime } p \\ &\Longleftrightarrow F \textit{ is a Borcherds lift with input } -\frac{F_{m_0+1}}{F_{m_0}}. \end{aligned}$$

After we presented a first version of these converse results some time ago, N. Scheithauer asked if there is any relation to J. Bruinier's converse results [10] via Heegner divisors. This leads us to the following results. From the Koecher principle it follows that Ⓐ $\Longleftrightarrow$ Ⓢ for $n \geq 1$. Since $\Delta \otimes \Delta$ is a Borcherds product, we can also include the case $n = 0$. Hence we also obtain

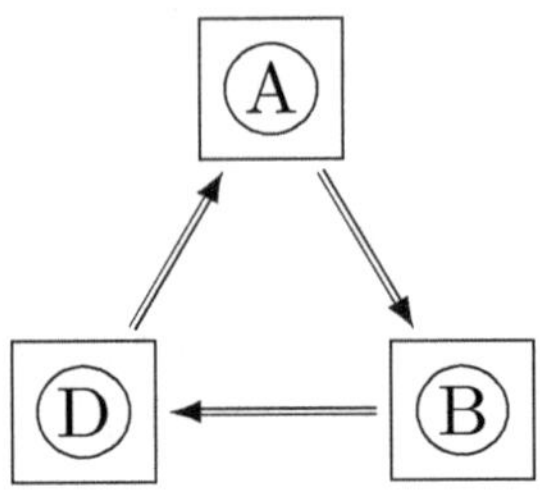

Theorem II. *Let $F \in M_k(\Gamma(L))$ be non-trivial. Let $\mathcal{X}$ be the divisor of F. Then we have*

$$\begin{aligned} \mathcal{X}|T_{\Sigma}^{\uparrow}(p) &= \mathcal{X}|T_{\Sigma}^{\downarrow}(p) \textit{ for all prime } p \\ &\Longleftrightarrow \mathcal{X} \textit{ is a special divisor} \Longleftrightarrow F \textit{ is a Borcherds lift}. \end{aligned}$$

In this paper we assume that the lattice L splits into to orthogonal hyperbolic planes over $\mathbb{Z}$ (see also Bruinier ([10], Theorem 5.12)). Recently [13] extended his results to converse theorems dropping the hypothesis that L splits hyperbolic planes of $\mathbb{Z}$.

Let $\mathcal{J}^{!}_{0,1}(\mathbb{Z})$ be the space of weakly Jacobi forms on $\mathfrak{H} \times \mathbb{C}$ of weight 0, index 1, and integral principal part. This space correspondence to weakly elliptic modular forms of weight $-1/2$ and integral principal part. The Fourier expansion is given by

$$\varphi(\tau, z) = \sum_{D \ll -\infty}^{\infty} a_\varphi(D) \sum_{r \in \mathbb{Z}, -r^2 \equiv D(4)} q^{\frac{D+r^2}{4}} \zeta^r. \tag{11}$$

Let further $M(\Gamma_2)$ be the set of all Siegel modular forms with respect Γ_2 (without character). Then we put

$$\mathbb{P}\left(M(\Gamma_2)\right) = \left(M(\Gamma_2)\,/\mathbb{C}^*. \tag{12}$$

We also put $\mathcal{D} = \{D \in \mathbb{N} | D \equiv 0, 1 \pmod 4\}$. Then we have the following concrete example of a reversing result:

Theorem III. *We have a correspondence between the multiplicative semigroup of all $F \in \mathbb{P}(M(\Gamma_2))$ with*

$$F|T^{\uparrow}_{\Pi}(p) = \varepsilon_F(p)\ F|T^{\downarrow}_{\Pi}(p) \text{ for all prime } p \tag{13}$$

and the additive semigroup of all $\varphi \in \mathcal{J}^{!}_{0,1}(\mathbb{Z})$ with

$$\sum_{l=1}^{\infty} a_\varphi(-l^2 D) \geq 0,\ D \in \mathcal{D} \text{ and } \sum_{l=1}^{\infty} \left(a_\varphi(-l^2) + a_\varphi(-4l^2)\right) \equiv 0 \pmod 2. \tag{14}$$

Let F_{m_0} be the first non-zero Fourier-Jacobi coefficient of F. Then the correspondence is given by $F \mapsto -\frac{F_{m_0+1}}{F_{m_0}}$.

Remark 1.1. The first condition in (14) is saying that F_φ is holomorphic and the second condition that the lift has trivial character.

Remark 1.2. (a) Let $G = F^{-1}$ with $F \in M_k(\Gamma)$. Then the converse theorem can also be applied.

(b) Let F be a holomorphic form on $O(2, n+2)$ with character of finite order. Then the converse theorem is still true. Take F to the power l, where l is the order of the character. Then $G = F^l$ satisfies also the symmetry conditions and hence is a Borcherds product and has a special divisor. This implies that F has a special divisor and hence a Borcherds lift (by the result of Bruinier [10]). With some more effort we think we can prove this also directly without Bruinier's result. Actually we would also reprove his result with the methods introduced in this paper.

$$\text{Ⓐ}\ \mathcal{X}|T^{\uparrow}_{\Sigma}(p) = \mathcal{X}|T^{\downarrow}_{\Sigma}(p) \text{ (for all prime } p) \Longleftrightarrow \text{Ⓢ}$$

$$F|T^{\uparrow}_{\Pi}(p) = \varepsilon_F(p)\ F|T^{\downarrow}_{\Pi}(p) \text{ (for all prime } p)$$

Summary: Putting everything together leads to the following equivalent statements:

$$\text{Ⓐ}\ \mathcal{X}|T^{\uparrow}_{\Sigma}(p) = \mathcal{X}|T^{\downarrow}_{\Sigma}(p) \text{ (for all prime } p) \Longleftrightarrow \text{Ⓢ}\ F|T^{\uparrow}_{\Pi}(p) = \varepsilon_F(p)\ F|T^{\downarrow}_{\Pi}(p) \text{ (for all prime } p)$$

$$\Longleftrightarrow \text{Ⓑ}\ F \text{ is Borcherds lift of } -f_1$$

$$\Longleftrightarrow \text{Ⓡ}\ \mathcal{G}(B_1(z), \cdots, B_n(z)) = \sum_{d|n} d^{-1} B_1(dz)$$

$$\mathcal{G}(f_1, \cdots, f_n) = -f_1|_{0,S'} V(n) \text{ (for } n \in \mathbb{N})$$

$$\Longleftrightarrow \text{Ⓓ}\ \mathcal{X} \text{ is a special divisor}$$

Hence Borcherds lifts are characterized by an additive symmetry of the divisor, or a multiplicative symmetry of the automorphic form, or recursion formulas of the Fourier-Jacobi coefficients (and the first non-trivial Fourier-Jacobi coefficient), or by special divisors. All of these properties are equivalent.

Acknowledgement The authors thank the referee for sharpening the structure of the survey.

References

1. D. Allcock, E. Freitag, Cubic surfaces and Borcherds products, Comment. Math. Helv. **77**, 270–296 (2002)
2. H. Aoki, T. Ibukiyama, Simple graded rings of Siegel modular forms, differential operators and Borcherds products. Int. J. Math. **16**, 249–279 (2005)
3. A. Barnard, The singular theta correspondence, lorentzian lattices and Borcherds-Kac-Moody algebras. Ph.D. Dissertation, U.C. Berkley (2003). arXiv:math/0307102v1 [math.GR]
4. R.E. Borcherds, Monstrous moonshine and monstrous Lie superalgebras. Invent. Math. **109**, 405–444 (1992)
5. R.E. Borcherds, Automorphic forms on $O_{s+2,2}(\mathbb{R})$ and infinite products, Invent. Math. **120**. 161–213 (1995)
6. R.E. Borcherds, The moduli space of Enriques surfaces and the fake monster Lie superalgebra. Topology **35**, 699–710 (1996)
7. R.E. Borcherds, Automorphic forms with singularities on grassmannians. Invent. Math. **132**, 491–562 (1998)
8. R.E. Borcherds, The Gross-Kohnen-Zagier theorem in higher dimensions. Duke Math. J. **97**, 219–233 (1999) Correction to "The Gross-Kohnen-Zagier theorem in higher dimensions". Duke Math. J. **105**, 183–184 (2000)
9. J.H. Bruinier, E. Freitag, Local Borcherds products. Ann. I. Fourier **51.1**, 1–26 (2001)
10. J.H. Bruinier, in *Borcherds Products on $O(2,l)$ and Chern Classes of Heegner Divisors*, Lecture Notes in Mathematics, vol. 1780 (Springer, New York, 2002)
11. J.H. Bruinier, J.I. Burgos Gil, U.Kühn, Borcherds products and arithmetic intersection theory on Hilbert modular surfaces. Duke Math. J. **139**, 1–88 (2007)
12. J.H. Bruinier, J. Funke, On the injectivity of the Kudla-Milson lift and surjectivity of the Borcherds lift, in *Moonshine: The First Quarter Century and Beyond*, ed. by J. Lepowski, J. McKay, M. Tuite (Cambridge University Press, Cambridge, 2010), pp. 12–39
13. J.H. Bruinier, On the converse theorem for Borcherds products. arXiv:1210.4821v1 [math.NT] 17 Oct 2012

14. M. Cheng, A. Dabholkar, Borcherds-Kac-Moody symmetry of $\mathcal{N} = 4$ dyons. Comm. Num. Theory Phys. **3**(1), 59–110 (2009)
15. A. Dabholkar, S. Murthy, D. Zagier, Quantum black holes, wall crossing, and mock modular forms. arXiv:1208.4074v1 [hep-th] 20 Aug (2012)
16. T. Dern, A. Krieg, Graded rings of Hermitian modular forms of degree 2. Manuscripta Math. **110**, 251–272 (2003)
17. R. Dijkgraff, E. Verlinde, H. Verlinde, Counting dyons in $N = 4$ string theory. Nucl. Phys. B. **484**, 543–561 (1997)
18. M. Eichler, D. Zagier, in *Theory of Jacobi Forms*, Progress in Mathematics, vol. 55, (Birkhäuser, Boston, 1985)
19. E. Freitag, R. Salvati Manni, Modular forms for the even modular lattice of signature $(2, 10)$. J. Algebraic Geom. **16**, 753–791 (2007)
20. V.A. Gritsenko, V.V. Nikulin, Igusa modular forms and 'the simplest' Lorentzian Kac-Moody algebras. Sb. Math. **187**(11), 1601–1641 (1996)
21. V.A. Gritsenko, V.V. Nikulin, Automorphic forms and Lorentzian Kac-Moody algebras. Part II. Int. J. Math. **9**, 201–275 (1998)
22. P. Guerzhoy, On the hecke equivariance of the borcherds isomorphism. Bull. London Math. Soc. **38**, 93–96 (2006)
23. J. Harvey, G. Moore, Algebras, BPS states, and strings. Nucl. Phys. B **463**, 315–368 (1996)
24. B. Heim, A. Murase, Symmetries for the Siegel theta functions, Borcherds lifts and automorphic Green functions. J. Number Theory **133**(10), 3485–3499. arXiv: 1003.2248
25. S. Kondo, The moduli space of Enriques surfaces and the Borcherds products. J. Algebraic Geom. **11**, 601–627 (2002)
26. M. Kontsevich, Product formulas for modular forms on $O(2, n)$. Séminaire Bourbaki, Vol. 1996/97. Astérisque No. 245, Exp. No. **821**(3), 41–56 (1997)
27. N. Scheithauer, On the classification of automorphic products and generalized Kac-Moody algebras. Invent. math. **164**, 641–678 (2006)
28. N. Scheithauer, in *Develompments and Trends in Infinite-Dimensional Lie Theory*, ed. by K.-H. Neeb, A. Pianzola. Lie Algebras, Vertex Algebras, and Automorphic Forms. Progess in Mathematics, vol. 288 (Brikhauser, Boston, 2011), pp. 151–168

Additive and Multiplicative Lifting Properties of the Igusa Modular Form

Bernhard Heim and Atsushi Murase

Mathematics Subject Classification (2010): 11F11, 11F25, 11F41

1 Introduction

The Igusa modular form χ_{10} has several remarkable properties. First of all it is a non-trivial Siegel modular cusp form of genus 2 of smallest weight for the full Siegel modular group. The Igusa modular form is a

- Saito–Kurokawa lift (proved by Maass [23–25, 27])
- Borcherds lift (proved by Gritsenko and Nikulin [1, 2, 13–15])
- Input data for a partition function for quarter-BPS dyons (see [4–6, 8, 9] for details)

The two lifting properties are related to the occurrence of the Igusa modular form in string theory. In this paper, we prove that every non-trivial Siegel modular form of genus 2 with the two lifting properties is already the Igusa modular form. This may serve as a key to determine all possible partition functions in more general settings relevant in string theory (congruence subgroups, modular forms

B. Heim (✉)
German University of Technology in Oman, Way No. 36, Building No. 331, North Ghubrah, Muscat, Sultanate of Oman
e-mail: bernhard.heim@gutech.edu.om

A. Murase
Department of Mathematics, Faculty of Science, Kyoto Sangyo University, Motoyama, Kamigamo, Kita-ku, Kyoto 603-8555, Japan
e-mail: murase@cc.kyoto-su.ac.jp

B. Heim et al. (eds.), *Automorphic Forms*, Springer Proceedings in Mathematics & Statistics 115, DOI 10.1007/978-3-319-11352-4_8

with multiplier, paramodular groups). See Cheng and Dabholkar [6] for candidates. Further Gritsenko and Nikulin have shown that the square root of the Igusa form is equal to the denominator of a Borcherds–Kac–Moody algebra [14], which has remarkable applications in the toroidally compactified heterotic string theory (see [6] for further results). To study a generalized Kac–Moody algebra one has to know the imaginary simple roots and the multiplicities of all positive roots. It is absolutely crucial that the underlying modular form has a degenerate Fourier expansion (Saito–Kurokawa lift) and an infinite product (Borcherds lift).

Let us briefly state how the Igusa form can be considered as a partition function. The dyonic $1/4$ states in 4D string theory with $\mathcal{N} \doteq 4$ space time supersymmetry are counted with modular forms [9]. In the $\mathbb{Z}$-CHL (Chaudhuri-Hockney-Lykken) model the dyons can be counted by a contour integral involving a Siegel modular form F of genus 2, the Igusa modular form of weight 10.

$$(-1)^{P\cdot Q+1} D(P,Q) = \oint d\Omega \frac{e^{\pi i (P^2\rho + Q^2\sigma - 2P\cdot PQ\nu)}}{F(\Omega)}, \quad \Omega = \left(\begin{smallmatrix} \sigma & \nu \\ \nu & \rho \end{smallmatrix}\right). \tag{1}$$

Here P and Q are two vectors in a certain Lorentzian charge lattice and P^2, Q^2, $P \cdot Q$ are related to the three T-duality invariants. This was first proposed in [9] and improved in [26] (we refer to [4] for a more detailed introduction).

In the theory of Siegel modular forms the Igusa modular form is well known. It appeared first in a famous theorem of Jun-ichi Igusa on the generators of the graded algebra of Siegel modular forms of even weight and degree 2 (see [19]). Let further E_k^2 be the normalized Siegel Eisenstein series of weight k (such that the Fourier coefficient related to 0-dim cusp at infinity is one), then the graded algebra is equal to

$$\mathbb{C}[E_4^2, E_6^2, \chi_{10}, E_{12}^2]. \tag{2}$$

We note that Igusa described χ_{10} in terms of Eisenstein series ([19], page 192):

$$\chi_{10} := -43867 \cdot 2^{-12} \cdot 3^{-5} \cdot 5^{-2} \cdot 7^{-1} \cdot 53^{-1} (E_4^2 E_6^2 - E_{10}^2).$$

In this paper we prove the following main theorem. The result is rather unexpected since a priori the additive and multiplicative lifts seem to be only randomly related.

Theorem. *Let F be a Siegel modular form of genus 2. If F is a Borcherds lift and a Saito–Kurokawa lift, then F is proportional to the Igusa modular form χ_{10} or $F \equiv 1$.*

By the converse theorem of Bruinier [3] on modular forms with Heegner divisors we obtain the following application which gives an infinite family of modular forms with divisors different from $\mathbb{Z}$-linear combinations of Humbert surfaces [11, 12].

Corollary. *Let F be a non-trivial Saito–Kurokawa lift. Let F be not proportional to the Igusa modular form. Then F has a divisor different from a Humbert surface (Heegner divisor).*

Finally, we want to note that there exist Eisenstein series in one variable which have a Heegner divisor. But this does not happen in the case of Siegel modular forms of genus 2. The results of this paper had been first announced at the RIMS conference (Japan) in January 2011 [17] and at the conference on explicit theory of automorphic forms, applications and computations in Luminy (France) May 2011. Gritsenko indicated to us that Theorem 3.3 in [15] may lead to a second, more natural proof. Up to our knowledge such a proof has not been given yet. We hope that in the future a more conceptional proof will be found to understand the uniqueness property better. Nevertheless, since our proof contains several new aspects of Borcherds and Saito–Kurokawa lifts (e.g. the Witt operator and higher Taylor coefficients of Borcherds lifts described by modular polynomials, a refined analysis of modular polynomials, cuspidality of Borcherds lifts, Siegel Φ-operator and Borcherds lifts, Hecke operators and Taylor expansions of Saito–Kurokawa lifts) the proof is interesting in its own right.

2 Siegel Modular Forms

As main reference for the basic facts and definitions from the theory of Siegel modular forms we refer to the book of Klingen [21]. We assume that the reader is familiar with the methods and results given in [16]. We denote by $M_k(\Gamma_n)$ the space of Siegel modular forms of weight k with respect to the modular group $\Gamma_n := Sp_n(\mathbb{Z})$. Modular forms are functions on the Siegel upper half-space $\mathfrak{H}_n$ of degree n. In this paper we are mainly interested in the cases $n = 1, 2$. To simplify notation we drop the index $n = 1$ and use the parametrization (τ_1, z, τ_2) for a point

$$Z = \begin{pmatrix} \tau_1 & z \\ z & \tau_2 \end{pmatrix} \in \mathfrak{H}_2.$$

We put $q_1 := \mathbf{e}(\tau_1)$, $\zeta := \mathbf{e}(z)$, $q_2 := \mathbf{e}(\tau_2)$, where $\mathbf{e}(w) := \exp(2\pi i w)$ for $w \in \mathbb{C}$.

In the following let $F \in M_k(\Gamma_2)$. To compare properties of Borcherds and Saito–Kurokawa lifts we study the Fourier expansion of F:

$$F(\tau_1, z, \tau_2) = \sum_{n,r,m \in \mathbb{Z}} A_F(n, r, m)\ \mathbf{e}(n\tau_1 + rz + m\tau_2).$$

Note that $A_F(n, r, m) = 0$ unless $n, m, 4nm - r^2 \geq 0$. The space of cusp forms $S_k(\Gamma_2)$ of degree 2 is given by all forms with $A_F(n, r, m) = 0$ unless $n, m, 4nm - r^2 > 0$. We denote by $S_k(\Gamma)$ the cusp forms in $M_k(\Gamma)$. Let $F \in M_k(\Gamma_2)$ be not identically zero. Then the order of F is defined by

$$\text{ord}(F) := \min \{n \in \mathbb{N}_0 \mid A_F(n,r,m) \neq 0\}.$$

Let G be a non-zero function on $\mathfrak{H}_2$. Assume there exists $l \in \mathbb{N}$, such that G^l is a Siegel modular form. Let m be the smallest $m \in \mathbb{N}$ with this property. Then the order of G is defined by

$$\text{ord}(G) := \frac{\text{ord}(G^m)}{m}.$$

Similarly we define the order of functions on $\mathfrak{H}$.

The **Siegel Φ-operator** maps modular forms of degree n to degree $n-1$ (see [21], Sect. 5). For $n = 2$,

$$\Phi(F)(\tau) := \lim_{y\to\infty} F(\tau, 0, iy) \in M_k(\Gamma) \qquad (\tau \in \mathfrak{H}). \tag{3}$$

Then $S_k(\Gamma_2) = \{F \in M_k(\Gamma_2) \mid \Phi(F) = 0\}$, since

$$\Phi(F)(\tau) = \sum_{n=0}^{\infty} A_F(n,0,0)\, q^n \qquad (q = e(\tau)).$$

Let k be even. Let E_k^n denote the Siegel-type Eisenstein series on Γ_n, normalized by $\Phi^n(E_k^n) = 1$, here $k > n+1$ and Φ^n denotes the nth iteration of the Φ operator. For simplification we put $e_k := E_k^1$. Let $E_k^n(f)$ denote the Klingen Eisenstein series [21] attached to $f \in S_k(\Gamma)$, $f \neq 0$ with $\Phi^{n-1}(E_k^n(f)) = f$ (here $k > n+2$). Then Theorem 2 in [21] leads to

$$M_k(\Gamma_2) = M_k^{2,0} \oplus M_k^{2,1} \oplus M_k^{2,2}, \tag{4}$$

where $M_k^{2,0} := \mathbb{C}E_k(\Gamma_2)$. The space $M_k^{2,1}$ is generated by all Klingen Eisenstein series $E_k^2(f)$ ($f \in S_k(\Gamma)$) and $M_k^{2,2} = S_k(\Gamma_2)$. The direct decomposition is given with respect to the Petersson scalar product and compatible with the Siegel Φ-operator. Let $F \in M_k(\Gamma_2)$ with decomposition $F_0 + F_1 + F_2$. Then

$$\begin{aligned}\Phi(F) &= \Phi(F_0) + \Phi(F_1) + \Phi(F_2) \\ &= c_1\, e_k + c_2\, f \quad (c_1, c_2 \in \mathbb{C},\ f \in S_k(\Gamma)).\end{aligned} \tag{5}$$

Moreover it is well known that the graded algebra of Siegel modular forms of even weight is given by

$$A_{\text{even}}(\Gamma_2) := \mathbb{C}[E_4^2, E_6^2, \chi_{10}, E_{12}^2]. \tag{6}$$

as stated in the introduction. Here χ_{10} is the Igusa modular form, the unique cusp form of weight 10 [19, 20]. It is known that there exists also a unique cusp form

χ_{35} of weight 35. Moreover every Siegel modular form of odd weight (actually only cusp forms of odd weight exist) is an element of $\chi_{35} \cdot A_{\text{even}}(\Gamma_2)$. We note that $M_0(\Gamma_2) = \mathbb{C}$ and $S_0(\Gamma_2) = \{0\}$.

Another important linear map is given by the **Witt operator**:

$$\mathcal{W}(F)(\tau_1, \tau_2) := F(\tau_1, 0, \tau_2) \qquad (\tau_1, \tau_2 \in \mathfrak{H}). \tag{7}$$

Here $\mathcal{W}(F) \in \text{Sym}^2(M_k(\Gamma))$. More generally let the Taylor expansion of $F \in M_k(\Gamma_2)$ around $z = 0$ be given by

$$F(\tau_1, z, \tau_2) = \sum_{l=0}^{\infty} \Psi_l(\tau_1, \tau_2)\, z^l. \tag{8}$$

Then $\mathcal{W}(F) = \Psi_0$. We note that $\Psi_{2l+1}(\tau_1, \tau_2) = 0$ if k is even. Moreover, if $\Psi_0 = 0$ and k is even, then $\Psi_2 \in \text{Sym}^2(S_{k+2}(\Gamma))$ (for more details see [16], Sect. 3).

Let $f_1, f_2, \ldots, f_d$ be a basis of newforms of S_k and $f_0 = e_k$. Here e_k denotes the elliptic Eisenstein series with constant term $a(0) = 1$.

Then we define

$$\text{Sym}^2(M_k(\Gamma))^D := \left\{ \sum_{i=0}^{d} \alpha_i\, f_i \otimes f_i \mid \alpha_i \in \mathbb{C} \right\}. \tag{9}$$

By $\text{Sym}^2(S_k(\Gamma))^D$ we denote the cuspidal part.

Remark 2.1. Let $F \in M_k(\Gamma_2)$ with $\mathcal{W}(F) \neq 0$. If $\text{ord}(F) \geq 2$, then $\mathcal{W}F \notin \text{Sym}^2(M_k(\Gamma))^D$.

Examine the q-expansion of $\mathcal{W}(F)$. The order of q is finite and larger than 1. Suppose that $\mathcal{W}(F)$ is an element of the diagonal space. This leads to a contradiction.

Remark 2.2. We denote by T_p the Hecke operator attached to a prime number p. Let

$$H(\tau_1, \tau_2) := \sum_{i,j=0}^{d} \alpha_{i,j}\; f_i(\tau_1) f_j(\tau_2) \qquad \left(\alpha_{i,j} \in \mathbb{C}\right).$$

We denote by $T_p^{\uparrow}$ and $T_p^{\downarrow}$ the Hecke operators applied to H as a function of τ_1 and τ_2. Here (f_j) is as above. Then $H \in \text{Sym}^2(M_k(\Gamma))^D \Leftrightarrow T_p^{\uparrow}(H) = T_p^{\downarrow}(H)$ for all prime number p. This was first discovered in [16].

3 Saito–Kurokawa Lifts

We give a short introduction to Saito–Kurokawa lifts (see also [10, 27]). We also state several results used later in the proof of the main theorem. Let k be even, since Saito–Kurokawa lifts related to $M_k(\Gamma_2)$ for k odd do not exist. There exists an injective linear map

$$\mathrm{SKL} : M_{2k-2}(\Gamma) \longrightarrow M_k(\Gamma_2), \tag{10}$$

where Hecke eigenforms f map to Hecke eigenforms $F = \mathrm{SKL}(f)$. For a Hecke eigenform f, the spinor L-function $Z(\mathrm{SKL}(f), s)$ is given by

$$Z(\mathrm{SKL}(f), s) = \zeta(s-k+1)\,\zeta(s-k+2)\,L(f,s),$$

where $L(f,s)$ is the Hecke L-function of f and $\zeta(s)$ denotes the Riemann zeta function. We are interested in the image of the lifting, which is given by the so-called Maass Spezialschar $M_k^{Spez} :=$

$$\left\{ F \in M_k(\Gamma_2) \,\middle|\, A_F(n,r,m) = \sum_{d\in\mathbb{N},\ d|(n,r,m)} d^{k-1} A_F\left(\frac{nm}{d^2}, \frac{r}{d}, 1\right) \text{ for } (n,r,m) \neq (0,0,0) \right\}. \tag{11}$$

Here (n,r,m) denotes the greatest common divisor of n, r, m. To prove our main result we use the following properties of the Maass Spezialschar (see [10, 16]):

Remark 3.1. Let $F \in M_k^{Spez}$. Then F is non-trivial iff Ψ_0^F or Ψ_2^F is not identically zero.

Remark 3.2. Let $F \in M_k^{Spez}$. Then $\Psi_0^F \in \mathrm{Sym}^2(M_k(\Gamma))^D$. If Ψ_0^F is identically zero then

$$\Psi_2^F \in \mathrm{Sym}^2(S_{k+2}(\Gamma))^D. \tag{12}$$

Remark 3.3. Let $F \in M_k(\Gamma_2)$ have the decomposition $F_0 + F_1 + F_2$ as described before. If F_1 is non-trivial, then F is not in the Spezialschar. Note that $\Psi_0^{F_0} = \alpha(e_k \otimes e_k)$. If Ψ_0^F contains a (non-vanishing) part of type $f \otimes e_k + e_k \otimes f$, then F is not in the Spezialschar.

4 Borcherds Lifts

Roughly speaking a Borcherds lift BL is a correspondence between weakly modular forms of weight $1 - \frac{m}{2}$ on $\mathfrak{H}$ with possible singularities at the cusps and certain meromorphic automorphic forms with possible characters on symmetric domains of type IV related to orthogonal groups $O(2,m)$ $(m \in \mathbb{N})$ [1–3, 15]. We note that Borcherds lifts are multiplicative lifts:

$$\mathrm{BL}(f+g) = \mathrm{BL}(f) \cdot \mathrm{BL}(g)$$

in contrast to the linear Saito–Kurokawa lifts.

Lifts to Siegel modular forms of degree 2 are related to the case $m = 3$, where the image is uniquely (up to a scalar) determined by the divisor

$$\operatorname{div}(\mathrm{BL}(f)) = \sum_{d \in \mathcal{D}} n_d \, H_d . \tag{13}$$

Here $\mathcal{D}$ is the set of all positive integers congruent to 0 or 1 modulo 4. The sum is finite and $n_d \in \mathbb{Z}$. The H_d are the Humbert surfaces (for general m they are called Heegner divisors due to Borcherds). In the following we assume that the reader is familiar with basic results of [18], which places special emphasis on the classification of Borcherds lifts on $\mathfrak{H}_2$ with respect to Γ_2. Let υ be the unique non-trivial character on Γ_2. Let $M_k(\Gamma_2, \upsilon)$ be the space of Siegel modular forms with this character. By $S_k(\Gamma_2, \upsilon)$ we denote the subspace of cusp forms. To unify notation we identify $M_k(\Gamma_2, \upsilon^0)$ with $M_k(\Gamma_2)$.

Recently [18] we found an explicit description of the Borcherds lifts related to single Heegner divisors. As a by-product one can see that the character is only related to the divisors H_1 and H_4.

Theorem 4.1 ([18]). *Classification of Borcherds products with respect to* Γ_2.

(i) For each positive integer d *with* $d \equiv 0$ *or* $1 \pmod 4$, *there exists an* $F_d \in M_{k_d}(\Gamma_2, \upsilon^{\alpha_d})$ *with* $\alpha_d \in \{0,1\}$ *satisfying* $\operatorname{div}(F_d) = H_d$.
(ii) We have $F_1 \in S_5(\Gamma_2, \upsilon)$, $F_4 \in S_{30}(\Gamma_2, \upsilon)$, *and* $F_d \in M_{k_d}(\Gamma_2)$ *if* $d > 4$.
(iii) A Borcherds lift $F \in M_k(\Gamma_2, \upsilon^{\alpha})$ $(\alpha \in \{0,1\})$ *is a constant multiple of* $\prod_d F_d^{A(d)}$, *where* d *runs through the positive integers with* $d \equiv 0$ *or* $1 \pmod 4$, *and* $A(d)$ *is a non-negative integer* ($A(d) = 0$ *except for a finite number of* d) *satisfying* $A(1) + A(4) \equiv \alpha \pmod 2$.

It is well known that $\dim(S_{10}(\Gamma_2)) = 1$ (see also [21]). Hence χ_{10} is proportional to F_1^2. Moreover, we have $\chi_{35} = F_1 \cdot F_4$.

Remark 4.2. The Borcherds lifts in $M_k(\Gamma_2)$ with $0 < k \leq 60$ are listed as follows:

Borcherds lift	Weight	Divisor
F_1^{2a} $(1 \leq a \leq 6)$	$10a$	$2aH_1$
$F_1^{2a+1} F_4$ $(0 \leq a \leq 2)$	$10a + 35$	$(2a+1)H_1 + H_4$
$F_1^{2a} F_5$ $(0 \leq a \leq 3)$	$10a + 24$	$2aH_1 + H_5$
F_4^2	60	$2H_4$
F_5^2	48	$2H_5$
F_8	60	H_8

The table shows that every Borcherds lift of weight less than or equal to 60 is a monomial of F_1, F_4, F_5 and F_8. (Here we used the following additional result: The weight k_d of F_d is divisible by 120 for $k_d \geq 9$ (see [18], Sect. 4).)

Remark 4.3. The Borcherds lift of the zero function $F_0 := \mathrm{BL}(0)$ is equal to 1. Hence we can consider $F_0 \in M_0(\Gamma_2)$ as the constant Borcherds product (see [18], Sect. 2 for an explicit version of Borcherds products).

Remark 4.4. The orders of F_1 and F_4 are equal to $\frac{1}{2}$ and $\frac{3}{2}$. Moreover, F_d is a cusp form iff d is a square. If d is a square and $d > 4$, then $\mathrm{ord}(F_d) > 1$.

Remark 4.5. Assume that $F \in M_k(\Gamma_2)$ is a Borcherds lift. Then $\Phi(F)$ is proportional $\Delta^{\frac{k}{12}}$. Here Δ is the modular discriminant given by $q \prod_{n=1}^{\infty}(1-q^n)^{24}$. This implies, if $\Phi(F)$ is non-trivial, then k is divisible by 12.

5 Proof of the Theorem

5.1 Pullbacks of Borcherds Lifts and the Modular Invariant

We utilize that the image of the Witt operator of a Borcherds lift is again a Borcherds lift. Borcherds lifts on $O(2,2)$ can be described in terms of the discriminant function, the modular invariant and modular polynomials. For detailed information on the modular invariant and modular polynomials we refer to [7, 22].

Let $j(\tau)$ be the modular invariant of Γ:

$$j(\tau) = q^{-1} + 744 + 196884\, q + 21493\, q^2 + \dots . \tag{14}$$

The modular invariant generates the space of weakly modular forms $M_0^!(\Gamma)$ of weight 0, i.e. $M_0^!(\Gamma) = \mathbb{C}[j]$. It induces a bijection $\Gamma\backslash\mathfrak{H} \longrightarrow \mathbb{C}$. Let $\Delta(\tau)$ be the discriminant function for Γ

$$\Delta(\tau) = q - 24q^2 + 252q^2 - 1472q^3 + \dots . \tag{15}$$

It is well known that $S_{12}(\Gamma) = \mathbb{C}\Delta$. Moreover, Δ is non-vanishing on $\mathfrak{H}$. The non-vanishing follows from

$$\Delta(\tau) = q \prod_{n=1}^{\infty} (1-q^n)^{24} = \sum_{n=1}^{\infty} \tau(n)\, q^n . \tag{16}$$

Here the map $n \mapsto \tau(n)$ is called the Ramanujan function. The discriminant function is a Hecke eigenform with eigenvalues $\tau(n)$.

Next we introduce the modular polynomial $\Phi_m(X,Y)$ and its irreducible factors, the primitive modular polynomials $\Phi_m^*(X,Y)$. Put

$$\mathcal{M}_m := \left\{M \in \mathbb{Z}^{2,2} \mid \det M = m\right\},$$
$$\mathcal{M}_m^* := \left\{M = \begin{pmatrix} a & b \\ c & d \end{pmatrix} \in \mathcal{M}_m \mid (a,b,c,d) = 1\right\}.$$

Then we have the double cosets decompositions

$$\mathcal{M}_m = \bigcup_{a,d>0,\, ad=m,\, d|a} \Gamma \begin{pmatrix} a & 0 \\ 0 & d \end{pmatrix} \Gamma \qquad \text{(disjoint union)},$$
$$\mathcal{M}_m^* = \Gamma \begin{pmatrix} m & 0 \\ 0 & 1 \end{pmatrix} \Gamma.$$

Let $M = \begin{pmatrix} a & b \\ c & d \end{pmatrix} \in \mathcal{M}_m$ and $\tau \in \mathfrak{H}$. Then $M\langle\tau\rangle := (a\tau+b)(c\tau+d)^{-1}$. It is well known that there exist polynomials $\Phi_m(X,Y)$ and $\Phi_m^*(X,Y)$ in $\mathbb{Z}[X,Y]$ such that

$$\Phi_m(j(\tau_1), j(\tau_2)) = \prod_{M \in \Gamma\backslash\mathcal{M}_m} \left(j(\tau_1) - j(M\langle\tau_2\rangle)\right), \tag{17}$$

$$\Phi_m^*(j(\tau_1), j(\tau_2)) = \prod_{M \in \Gamma\backslash\mathcal{M}_m^*} \left(j(\tau_1) - j(M\langle\tau_2\rangle)\right), \tag{18}$$

$$\Phi_m(X,Y) = \prod_{d>0,\, d^2|m} \Phi_{d^{-2}m}^*(X,Y). \tag{19}$$

Remark 5.1. In the following we list several properties of the modular polynomials. All properties are well known. Note that (iv) follows from (18).

(i) $\Phi_1(X,Y) = \Phi_1^*(X,Y) = X - Y$. Otherwise $\Phi_m(X,Y)$ and $\Phi_m^*(X,Y)$ are symmetric in X, Y.
(ii) The degree of $\Phi_m(X,Y)$ in X is equal to $\sigma(m) := \sum_{d|m} d$.
(iii) The degree of $\Phi_m^*(X,Y)$ in X is equal to $\sigma^*(m) := m\prod_{p|m}(1+\frac{1}{p})$. Here p are prime numbers. Hence

$$\sigma(m) \geq \sigma^*(m) > m \quad (m > 1). \tag{20}$$

(iv) Let $m > 1$. Then the expansion of $\Phi_m^*(X,Y)$ with respect to X is given by

$$X^d - (Y^m + q_m(Y))\, X^{d-1} + \ldots + Y^d, \tag{21}$$

where $q_m(Y) \in \mathbb{Z}[Y]$ of degree smaller than m and $d = \sigma^*(m)$.

The image of the Witt operator of a Borcherds lift can be described in terms of the input data of the Borcherds lift itself. For our purpose we need a slightly weaker result which we state here.

Proposition 5.2. *Let $F \in M_k(\Gamma_2)$ be a Borcherds lift. Let the image of the Witt operator be not identically zero. Then the image of the Witt operator of F is proportional to*

$$\Big(\Delta(\tau_1)\Delta(\tau_2)\Big)^{\frac{k}{12}} \prod_{n=1}^{\infty} \Phi_n^* \left(j(\tau_1), j(\tau_2)\right)^{a_n}, \tag{22}$$

where $a_n \in \mathbb{N}_0$ and almost all a_n are zero.

This follows from two observations. Borcherds products in two variables have Heegner divisors and are characterized up to powers of the discriminant function by this property. The divisor of the function

$$\Phi_n^* \left(j(\tau_1), j(\tau_2)\right)$$

has an irreducible Heegner divisor. Every irreducible Heegner divisor can be realized in this way. Secondly, the image of the Witt operator (if non-trivial) is a Borcherds product on $\mathfrak{H} \times \mathfrak{H}$. We note that it is essential to have the primitive modular polynomial instead of just the modular polynomial in the proposition. Since otherwise the exponents could be negative and hence virtual poles occur, which is more difficult to handle.

5.2 Final Steps

Let $F \in M_k(\Gamma_2)$ be of weight $k \in \mathbb{N}_0$. Let F be a Borcherds lift (BL) and a Saito–Kurokawa lift (SKL). Then $F = 1$ for $k = 0$, since F is a BL. Moreover, k is always even, since F is a ZKL. Hence we can assume that k is positive and even.

We apply the structure theorem of Borcherds lifts (Theorem 4.1) and obtain that every holomorphic Borcherds lift F can be decomposed in a finite product:

$$F \sim \prod_{d \in \mathcal{D}} F_d^{n_d} \qquad (n_d \in \mathbb{N}_0), \tag{23}$$

where $n_1 + n_4 \equiv 0 \pmod 2$ ($\sim$ means proportional up to non-zero constant). Since F is also a SKL we have $ord(F) \leq 1$. If $\mathcal{W}(F) \neq 0$ this would follow from the remarks 2.1, 3.1 and 3.2. Let $\mathcal{W}(F) = 0$. Then F is divisible by χ_{10}. And it is easy to see that $F \sim \chi_{10} F'$, where $\mathcal{W}(F') \neq 0$. From this it follows that $\psi_2(F) \sim (\Delta \otimes \Delta)\, \mathcal{W}(F')$. Hence the order of F' has to be zero; otherwise the order of $\psi_2(F)$ would be larger than one, and hence not in the diagonal. Hence we can assume that $ord(F) \leq 1$. This leads to

$$F \sim F_1^{\alpha} \cdot \prod_{d \geq 5,\ d \text{ not a square}} F_d^{n_d} \quad (\alpha \in \{0, 2\}). \tag{24}$$

Put $G := F/F_1^{\alpha}$. Since G is a BL and not a cusp form we have [18]

$$\Phi(G) \sim \Delta^r \quad (r = \frac{k - 5\alpha}{12} \in \mathbb{N}).$$

Then it is easy to see that

$$\mathcal{W}(G) \sim \Delta^r \otimes E_k + E_k \otimes \Delta^r + \text{ cuspidal}.$$

This shows that, if $\alpha = 0$, then $G = F$ is not a SKL, a contradiction. Thus we have $\alpha = 2$. Finally the case $\alpha = 2$ remains. We show that $r \geq 1$ is not possible (then the theorem is proven).

Let us assume that in the following $F \sim F_1^2 \dot{G}$, with $\Phi(G) = \Delta^r$ $(r \geq 1)$. Then Ψ_0^F is identically 0. Since F is a SKL and not identically zero, we can assume that $\Psi_2^F \neq 0$ and that

$$\Psi_2^F \in \operatorname{Sym}^2(S_{k+2}(\Gamma))^D. \tag{25}$$

Since the second Taylor coefficient of F_1^2 is proportional to $\Delta \otimes \Delta$ we obtain

$$\Psi_2^F \sim (\Delta \otimes \Delta) \cdot \mathcal{W}(G). \tag{26}$$

On the other hand as we have already observed (see Proposition 5.2) $\mathcal{W}(G)$ can be expressed in terms of the modular function j and the primitive modular polynomials. This leads to

$$\mathcal{W}(G)(\tau_1, \tau_2) \sim (\Delta^r(\tau_1) \otimes \Delta^r(\tau_2)) \prod_{n>0} \Phi_n^*(j(\tau_1), j(\tau_2))^{a(n)}, \tag{27}$$

where $a(n) \in \mathbb{N}_0$. Here $r := \frac{k-10}{12}$. Let $\overline{n}$ be the largest n such that $a(n) > 0$. Note that $a(1) = 0$. Moreover let

$$t_0 := r + 1 \quad \text{and} \quad s_0 := t_0 - \overline{n}. \tag{28}$$

Then we obtain

$$\mathcal{W}(G)(\tau_1, \tau_2) \sim \left(\Delta_x^r\ \Delta_y^r\right) \left(X^r - \left(a(\overline{n}) Y^{\overline{n}} + \dots Y^{n-1} + \dots\right) X^{r-1} + \dots X^0\right) \tag{29}$$

$$\Psi_2^F(\tau_1, \tau_2) \sim \Delta_y^{t_0} q_x + \left(A q_y^{s_0} + B q_y^{s_0+1} + \dots\right) q_x^2 + \dots. \tag{30}$$

Here we introduced the shortcuts $X = j(\tau_1)$, $Y = j(\tau_2)$, $q_x := q_1$, $\Delta_x := \Delta(\tau_1)$, $q_y := q_2$, $and \Delta_y := \Delta(\tau_2)$. Moreover, $A = -a(\overline{n})$ and $B \in \mathbb{C}$. We note that $\mathcal{W}(G)$ is not a cusp form. In the following we put $\tilde{k} = k + 2$.

Lemma 5.3. *Let H be equal to Ψ_2^F. Then $H \in Sym^2(S_{\tilde{k}}(\Gamma))^D$ and*

$$H(\tau_1, \tau_2) = \sum_{j=1}^{d} \alpha_j \, f_j(\tau_1) \, f_j(\tau_2) \qquad (\alpha_j \in \mathbb{C}) \tag{31}$$

$$= \sum_{n,m \in \mathrm{N}} a(n,m) \, q_x^n q_y^m \tag{32}$$

$$= \Delta_y^{t_0} \, q_x + \left(A q_y^{s_o} + B q_y^{s_0+1} + \ldots \right) q_x^2 + \ldots, \tag{33}$$

where $f_1, f_2, \ldots f_d$ is a basis of newforms of $S_{\tilde{k}}(\Gamma)$. We denote by $\lambda_j(n)$ the nth eigenvalue of the form f_j. Then

$$a(1,n) = \sum_j \alpha_j \, \lambda_j(n) = \begin{cases} 0 & n < t_0 \\ 1 & n = t_0 \\ t_0 \tau(2) & n = t_0 + 1 \\ a_n & \textit{otherwise}. \end{cases} \tag{34}$$

$$a(2,n) = \sum_j \alpha_j \, \lambda_j(2) \lambda_j(n) = \begin{cases} 0 & n < s_0 \\ A & n = s_0 \\ b_n & \textit{otherwise}\ . \end{cases} \tag{35}$$

Here $a_n, b_n \in \mathbb{C}$ and $\Delta(\tau) = \sum_{n=1}^{\infty} \tau(n) \, q^n$.

To obtain a contradiction we decompose the task into two cases.

I) $t_0 \geq 2s_0$.
II) $t_0 < 2s_0$ (we will distinguish between t_0 even or odd).

Case I We consider the expression

$$\sum_i \alpha_i \lambda_i(2) \lambda_i(s_0).$$

By applying (35) it is equal to A, and hence a negative integer. On the other side since the λ_i are eigenvalues we obtain

$$\sum_i \alpha_i \left(\lambda_i(2s_0) + \text{terms related to } \lambda_i(l), \ l < t_0 \right). \tag{36}$$

Hence by applying (34) only the term $\sum_i \alpha_i \lambda_i(2s_0)$ remains. Applying again (34) for $n \leq t_0$ shows that

$$\sum_i \alpha_i \lambda_i(2s_0) \geq 0.$$

But this is not possible since it is also equal to A which is negative.

Case II We first assume that t_0 is even. It is obvious that

$$\sum_i \alpha_i \lambda_i(t_0) = 1.$$

Since t_0 is even we can also consider this as $\sum \alpha_i \lambda_i(2 \cdot \frac{t_0}{2})$ and use again the fact that the $\lambda_i(n)$ are Hecke eigenvalues. Then we obtain

$$\sum_i \alpha_i \left(\lambda_i(2)\lambda_i(\frac{t_0}{2}) + \text{terms related to } \lambda_i(l),\ l < t_0 \right). \tag{37}$$

This is equal to

$$\sum_i \alpha_i \left(\lambda_i(2)\lambda_i(\frac{t_0}{2}) \right).$$

Now we can apply (35) and see that the value is equal to 0, which is a contradiction.

It remains to study the case t_0 odd. Note that $\frac{t_0+1}{2} \leq s_0$. Applying (35) leads with similar arguments to

$$\sum \alpha_i \lambda_i \left(2 \cdot \frac{t_0+1}{2} \right) = \begin{cases} 0 & \frac{t_0+1}{2} < s_0 \\ A & \frac{t_0+1}{2} = s_0 \end{cases}. \tag{38}$$

On the other hand it is also equal to $(t_0+1)\tau(2)$, since it is equal to $a(1, t_0+1)$. It is well known that $\tau(2) < 0$ ($\tau(2) = -24$). Hence, the case

$$\frac{t_0+1}{2} = s_0 \tag{39}$$

remains, since A is also negative. This leads to a contradiction. It is easy to see that (39) leads to $r = 2\overline{n}$. We recall that

$$\begin{aligned} r &= \sum_l \sigma^*(n_l)\, a(n_l) \\ &= \sigma^*(\overline{n})\, a(\overline{n}) + C \\ &= 2\overline{n}. \end{aligned}$$

Here $C \geq 0$. Since $\sigma^*(\overline{n}) > \overline{n}$. We use the fact that $\overline{n} \neq 1$. Since $a(\overline{n})$ is an integer, it has to be equal to 1. This translates into

$$t_0\, \tau(2) = -1,$$

which is not possible since $t_0 = r + 1$ and $\tau(2)$ is an integer. Hence, $r = 0$. This proves our theorem.

Acknowledgement The authors thank the referee for many helpful comments.

References

1. R.E. Borcherds, Automorphic forms on $O_{s+2,2}(\mathbb{R})$ and infinite products, Invent. Math. **120**, 161–213 (1995)
2. R.E. Borcherds, Automorphic forms with singularities on Grassmannians. Invent. Math. **132**, 491–562 (1998)
3. J.H. Bruinier, in *Borcherds Products on $O(2,l)$ and Chern Classes of Heegner Divisors*, Lecture Notes in Mathematics, vol. 1780 (Springer, New York, 2002)
4. M. Cheng, E. Verlinde, Wall crossing, discrete attractor flow, and borcherds algebra. SIGMA 4, (2008)
5. M. Cheng, E. Verlinde, Dying dyons don't count. arXiv:0706.2363. JHEP **0709** (2007)
6. M. Cheng, A. Dabholkar, Borcherds-Kac-Moody symmetry of $\mathcal{N} = 4$ dyons. Comm. Num. Theory Phys. **3**(1), 59–110 (2009)
7. D. Cox, *Primes of the Form $x^2 + ny^2$* (Wiley, New York, 1989)
8. A. Dabholkar, S. Murthy, D. Zagier, Quantum black holes, wall crossing, and mock modular forms. arXiv:1208.4074v1 [hep-th] 20 Aug (2012)
9. R. Dijkgraff, E. Verlinde, H. Verlinde, Counting dyons in $N = 4$ string theory. Nucl. Phys. B. **484**, 543–561 (1997)
10. M. Eichler, D. Zagier, in *Theory of Jacobi Forms*, Progress in Mathematics, vol. 55 (Birkhäuser, Boston, 1985)
11. G. van der Geer, On the geometry of a Siegel modular threefold. Math. Ann. **260**, 317–350 (1982)
12. G. van der Geer, in *Hilbert Modular Surfaces* (Springer, New York, 1988)
13. V. A. Gritsenko, V. V. Nikulin, The Igusa modular forms and "the simplest" Lorentzian Kac–Moody algebra. Sb. Math. **187**, 1601–1643 (1996)
14. V.A. Gritsenko, V.V. Nikulin, Siegel automorphic form corrections of some Lorentzian Kac-Moody Lie algebras. Am. J. Math. **119**, 181–224 (1997)
15. V.A. Gritsenko, V.V. Nikulin, Automorphic forms and Lorentzian Kac-Moody algebra, part II. Int. J. Math. **9**, 201–275 (1998)
16. B. Heim, On the spezialschar of maass. Int. J. Math. Math. Sci. **2010** (2010)
17. B. Heim, A. Murase, On the Igusa modular form of weight 10, RIMS Proceeding on Automorphic forms, trace formulas and zeta functions. RIMS Kokyuroku **1767**, 179–187 (2011)
18. B. Heim, A. Murase, Borcherds lifts on $Sp_2(\mathbb{Z})$, in *Geometry and Analysis of Automorphic Forms of Several Variables*, Proceedings of the international symposium in honor of Takayuki Oda on the Occasion of his 60th birthday (World Scientific, Singapore, 2012), pp. 56–76
19. J. Igusa, On Siegel modular forms of genus two. Am. J. Math. **84**, 175–200 (1962)
20. J. Igusa, On Siegel modular forms of genus two (II). Am. J. Math. **86**, 392–412 (1964)

21. H. Klingen, *Introductory Lectures on Siegel Modular Forms* (Cambridge University Press, Cambridge, 1990)
22. S. Lang, *Elliptic Functions* (Springer, New York/Berlin, 1987)
23. H. Maass, Über eine spezialschar von modulformen zweiten grades I. Invent. Math. **52**, 95–104 (1979)
24. H. Maass, Über eine spezialschar von modulformen zweiten grades II. Invent. Math. **53**, 249–253 (1979)
25. H. Maass, Über eine Spezialschar von Modulformen zweiten grades III. Invent. Math. **53** 255–265 (1979)
26. D. Shih, A. Strominger, X. Yin, Recounting dyons in $N = 4$ string theory. J. High Energy Phys. **10** (2006)
27. D. Zagier, Sur la conjecture de Saito-Kurokawa (d'aprè H. Maaß), in *Séminaire Delange-Pisot-Poitou*, Progress in Mathematics. vol. 12 (Birkhäuser, Boston, 1980), pp. 371–394

On the Special Values of Certain L-Series Related to Half-Integral Weight Modular Forms

Hidenori Katsurada*

Mathematics Subject Classification (2000): Primary 11F67,11F46, 11F66.

1 Introduction

For two modular forms $h_1(z)$ and $h_2(z)$ of half-integral weights $k_1 + 1/2$ and $k_2 + 1/2$, respectively, for $\Gamma_0(4)$, and a primitive character χ we define the Rankin–Selberg convolution product $\tilde{R}(s, h_1, h_2, \chi)$ twisted by χ as

$$\tilde{R}(s, h_1, h_2, \chi) = L(2s - k_1 - k_2 + 1, \chi_{-1}^{k_1 - k_2} \chi^2) \sum_{m=1}^{\infty} \frac{c_1(m) c_2(m) \chi(m)}{m^s},$$

where $c_1(m)$ and $c_2(m)$ denote the mth Fourier coefficients of h_1 and h_2, respectively, and $L(s, \chi_{-1}^{k_1-k_2} \chi^2)$ is the Dirichlet L-function for $\chi_{-1}^{k_1-k_2} \chi^2$ (for the precise definition of χ_{-1} see Sect. 2.)

The analytic properties of this Dirichlet series were investigated by Shimura [9]. Furthermore the algebraicity of the values of this Dirichlet series at half-integers was deeply investigated by Shimura [9]. However, as far as we know, there is no

*The author was partly supported by Grant-in-Aid for Scientific Research, JSPS

H. Katsurada (✉)
Muroran Institute of Technology,
27-1 Mizumoto, Muroran, 050-8585, Japan
e-mail: hidenori@mmm.muroran-it.ac.jp

B. Heim et al. (eds.), *Automorphic Forms*, Springer Proceedings
in Mathematics & Statistics 115, DOI 10.1007/978-3-319-11352-4_9

literature on the algebraicity of its special values at integers except for [5]. Therefore we naturally ask the following question:

Question What can one say about the algebraicity of $\tilde{R}(m, h_1, h_2, \chi)$ with m an integer?

In [5], we gave a partial answer to the above question in the case h_1 is a cuspidal Hecke eigenform in Kohnen's plus subspace for $\Gamma_0(4)$ and h_2 is Zagier's Eisenstein series of weight $3/2$. In this paper, we consider the above question in the case h_1 is a cuspidal Hecke eigenform in Kohnen's plus subspace for $\Gamma_0(4)$ and h_2 is Cohen's Eisenstein series. This paper is a summary of our paper [4]. To state our main result more explicitly, we define another Dirichlet series $R(s, h_1, h_2, \chi)$ by

$$R(s, h_1, h_2, \chi) = L(2s - k_1 - k_2 + 1, \chi^2) \sum_{m=1}^{\infty} c_{h_1}(m) c_{h_2}(m) \chi(m) m^{-s}.$$

Assume that $k_1 + k_2$ is even and that the conductor of χ is odd. Then, as will be explained in Sect. 2, it suffices to consider the above question for $R(m, h_1, h_2, \chi)$ with integer m. Now let k and n be even integers such that $n \geq 4$ and $2k - n \geq 12$. Let h be a Hecke eigenform of weight $k - n/2 + 1/2$ for $\Gamma_0(4)$ belonging to Kohnen's plus subspace and $S(h)$ the normalized Hecke eigenform of weight $2k - n$ for $SL_2(\mathbf{Z})$ corresponding to h under the Shimura correspondence. Moreover let $E_{n/2+1/2}$ be Cohen's Eisenstein series of weight $n/2+1/2$ (for the precise definition of $E_{n/2+1/2}$, see Sect. 3). Let χ be a primitive character of conductor N. We assume that N is square free and let $N = p_1 \cdots p_r$ be the prime decomposition of N. Put $l_j = l_{n,p_j} = \text{G.C.D}(n, p_j - 1)$. For an r-tuple $(i_1, i_2, \cdots, i_r)$ of integers put

$$\chi_{(i_1, \cdots, i_r)} = \chi \prod_{j=1}^{r} \left(\frac{*}{p_j}\right)_{l_j}^{i_j},$$

where $\left(\frac{*}{p_j}\right)_{l_j}$ denotes the l_jth power residue symbol mod p_j. For two Dirichlet characters η_1 and η_2 mod N, we define $J_m(\eta_1, \eta_2)$ by

$$J_m(\eta_1, \eta_2) = \sum_{Z} \eta_1(\det Z) \eta_2(1 - \text{tr}(Z)),$$

where Z runs over all symmetric matrices of degree m with entries in $\mathbf{Z}/N\mathbf{Z}$ and $\text{tr}(Z)$ denotes the trace of a matrix Z. We note that $J_1(\eta_1, \eta_2)$ is the Jacobi sum $J(\eta_1, \eta_2)$ associated with η_1 and η_2. We also put $J_m(\eta_1) = J_m(\eta_1 \left(\frac{*}{N}\right)^{m-1}, \eta_1)$, where $\left(\frac{*}{N}\right)$ is the Jacobi symbol. We then define

$$R^{(\chi)}(s,h,E_{n/2+1/2}) = \sum_{i_1=0}^{l_1-1}\cdots\sum_{i_r=0}^{l_r-1}\overline{\chi_{(i_1,\cdots,i_r)}(2^n)R(s,h,E_{n/2+1/2},\chi_{(i_1,\ldots,i_r)})}$$

$$\times \overline{J(\chi_{(i_1,\cdots,i_r)},\left(\frac{*}{N}\right))J_{n-1}(\chi_{(i_1,\cdots,i_r)})}\prod_{j=1}^{n/2-1}L(2s-2j,S(h),\chi^2_{(i_1,\cdots,i_r)}),$$

where $L(s,S(h),\chi^2_{(i_1,\cdots,i_r)})$ is Hecke's L-function of $S(h)$ twisted by $\chi^2_{(i_1,\cdots,i_r)}$. Then our main result (Theorem 3.1) can be stated as follows:

There exists a finite dimensional $\overline{\mathbf{Q}}$*-vector space* $W_{h,E_{n/2+1/2}}$ *in* $\mathbf{C}$ *such that*

$$\frac{R^{(\chi)}(m,h,E_{n/2+1/2})}{\pi^{mn}} \in W_{h,E_{n/2+1/2}}$$

for any integer m *such that* $n/2+1 \le m \le k-n/2-1$ *and a character* χ *of odd square free conductor such that* χ^n *is primitive.*

From the above result we easily obtain the following (cf. Theorem 3.2):

Let $r > \dim_{\overline{\mathbf{Q}}} W_{h,E_{n/2+1/2}}$*. Let* $m_1,m_2,\cdots,m_r$ *be integers such that* $n/2+1 \le m_1,m_2,\cdots,m_r \le k-n/2-1$ *and* $\chi_1,\chi_2,\cdots,\chi_r$ *be Dirichlet characters of odd square-free conductors* $N_1,N_2,\cdots,N_r$*, respectively, such that* χ_i^n *is primitive for any* $i=1,2,\cdots r$*. Then the values*

$$\frac{R^{(\chi_1)}(m_1,h,E_{n/2+1/2})}{\pi^{m_1n}},\cdots,\frac{R^{(\chi_r)}(m_r,h,E_{n/2+1/2})}{\pi^{m_rn}}$$

are linearly dependent over $\overline{\mathbf{Q}}$*.*

This is a certain generalization of a main result in [5] as will be explained later.

A main tool for proving Theorem 3.1 is the twisted Koecher–Maaß series of the Duke-Imamoglu-Ikeda lift of h. To explain this, we define the twisted Koecher–Maaß series of a Siegel modular form in a more general setting. Let $F(Z)$ be a modular form of weight k with respect to the symplectic group $Sp_n(\mathbf{Z})$. For a positive integer N let $SL_{n,N}(\mathbf{Z}) = \{U \in SL_n(\mathbf{Z}) \mid U \equiv 1_n \bmod N\}$ and $e_N(T) = \#\{U \in SL_{n,N}(\mathbf{Z}) \mid T[U] = T\}$. For a primitive Dirichlet character χ mod N we define the Koecher–Maaß series $L(s,F,\chi)$ of F twisted by χ as

$$L(s,F,\chi) = \sum_T \frac{\chi(\mathrm{tr}(T))c_F(T)}{e_N(T)(\det T)^s},$$

where T runs over a complete set of representatives of $SL_{n,N}(\mathbf{Z})$-equivalence classes of positive definite half-integral matrices of degree n and $c_F(T)$ denotes the Tth Fourier coefficient of F. We note that this Dirichlet series coincides with the Hecke L-function associated to F twisted by χ in case $n=1$. Though we are mainly concerned with $L(s, F, \chi)$ in this paper, we also define another type of twisted Koecher–Maaß series $L^*(s, F, \chi)$ as

$$L^*(s, F, \chi) = \sum_T \frac{\chi(\det(2T))c_F(T)}{e(T)(\det T)^s},$$

where T runs over a complete set of representatives of $SL_n(\mathbf{Z})$-equivalence classes of positive definite half-integral matrices of degree n and $e(T) = e_1(T)$. These two Dirichlet series $L(s, F, \chi)$ and $L^*(s, F, \chi)$ essentially coincide with each other in case $n = 1$, but they don't in general. To distinguish these two Dirichlet series, we sometimes call $L(s, F, \chi)$ and $L^*(s, F, \chi)$ the twisted Koecher–Maaß series of the first and second kind, respectively. In Sect. 4, we will discuss a relation between these two Dirichlet series (cf. Theorem 4.5). Now for the integers k and n stated above, let h a cuspidal Hecke eigenform h in Kohnen's plus subspace of weight $k - n/2 + 1/2$ for $\Gamma_0(4)$. Let $I_n(h)$ be the Duke-Imamoglu-Ikeda lift of h to the space of Siegel cusp forms of degree n. Then, in Sect. 5, first we give an explicit formula of $L^*(s, I_n(h), \eta)$ in terms of the Rankin–Selberg series $R(s, h, E_{n/2+1/2}, \eta)$ and shifted products of Hecke's L-functions of $S(h)$ twisted by η^2 in the case η is a primitive character (cf. Theorem 5.1.) Next, by this result combined with Theorem 4.5, we give an explicit formula of $L(s, I_n(h), \chi^n)$ in terms of $R^{(\chi)}(s, h, E_{n/2+1/2})$ and a sum of the shifted products $\prod_{j=1}^{n/2-1} L(2s - 2j + 1, S(h), \chi^2_{(i_1,\cdots,i_r)})$ (cf. Theorem 5.2 and its corollary). This implies that $R^{(\chi)}(s, h, E_{n/2+1/2})$ can be expressed in terms of $L(s, I_n(h), \chi^n)$ and the sum of the shifted products. Thus we can prove our main result using the algebraicity of Hecke's L-function of $S(h)$ (cf. Theorem 2.1) combined with the arithmetic properties of $L(s, I_n(h), \chi^n)$, which were investigated by Choie and Kohnen [1] in a more general setting (cf. Theorem 4.2). We can also prove a functional equation for $R^{(\chi)}(s, h, E_{n/2+1/2})$ in case $n \equiv 2 \bmod 4$ using the functional equation for $L(s, F, \chi^n)$ (cf. Theorem 3.3).

Notation We denote by $\mathbf{e}(x) = \exp(2\pi\sqrt{-1}x)$ for a complex number x. For a commutative ring R, we denote by $M_{mn}(R)$ the set of (m, n)-matrices with entries in R. For an (m, n)-matrix X and an (m, m)-matrix A, we write $A[X] = {}^tXAX$, where ${}^tX$ denotes the transpose of X. Let a be an element of R. Then for an element X of $M_{mn}(R)$ we often use the same symbol X to denote the coset $X \bmod aM_{mn}(R)$. Put $GL_m(R) = \{A \in M_m(R) \mid \det A \in R^*\}$ and $SL_m(R) = \{A \in M_m(R) \mid \det A = 1\}$, where $\det A$ denotes the determinant of a square matrix A and R^* is the unit group of R. We denote by $S_n(R)$ the set of symmetric matrices of degree n with entries in R. In particular, if S is a subset of $S_n(\mathbf{R})$ with $\mathbf{R}$ the field of real numbers, we denote by $S_{>0}$ (resp. $S_{\geq 0}$) the subset of S consisting of positive definite (resp. semi-positive definite) matrices. The group $SL_n(\mathbf{Z})$ acts on the set $S_n(\mathbf{R})$ in the following way:

$$SL_n(\mathbf{Z}) \times S_n(\mathbf{R}) \ni (g, A) \longrightarrow {}^tgAg \in S_n(\mathbf{R}).$$

Let G be a subgroup of $GL_n(\mathbf{Z})$. For a subset $\mathcal{B}$ of $S_n(\mathbf{R})$ stable under the action of G we denote by $\mathcal{B}/G$ the set of equivalence classes of $\mathcal{B}$ with respect to G. We sometimes identify $\mathcal{B}/G$ with a complete set of representatives of $\mathcal{B}/G$. Two symmetric matrices A and A' with entries in R are said to be equivalent with each other with respect to G and write $A \sim_G A'$ if there is an element X of G such that $A' = A[X]$. Let $\mathcal{L}_n$ denote the set of half-integral matrices of degree n over $\mathbf{Z}$, that is, $\mathcal{L}_n$ is the set of symmetric matrices of degree n whose (i, j)-component belongs to $\mathbf{Z}$ or $\frac{1}{2}\mathbf{Z}$ according as $i = j$ or not.

2 Review on the Algebraicity of L-Values of Elliptic Modular Forms of Integral and Half-Integral Weights

Before stating our main results, we review on the special values of L-functions of elliptic modular forms of integral and half-integral weights. Put $J_n = \begin{pmatrix} O_n & -1_n \\ 1_n & O_n \end{pmatrix}$, where 1_n and O_n denotes the unit matrix and the zero matrix of degree n, respectively. Furthermore, put

$$Sp_n(\mathbf{Z}) = \{M \in GL_{2n}(\mathbf{Z}) \mid J_n[M] = J_n\}.$$

Let l be an integer or a half-integer, and let Γ be a congruence subgroup of $Sp_n(\mathbf{Z})$. We then denote by $M_l(\Gamma)$ the space of modular forms of weight l with respect to Γ and by $S_l(\Gamma)$ the subspace of $M_l(\Gamma)$ consisting of cusp forms. We also denote by $\Gamma_0(4)$ the subgroup of $SL_2(\mathbf{Z})$ consisting of matrices whose left lower entries are congruent to 0 mod N. Let

$$f(z) = \sum_{m=1}^{\infty} c_f(m)\mathbf{e}(mz)$$

be a normalized Hecke eigenform in $S_k(SL_2(\mathbf{Z}))$ and χ be a primitive Dirichlet character. Then let us define Hecke's L-function $L(s, f, \chi)$ of f twisted by χ as

$$L(s, f, \chi) = \sum_{m=1}^{\infty} c_f(m)\chi(m)m^{-s}.$$

Then we have the following result (cf. [8]):

Theorem 2.1. *There exist complex numbers $u_{\pm}(f)$ uniquely determined up to $\overline{\mathbf{Q}}^{\times}$ multiple such that*

$$L(m, f, \chi)(\pi^m u_j(f))^{-1} \in \overline{\mathbf{Q}}$$

for any integer $0 < m \leq k-1$ and a primitive character χ, where $j = +$ or $-$ according as $(-1)^m\chi(-1) = 1$ or -1.

We remark that we have $L(m, f, \chi) \neq 0$ if $m \neq k/2$ and $L(k/2, f, \chi) \neq 0$ for infinitely many χ.

Next let us consider the half-integral weight case. Let

$$h_1(z) = \sum_{m=1}^{\infty} c_{h_1}(m)\mathbf{e}(mz)$$

be a Hecke eigenform in $S_{k_1+1/2}(\Gamma_0(4))$ and

$$h_2(z) = \sum_{m=0}^{\infty} c_{h_2}(m)\mathbf{e}(mz)$$

be an element of $M_{k_2+1/2}(\Gamma_0(4))$. For positive integers e and l, let $\chi_{(-1)^l e}$ be the Dirichlet character corresponding to the extension $\mathbf{Q}(\sqrt{(-1)^l e}/\mathbf{Q})$. Let χ be a primitive character mod N. Then we define

$$\tilde{R}(s, h_1, h_2, \chi) = L(2s - k_1 - k_2 + 1, \omega) \sum_{m=1}^{\infty} c_{h_1}(m)c_{h_2}(m)\chi(m)m^{-s},$$

where $\omega(d) = \chi_{-1}^{k_1-k_2}\chi^2(d)$. Now let $S(h_1)$ be the normalized Hecke eigenform in $S_{2k_1}(SL_2(\mathbf{Z}))$ corresponding to h_1 under the Shimura correspondence. Then the following result is due to Shimura [9].

Theorem 2.2. *Assume that $k_1 > k_2$. Under the above notation we have*

$$\tilde{R}(m + 1/2, h_1, h_2, \chi)(u_-(S(h_1))\pi^{-k_2+1+2m})^{-1} \in \overline{\mathbf{Q}}(h_1)\overline{\mathbf{Q}}(h_2)$$

for any integer $k_2 \leq m \leq k_1 - 1$ and a primitive character χ, where $\overline{\mathbf{Q}}(h_i)$ is the field, generated over $\overline{\mathbf{Q}}$, by all the Fourier coefficients of h_i.

Corollary. *Let the notation be as above. Assume that $k_1 > k_2$ and that $c_{h_1}(n), c_{h_2}(n) \in \overline{\mathbf{Q}}$ for any $n \in \mathbf{Z}_{\geq 0}$. Then there exists a one-dimensional $\overline{\mathbf{Q}}$-vector space U_{h_1,h_2} in $\mathbf{C}$ such that*

$$\tilde{R}(m + 1/2, h_1, h_2, \chi)\pi^{-2m} \in U_{h_1,h_2}$$

for any integer $k_2 \leq m \leq k_1 - 1$ and a primitive character χ.

Now we consider the values of $\tilde{R}(s, h_1, h_2, \chi)$ at integers. Let

$$R(s, h_1, h_2, \chi) = L(2s - k_1 - k_2 + 1, \chi^2) \sum_{m=1}^{\infty} c_{h_1}(m)c_{h_2}(m)\chi(m)m^{-s}.$$

be the Dirichlet series defined in Sect. 1. Assume that $k_1 + k_2$ is even and that the conductor of χ is odd. Then we have

$$R(s, h_1, h_2, \chi) = (1 - 2^{-2s+k_1+k_2-1}\chi^2(2))^{-1}\tilde{R}(s, h_1, h_2, \chi).$$

Hence it suffices to consider the question in Sect. 1 for $R(m, h_1, h_2, \chi)$ with integer m.

3 Main Results

For a non-negative integer m and a positive integer l, Cohen's function $H(l, m)$ is given by $H(l, m) = L_{-m}(1 - l)$. Here

$$L_D(s) = \begin{cases} \zeta(2s-1), & D = 0 \\ L(s, \chi_{D_K}) \sum_{a|f} \mu(a)\chi_{D_K}(a)a^{-s}\sigma_{1-2s}(f/a), & D \neq 0,\ D \equiv 0, 1 \bmod 4 \\ 0, & D \equiv 2, 3 \bmod 4, \end{cases}$$

where the positive integer f is defined by $D = D_K f^2$ with the discriminant D_K of $K = \mathbf{Q}(\sqrt{D})$, χ_{D_K} is the Kronecker symbol, μ is the Möbius function, and $\sigma_s(n) = \sum_{d|n} d^s$. Furthermore we define Cohen's Eisenstein series $E_{l+1/2}(z)$ by

$$E_{l+1/2}(z) = \sum_{m=0}^{\infty} H(l, m)\mathbf{e}(mz).$$

It is known that $E_{l+1/2}(z)$ is a modular form of weight $l + 1/2$ belonging to Kohnen's plus space. Let k and n be positive even integers such that $n \geq 4,\ 2k - n \geq 12$. Let $h(z)$ be a Hecke eigenform in Kohnen's plus subspace $S^+_{k-n/2+1/2}(\Gamma_0(4))$ (cf. [7]) and $S(h)$ be the normalized Hecke eigenform in $S_{2k-n}(SL_2(\mathbf{Z}))$ corresponding to h under the Shimura correspondence. Let p be a prime number and l be a positive integer dividing $p - 1$. Take an lth root of unity ζ_l and a prime ideal $\mathfrak{p}$ of $\mathbf{Q}(\zeta_l)$ lying above p. Let a be an integer prime to p. Then we have $a^{(p-1)/l} \equiv \zeta_l^i \bmod \mathfrak{p}$ with some $i \in \mathbf{Z}$. We then put $\left(\frac{a}{p}\right)_l = \zeta^i$. We call $\left(\frac{*}{p}\right)_l$ the lth power residue symbol mod p. In the case $l = 2$, this is the Legendre symbol, and we write it as $\left(\frac{*}{p}\right)$ as usual. We note that this definition of the power residue symbol is different from the usual one and depends on the choice of $\mathfrak{p}$ and ζ_l except the case $l = 2$. We denote by $\left(\frac{*}{N}\right)$ the Jacobi symbol for a positive odd integer M. Let χ be a primitive Dirichlet character of conductor N. We assume that N is a square-free odd integer and write $N = p_1 \cdots p_r$ with $p_1, \cdots, p_r$ prime numbers. Put $l_j = l_{n,p_j} = \mathrm{G.C.D}(n, p_j - 1)$. For an r-tuple $(i_1, i_2, \cdots, i_r)$ of integers put

$$\chi_{(i_1,\cdots,i_r)} = \chi \prod_{j=1}^{r} \left(\frac{*}{p_j}\right)_{l_j}^{i_j}.$$

For two Dirichlet characters η_1 and η_2 mod N, let $J_m(\eta_1, \eta_2)$ and $J_m(\eta_1)$ be as those defined in Sect. 1. By definition, $J_m(\eta_1, \eta_2)$ is an algebraic number. As in Sect. 1, we define

$$R^{(\chi)}(s, h, E_{n/2+1/2})$$
$$= \sum_{i_1=0}^{l_1-1} \cdots \sum_{i_r=0}^{l_r-1} \overline{\chi_{(i_1,\cdots,i_r)}(2^n)}\, \overline{J(\chi_{(i_1,\cdots,i_r)}, \left(\frac{*}{N}\right))}\, \overline{J_{n-1}(\chi_{(i_1,\cdots,i_r)})}$$
$$\times R(s, h, E_{n/2+1/2}, \chi_{(i_1,\ldots,i_r)}) \mathbf{L}_n(s, S(h), \chi_{(i_1,\cdots,i_r)}),$$

where

$$\mathbf{L}_n(s, S(h), \eta) = \prod_{j=1}^{n/2-1} L(2s - 2j, S(h), \eta^2)$$

for a primitive character η. We note that $R^{(\chi)}(s, h, E_{n/2+1/2})$ does not depend on the choice of an l_ith root of unity ζ_{l_i} and an prime ideal $\mathfrak{p}_i$ of $\mathbf{Q}(\zeta_{l_i})$ lying above p_i.

Remark. (1) Let m be an integer s.t. $n/2 + 1 \le m \le k - n/2 - 1$. Then the value $\dfrac{\mathbf{L}_n(m, S(h), \chi^2_{(i_1,\cdots,i_r)})}{\pi^{m(n-2)}}$ belongs to $\overline{\mathbf{Q}}u_+(S(h))^{n/2-1}\pi^{-n^2/4+n/2}$ for any χ. In particular if $n \equiv 2 \bmod 4$, then it is non-zero for any χ, and if $n \equiv 0 \bmod 4$, then it is non-zero for infinitely many χ.

(2) As will be stated in Sect. 4, $J_{n-1}(\chi_{(i_1,\cdots,i_r)})$ is expressed as a product of Jacobi sums, and it is non-zero algebraic number if χ^n is primitive.

Theorem 3.1. *There exists a finite dimensional* $\overline{\mathbf{Q}}$*-vector space* $W_{h,E_{n/2+1/2}}$ *in* $\mathbf{C}$ *such that*

$$\frac{R^{(\chi)}(m, h, E_{n/2+1/2})}{\pi^{mn}} \in W_{h,E_{n/2+1/2}}$$

for any integer $n/2 + 1 \le m \le k - n/2 - 1$ *and a character* χ *of odd square-free conductor such that* χ^n *is primitive.*

Theorem 3.2. *Let* $r > \dim_{\overline{\mathbf{Q}}} W_{h,E_{n/2+1/2}}$. *Let* $m_1, m_2, \cdots, m_r$ *be integers such that* $n/2 + 1 \le m_1, m_2, \cdots, m_r \le k - n/2 - 1$ *and* $\chi_1, \chi_2, \cdots, \chi_r$ *be Dirichlet characters of odd square-free conductors* $N_1, N_2, \cdots, N_r$*, respectively, such that*

χ_i^n is primitive for any $i = 1, 2, \cdots r$. Then the values $\dfrac{R^{(\chi_1)}(m_1, h, E_{n/2+1/2})}{\pi^{m_1 n}}, \cdots,$ $\dfrac{R^{(\chi_r)}(m_r, h, E_{n/2+1/2})}{\pi^{m_r n}}$ *are linearly dependent over $\overline{\mathbf{Q}}$.*

Corollary. *Assume that $n \equiv 2 \bmod 4$. Let r and $m_1, m_2, \cdots, m_r$ be as above. Let $\chi_1, \chi_2, \cdots, \chi_r$ be Dirichlet characters of odd prime conductors $p_1, p_2, \cdots, p_r$, respectively, such that χ_i^n is non-trivial for any $i = 1, 2, \cdots r$. Put $l_i = \mathrm{GCD}(n, p_i - 1)$. Then the values* $\left\{\dfrac{R(m_i, h, E_{n/2+1/2}, \chi_{i\,(j)})}{\pi^{2m_i}}\right\}_{1 \le i \le r, 0 \le j \le l_i - 1}$ *are linearly dependent over $\overline{\mathbf{Q}}$.*

We also have a functional equation for $R^{(\chi)}(s, h, E_{n/2+1/2})$:

Theorem 3.3. *Let h be as above. Let χ be a primitive character of odd square-free conductor N. Assume that $n \equiv 2 \bmod 4$ and that χ^n is primitive. Put*

$$\mathcal{R}^{(\chi)}(s, h, E_{n/2+1/2}) = N^{2s}\tau(\chi^n)^{-1}\gamma_n(s) R^{(\chi)}(s, h, E_{n/2+1/2}),$$

where $\tau(\chi^n)$ is the Gauss sum of χ^n, and

$$\gamma_n(s) = (2\pi)^{-ns}\prod_{i=1}^{n} \pi^{(i-1)/2}\Gamma(s - (i-1)/2).$$

Then $\mathcal{R}^{(\chi)}(s, h, E_{n/2+1/2})$ has an analytic continuation to the whole s-plane and has the following functional equation:

$$\mathcal{R}^{(\chi)}(k - s, h, E_{n/2+1/2}) = \mathcal{R}^{(\chi)}(s, h, E_{n/2+1/2}).$$

Remark. (1) The series $\{R(s, h, E_{n/2+1/2}, \chi_{i\,(j)})\}_{1 \le i \le r, 0 \le j \le l_i - 1}$ are linearly independent over $\mathbf{C}$ as functions of s.

(2) In the case of $n = 2$, this type of result was given for $R(m, h, E_{3/2})$ with $E_{3/2}$ Zagier's Eisenstein series of weight $3/2$ by [5]. Cohen's Eisenstein series is a holomorphic modular form, where as Zagier's Eisenstein series is not. Nevertheless, the former can be regarded as a generalization of the latter. Therefore, our present result can be regarded as a generalization of [5]. (3) The meromorphy of this type of series was derived in [9] by using so called the Rankin–Selberg integral expression in a more general setting, but we don't know whether the functional equation of the above type can be directly proved without using the above method.

4 Twisted Koecher–Maaß Series

To prove the main results, in this section and the next, we consider the twisted Koecher–Maaß series of a Siegel modular form. Let $F(Z) \in M_k(Sp_n(\mathbf{Z}))$. Then $F(Z)$ has the Fourier expansion:

$$F(Z) = \sum_{T \in \mathcal{L}_{n \geq 0}} c_F(T)\mathbf{e}(\mathrm{tr}(TZ)),$$

where $\mathrm{tr}(X)$ denotes the trace of a matrix X. For $N \in \mathbf{Z}_{>0}$, put $SL_{n,N}(\mathbf{Z}) = \{U \in SL_n(\mathbf{Z}) \mid U \equiv 1_n \bmod N\}$, and for $T \in \mathcal{L}_{n>0}$ put $e_N(T) = \#\{U \in SL_{n,N}(\mathbf{Z}) \mid T[U] = T\}$. For a primitive Dirichlet character χ mod N let

$$L(s, F, \chi) = \sum_{T \in \mathcal{L}_{n>0}/SL_{n,N}(\mathbf{Z})} \frac{\chi(\mathrm{tr}(T))c_F(T)}{e_N(T)(\det T)^s}$$

be the twisted Koecher–Maaß series of F of the first kind as in Sect. 1. The following two theorems are due to Choie and Kohnen [1].

Theorem 4.1. *Let $F \in S_k(Sp_n(\mathbf{Z}))$ and χ a primitive character of conductor N. Put*

$$\gamma_n(s) = (2\pi)^{-ns} \prod_{i=1}^{n} \pi^{(i-1)/2}\Gamma(s - (i-1)/2),$$

and

$$\Lambda(s, F, \chi) = N^{2s}\tau(\chi)^{-1}\gamma_n(s)L(s, F, \chi) \quad (\mathrm{Re}(s) >> 0),$$

where $\tau(\chi)$ is the Gauss sum of χ. Then $\Lambda(s, F, \chi)$ has an analytic continuation to the whole s-plane and has the following functional equation:

$$\Lambda(k - s, F, \chi) = (-1)^{nk/2}\chi(-1)\Lambda(s, F, \overline{\chi}).$$

Theorem 4.2. *Let F and χ be as above. Then there exists a finite dimensional $\overline{\mathbf{Q}}$-vector space V_F in $\mathbf{C}$ such that*

$$L(m, F, \chi)\pi^{-nm} \in V_F$$

for any primitive character χ and any integer m such that $(n+1)/2 \leq m \leq k - (n+1)/2$.

Example. Let $n = 1$. Take a basis $\{f_1, \cdots, f_d\}$ of $S_k(SL_2(\mathbf{Z}))$ consisting of normalized Hecke eigenforms. Write $f \in S_k(SL_2(\mathbf{Z}))$ as

$$f = a_1 f_1 + \cdots + a_d f_d$$

with $a_1, \cdots, a_d \in \mathbf{C}$. Then put $w_i = a_i u_+(f_i), w_{d+i} = a_i u_-(f_i)$ $(i = 1, \cdots, d)$, and $V_f = \sum_{i=1}^{2d} \overline{\mathbf{Q}} w_i$. Then V_f satisfies the required property for f.

Now let

$$L^*(s, F, \chi) = \sum_{T \in \mathcal{L}_{n>0}/SL_n(\mathbf{Z})} \frac{\chi(\det(2T)) c_F(T)}{e(T)(\det T)^s}$$

be the twisted Koecher–Maaß series of F of the second kind as in Sect. 1. We will discuss a relation between these two Dirichlet series. Let N be a positive integer. Let g be a periodic function on $\mathbf{Z}$ with a period N and ϕ a polynomial in $t_1, \ldots, t_r$. Then for an element $u = (a_1 \bmod N, \ldots, a_r \bmod N) \in (\mathbf{Z}/N\mathbf{Z})^r$, the value $g(\phi(a_1, \ldots, a_r))$ does not depend on the choice of the representative u. Therefore we denote this value by $g(\phi(u))$. Now let χ be a primitive character mod N. For $A \in \mathcal{L}_{n>0}$, put

$$h(A, \chi) = \sum_{U \in SL_n(\mathbf{Z}/N\mathbf{Z})} \chi(\mathrm{tr}(A[U])).$$

The following proposition is due to [[5], Proposition 3.1].

Proposition 4.3. *Let*

$$F(Z) = \sum_{A \in \mathcal{L}_{n \geq 0}} c_F(A) \mathbf{e}(\mathrm{tr}(AZ))$$

be an element of $M_k(Sp_n(\mathbf{Z}))$. Let χ be a Dirichlet character mod N. Assume $N \neq 2$. Then we have

$$L(s, F, \chi) = \sum_{A \in \mathcal{L}_{n>0}/SL_n(\mathbf{Z})} \frac{c_F(A) h(A, \chi)}{e(A)(\det A)^s}.$$

For a Dirichlet character χ mod N, let $\chi^{(p)}$ be the p-factor of χ so that $\chi = \prod_{p|N} \chi^{(p)}$. For a prime number p put

$$\gamma_{n,p} = p^{n^2 - n(n+1)/2}(1 - p^{-n/2}) \prod_{e=1}^{(n-2)/2} (1 - p^{-2e})$$

or

$$\gamma_{n,p} = p^{n^2-n(n+1)/2} \prod_{e=1}^{(n-1)/2} (1-p^{-2e})$$

according as n is even or odd. The following result is a technical tool for proving our main result.

Theorem 4.4. *Let $A \in \mathcal{L}_{n>0}$. Let N be a square-free odd integer, and let $N = \prod_{i=1}^r p_i$ be the prime decomposition of N. Let χ be a primitive Dirichlet character mod N. For each positive integer $i \leq r$, put $l_i = \mathrm{G.C.D}(n, p_i - 1)$ and let $u_{0,i}$ be a primitive l_ith root of unity mod p_i.*

(1) *If $\chi^{(p_i)}(u_{0,i}) \neq 1$ for some i, then we have $h(A, \chi) = 0$.*
(2) *Assume that $\chi^{(p_i)}(u_{0,i}) = 1$ for any i. Fix a character $\tilde{\chi}$ such that $\tilde{\chi}^n = \chi$.*

(2.1) *Let n be even. Then we have*

$$h(A,\chi) = \prod_{i=1}^{r} (-1)^{n(p_i-1)/4} \gamma_{n,p_i}$$

$$\times \sum_{i_1=0}^{l_1-1} \cdots \sum_{i_r=0}^{l_r-1} \overline{\tilde{\chi}}_{(i_1,\cdots,i_r)}(2^n)\tilde{\chi}_{(i_1,\cdots,i_r)}(\det(2A))\overline{J(\tilde{\chi}_{(i_1,\cdots,i_r)}, \left(\frac{*}{N}\right))}\overline{J_{n-1}(\tilde{\chi}_{(i_1,\cdots,i_r)})}.$$

(2.2) *Let n be odd, and assume that χ^2 is primitive. Then we have*

$$h(A,\chi) = \prod_{i=1}^{r} (-1)^{(n-1)(p_i-1)/4} \gamma_{n,p_i}$$

$$\times \sum_{i_1=0}^{l_1-1} \cdots \sum_{i_r=0}^{l_r-1} \overline{\tilde{\chi}}_{(i_1,\cdots,i_r)}(2^n)\tilde{\chi}_{(i_1,\cdots,i_r)}(\det(2A))\overline{J_{n-1}(\tilde{\chi}_{(i_1,\cdots,i_r)})}.$$

The proof of the above theorem is elementary but is rather lengthy. The details will be given in [4].

Remark. Let η be a primitive Dirichlet character of odd prime conductor p. Assume that $\eta^2 \neq 1$. Then we can prove that we have

$$J(\eta, \left(\frac{*}{p}\right))J(\eta\left(\frac{*}{p}\right), \eta\left(\frac{*}{p}\right)) = \left(\frac{-1}{p}\right)\bar{\eta}(4)p.$$

(This is not so trivial. For the details, see [4].) Hence for $A \in \mathcal{L}_{2>0}$ and a primitive character χ of odd square-free conductor N such that $\chi^{(p)}(-1) = 1$ for any prime divisor p of N, we have

$$h(A,\chi)=\prod_{p|N}\left\{\left(1+\left(\frac{4\det A}{p}\right)\right)\left(1-\left(\frac{-1}{p}\right)p^{-1}\right)\right\}N^2\left(\frac{-1}{N}\right)\tilde{\chi}(4\det A)),$$

where $\tilde{\chi}$ is a character such that $\tilde{\chi}^2=\chi$. This coincides with (2) of Theorem 3.8 in [5].

By Theorem 4.4 and Proposition 4.3 we easily obtain:

Theorem 4.5. *Let $N, p_i, l_i, u_{0,i}$ $(i=1,\cdots,r)$ and χ be as in Theorem 4.4, and let F be an element of $M_k(Sp_n(\mathbf{Z}))$.*

(1) *If $\chi^{(p_i)}(u_{0,i})\neq 1$ for some i, then we have $L(s,F,\chi)=0$.*
(2) *Assume that $\chi^{(p_i)}(u_{0,i})=1$ for any i. Fix a character $\tilde{\chi}$ such that $\tilde{\chi}^n=\chi$.*

(2.1) *Let n be even. Then we have*

$$L(s,F,\chi)=\prod_{i=1}^{r}(-1)^{n(p_i-1)/4}\gamma_{n,p_i}$$

$$\times\sum_{i_1=0}^{l_1-1}\cdots\sum_{i_r=0}^{l_r-1}\overline{\tilde{\chi}}_{(i_1,\cdots,i_r)}(2^n)\overline{J(\tilde{\chi}_{(i_1,\cdots,i_r)},\left(\frac{*}{N}\right))}\overline{J_{n-1}(\tilde{\chi}_{(i_1,\cdots,i_r)})}L^*(s,F,\tilde{\chi}_{(i_1,i_2,\cdots,i_r)}).$$

(2.2) *Let n be odd, and assume that χ^2 is primitive. Then we have*

$$L(s,F,\chi)=\prod_{i=1}^{r}(-1)^{(n-1)(p_i-1)/4}\gamma_{n,p_i}$$

$$\times\sum_{i_1=0}^{l_1-1}\cdots\sum_{i_r=0}^{l_r-1}\overline{\tilde{\chi}}_{(i_1,\cdots,i_r)}(2^n)\overline{J_{n-1}(\tilde{\chi}_{(i_1,i_2,\cdots,i_r)})}L^*(s,F,\tilde{\chi}_{(i_1,i_2,\cdots,i_r)}).$$

To give an explicit formula of $J_m(\chi,\eta)$ for primitive characters χ,η mod N, we define $I_m(\chi,\eta)$ as

$$I_m(\chi,\eta)=\sum_{Z\in S_m(\mathbf{Z}/N\mathbf{Z})}\chi(\det Z)\eta(\mathrm{tr}(Z)).$$

Then we have the following two propositions, whose proof will be given precisely in [4].

Proposition 4.6. *Let χ and η be primitive character mod an odd prime number p. Assume that $\chi^2\neq 1$ and that η is non-trivial. Put $c_m(\chi,\eta)=1$ or 0 according as $\chi^{m-1}\eta=1$ or not.*

(1) *Assume that m is odd. Then*

$$I_m(\chi,\eta)=c_m(\chi,\eta)\left(\frac{-1}{p}\right)^{(m-1)/2}p^{(m-1)/2}(p-1)J_{m-1}(\chi\left(\frac{*}{p}\right),\eta).$$

(2) *Assume that m is even. Then*

$$I_m(\chi,\eta) = c_m(\chi,\eta)\left(\frac{-1}{p}\right)^{m/2} p^{(m-2)/2}(p-1)\chi(-1)J(\chi,\left(\frac{*}{p}\right))J_{m-1}(\chi\left(\frac{*}{p}\right),\eta).$$

Proposition 4.7. *Let χ, η, and p be as in Proposition 4.6.*

(1) *Assume that m is odd. Then*

$$J_m(\chi,\eta) = \left(\frac{-1}{p}\right)^{(m-1)/2} p^{(m-1)/2}$$
$$\times\{J(\chi,\chi^{m-1}\eta)J_{m-1}(\chi\left(\frac{*}{p}\right),\eta) + \eta(-1)I_{m-1}(\chi\left(\frac{*}{p}\right),\eta)\}.$$

(2) *Assume that m is even. Then*

$$J_m(\chi,\eta) = \left(\frac{-1}{p}\right)^{m/2} p^{(m-2)/2}J(\chi,\left(\frac{*}{p}\right))$$
$$\times\{J(\chi,\chi^{m-1}\left(\frac{*}{p}\right)\eta)J_{m-1}(\chi\left(\frac{*}{p}\right),\eta) + \eta(-1)I_{m-1}(\chi\left(\frac{*}{p}\right),\eta)\}.$$

From the above two propositions we have the following:

Theorem 4.8. *Let χ be a primitive character with a prime conductor p such that $\chi^2 \neq 1$.*

(1) *Let m be odd.*

(1.1) *Assume that $\chi^m \neq 1$. Then*

$$J_m(\chi\left(\frac{*}{p}\right)^i,\chi) = \left(\frac{-1}{p}\right)^{(m-1)/2} p^{(m-1)/2}J(\chi\left(\frac{*}{p}\right)^i,\chi^m)J_{m-1}(\chi\left(\frac{*}{p}\right)^{i+1},\chi).$$

(1.2) *Assume that $\chi^m = 1$. Then*

$$J_m(\chi\left(\frac{*}{p}\right)^i,\chi) = p^{m-1}\left(\frac{-1}{p}\right)^{i+1} J(\chi\left(\frac{*}{p}\right)^{i+1},\left(\frac{*}{p}\right))J_{m-2}(\chi\left(\frac{*}{p}\right)^i,\chi).$$

(2) *Let m be even.*

(2.1) *Assume that $\chi^m\left(\frac{*}{p}\right)^{i+1} \neq 1$. Then*

$$J_m(\chi\left(\frac{*}{p}\right)^i,\chi)$$
$$= \left(\frac{-1}{p}\right)^{m/2-1} J(\chi\left(\frac{*}{p}\right)^i,\left(\frac{*}{p}\right))J(\chi\left(\frac{*}{p}\right)^{i+1},\chi^m\left(\frac{*}{p}\right)^{i+1})J_{m-1}(\chi\left(\frac{*}{p}\right)^{i+1},\chi).$$

(2.2) *Assume that* $\chi^m \left(\frac{*}{p}\right)^{i+1} = 1$. *Then*

$$J_m(\chi \left(\frac{*}{p}\right)^i, \chi) = \chi(-1)p^{m-1}J(\chi \left(\frac{*}{p}\right)^i, \left(\frac{*}{p}\right))J_{m-2}(\chi \left(\frac{*}{p}\right)^i, \chi).$$

Corollary. *Let* χ *be a primitive character with an odd square-free conductor* N. *Assume that* χ^2 *is primitive. Then the value* $J_m(\chi)$ *is non-zero.*

5 An Explicit Formula for the Twisted Koecher–Maaß Series of the D–I–I Lift

Throughout this section and the next, we assume that n and k are even positive integers. Let h be a Hecke eigenform of weight $k-n/2+1/2$ belonging to Kohnen's plus space. Then h has the following Fourier expansion:

$$h(z) = \sum_e c_h(e)\mathbf{e}(ez),$$

where e runs over all positive integers such that $(-1)^{k-n/2}e \equiv 0, 1 \bmod 4$. Let

$$S(h)(z) = \sum_{m=1}^{\infty} c_{S(h)}(m)\mathbf{e}(mz)$$

be the normalized Hecke eigenform of weight $2k - n$ with respect to $SL_2(\mathbf{Z})$ corresponding to h under the Shimura correspondence. For a prime number p, let β_p be a non-zero complex number such that $\beta_p + \beta_p^{-1} = p^{-k+n/2+1/2}c_{S(h)}(p)$. For a prime number p, let $\mathbf{Q}_p$ and $\mathbf{Z}_p$ be the field of p-adic numbers and the ring of p-adic integers, respectively. We denote by ν_p the additive valuation on $\mathbf{Q}_p$ normalized so that $\nu_p(p) = 1$ and by $\mathbf{e}_p$ the continuous homomorphism from the additive group $\mathbf{Q}_p$ to $\mathbf{C}^\times$ such that $\mathbf{e}_p(x) = \mathbf{e}(x)$ for $x \in \mathbf{Z}[p^{-1}]$. For a positive definite half integral matrix T of degree n write $(-1)^{n/2}\det(2T)$ as $(-1)^{n/2}\det(2T) = \mathfrak{d}_T\mathfrak{f}_T^2$ with $\mathfrak{d}_T$ a fundamental discriminant and $\mathfrak{f}_T$ a positive integer. We then define the local Siegel series $b_p(T, s)$ by

$$b_p(T, s) = \sum_{R \in S_n(\mathbf{Q}_p)/S_n(\mathbf{Z}_p)} \mathbf{e}_p(\mathrm{tr}(TR))p^{-\nu_p(\mu_p(R))s} \quad (s \in \mathbf{C})$$

for each prime number p, where $\mu_p(R) = [R\mathbf{Z}_p^n + \mathbf{Z}_p^n : \mathbf{Z}_p^n]$. Then there exists a polynomial $F_p(T, X)$ in X such that

$$b_p(T, s) = F_p(T, p^{-s})(1 - p^{-s})(1 - \left(\frac{\mathfrak{d}_T}{p}\right) p^{n/2-s})^{-1} \prod_{i=1}^{n/2}(1 - p^{2i-2s})$$

(cf. [6].) We then put

$$c_{I_n(h)}(T) = c_h(|\mathfrak{d}_T|)\prod_p (p^{k-n/2-1/2}\beta_p)^{\nu_p(\mathfrak{f}_T)} F_p(T, p^{-(n+1)/2}\beta_p^{-1}).$$

We note that $c_{I_n(h)}(T)$ does not depend on the choice of β_p. Define a Fourier series $I_n(h)(Z)$ by

$$I_n(h)(Z) = \sum_{T\in\mathcal{L}_{n>0}} c_{I_n(h)}(T)\mathbf{e}(\mathrm{tr}(TZ)).$$

In [2] Ikeda showed that $I_n(h)(Z)$ is a cuspidal Hecke eigenform in $S_k(Sp_n(\mathbf{Z}))$ and its standard L-function $L(s, I_n(h), \mathrm{St})$ is given by

$$L(s, I_n(h), \mathrm{St}) = \zeta(s)\prod_{i=1}^{n} L(s+k-i, S(h)).$$

We call $I_n(h)$ the Duke-Imamoglu-Ikeda lift (D-I-I lift) of h. Now using the same argument as in the proof of Theorem 1 of [3] we obtain the following. For the details see [4].

Theorem 5.1. *Let χ be a primitive Dirichlet character mod N. Then we have*

$$L^*(s, F, \chi) = 2^{ns}\{c_n R(s, h, E_{n/2+1/2}, \chi)\prod_{j=1}^{n/2-1} L(2s-2j, S(h), \chi^2)$$

$$+d_n c_h(1)\prod_{j=1}^{n/2} L(2s-2j+1, S(h), \chi^2)\},$$

where c_n and d_n are non-zero rational numbers depending only on n.

Now by the above theorem combined with Theorem 4.5 we obtain:

Theorem 5.2. *Let N be a square-free odd integer and $N = p_1\cdots p_r$ be the prime decomposition of N. For each $i = 1,\cdots,r$ let $l_i = \mathrm{G.C.D}(n, p_i-1)$ and $u_0\in\mathbf{Z}$ be a primitive l_ith root of unity mod p_i.*

(1) *Assume $\chi^{(p_i)}(u_i)\neq 1$ for some i. Then $L(s, I_n(h), \chi) = 0$.*
(2) *Assume $\chi^{(p_i)}(u_i) = 1$ for any i. Then*

$$L(s, I_n(h), \chi) = 2^{ns}\sum_{i_1=0}^{l_1-1}\cdots\sum_{i_r=0}^{l_r-1}\overline{\tilde{\chi}_{(i_1,\cdots,i_r)}(2^n)\overline{J(\tilde{\chi}_{(i_1,\cdots,i_r)},\left(\frac{*}{N}\right))}J_{n-1}(\tilde{\chi}_{(i_1,\cdots,i_r)})}$$

$$\times\{c_{n,N}R(s,h,E_{n/2+1/2},\tilde{\chi}_{(i_1,\cdots,i_r)})\prod_{j=1}^{n/2-1} L(2s-2j, S(h), \tilde{\chi}^2_{(i_1,\cdots,i_r)})$$

$$+d_{n,N}c_h(1)\prod_{j=1}^{n/2} L(2s-2j+1, S(h), \tilde{\chi}^2_{(i_1,\cdots,i_r)})\},$$

where $c_{n,N}$ and $d_{n,N}$ are non-zero rational numbers depending only on n and N and $\tilde{\chi}$ is a character s.t. $\tilde{\chi}^n = \chi$.

Remark. In the case $n = 2$, an explicit formula for $L(s, I_2(h), \chi)$ was given by Katsurada-Mizuno [5].

Corollary. *Let χ be a Dirichlet character of odd square-free conductor N such that χ^n is primitive. Then for any integer $n/2+1 \le m \le k-n/2-1$*

$$\frac{L(m, I_n(h), \chi^n)}{\pi^{mn}} = \left\{ \gamma_{n,N} \frac{R^{(\chi)}(m,h,E_{n/2+1/2})}{\pi^{mn}} + \delta_{n,N} c_h(1) \frac{\mathbf{M}^{(\chi)}(m, S(h))}{\pi^{mn}} \right\},$$

where $\gamma_{n,N}$ and $\delta_{n,N}$ are non-zero numbers, and

$$\mathbf{M}^{(\chi)}(m, S(h)) = \sum_{i_1=0}^{l_1-1} \cdots \sum_{i_r=0}^{l_r-1} \overline{\chi_{(i_1,\cdots,i_r)}(2^n) J(\chi_{(i_1,\cdots,i_r)}, \left(\frac{*}{N}\right))} \overline{J_{n-1}(\chi_{(i_1,\cdots,i_r)})}$$
$$\times \prod_{j=1}^{n/2} L(2m-2j+1, S(h), (\chi_{(i_1,\cdots,i_r)})^2).$$

6 Proof of Main Results and Some Comments

We prove the results in Sect. 3.

Proof of Theorem 3.1. Assume that $n \equiv 2 \bmod 4$. Then we have $c_h(1) = 0$, and by Theorem 4.1 and Corollary to Theorem 5.2, we have

$$\frac{R^{(\chi)}(m,h,E_{n/2+1/2})}{\pi^{mn}} \in \overline{\mathbf{Q}}u_1 \otimes_{\overline{\mathbf{Q}}} V_{I_n(h)}$$

with some complex number u_1, where $V_{I_n(h)}$ is the $\overline{\mathbf{Q}}$-vector space associated with $I_n(h)$ in Theorem 4.1. Assume that $n \equiv 0 \bmod 4$. By Theorem 2.1 we have

$$\frac{\mathbf{M}^{(\chi)}(m, S(h))}{\pi^{mn}} \in \overline{\mathbf{Q}}u_-(S(h))^{n/2}\pi^{-n^2/4}.$$

Hence, again by Theorem 4.1 and Corollary to Theorem 5.2,

$$\frac{R^{(\chi)}(m,h,E_{n/2+1/2})}{\pi^{mn}} \in \overline{\mathbf{Q}}u_1 \otimes_{\overline{\mathbf{Q}}} V_{I_n(h)} + \overline{\mathbf{Q}}u_2$$

with complex numbers u_1 and u_2. This proves the assertion. □

Proof of Theorem 3.2 and Its Corollary. Theorem 3.2 follows directly from Theorem 3.1. We note that $J_{n-1}(\chi_{(i_1,\cdots,i_r)})$ is a non-zero algebraic number by virtue of Corollary to Theorem 4.8 We also note that $\dfrac{\mathbf{L}_n(m, S(h), \eta)}{\pi^{m(n-2)}}$ belongs to $\overline{\mathbf{Q}}u_+(S(h))^{n/2-1}\pi^{-n^2/4+n/2}$ and non-zero for any integer $n/2+1 \le m \le k-n/2-1$ and primitive character η. This proves the corollary. □

Proof of Theorem 3.3. The assertion follows from Theorem 4.2. □

Now we give some comments. First we are interested in the dimension of $W_{h,E_{n/2+1/2}}$ over $\overline{\mathbf{Q}}$. Therefore we propose the following problem.

Problem 1. *Give* $\dim_{\overline{\mathbf{Q}}} W_{h,E_{n/2+1/2}}$ *explicitly or estimate it.*

This problem is reduced to the following problem:

Problem 2. *Give* $\dim_{\overline{\mathbf{Q}}} V_{I_n(h)}$ *explicitly or estimate it.*

Next we consider a generalization or a refinement of Theorem 3.1. Namely we propose the following conjecture.

Conjecture. Let $h_1(z)$ be a Hecke eigenform in $S^+_{k_1+1/2}(\Gamma_0(4))$ and $h_2(z) \in M_{k_2+1/2}(\Gamma_0(4))$ with $k_1 \geq k_2 + 2$. Assume that $c_{h_2}(m) \in \overline{\mathbf{Q}}$ for any $m \in \mathbf{Z}_{\geq 0}$. Then there exists a finite dimensional $\overline{\mathbf{Q}}$-vector space $W_{h_1,h_2} \subset \mathbf{C}$ such that

$$R(m, h_1, h_2, \chi)\pi^{-2m} \in W_{h_1,h_2}$$

for any $k_2 + 1 \leq m \leq k_1 - 1$ and any primitive character χ.

Problem 3. Prove Theorem 3.1 without using the relation between the twisted Koecher–Maaß series of the Duke–Imamoglu–Ikeda lift and the twisted Rankin–Selberg series of modular forms of half-integral weight.

References

1. Y. Choie, W. Kohnen, Special values of Koecher–Maaß series of Siegel cusp forms. Pacific J. Math. **198**, 373–383 (2001)
2. T. Ikeda, On the lifting of elliptic modular forms to Siegel cusp forms of degree $2n$. Ann. Math. *154*, 641–681 (2001)
3. T. Ibukiyama, H. Katsurada, An explicit formula for Koecher–Maaß Dirichlet series for the Ikeda lifting. Abh. Math. Sem. Univ. Hamburg **74**, 101–121 (2004)
4. H. Katsurada, Explicit formulas of twisted Koecher–Maaß series of the Duke-Imamoglu-Ikeda lift and their applications. Math. Z. **276**, 1049–1075 (2014)
5. H. Katsurada, Y. Mizuno, Linear dependence of certain L-values of half-integral weight modular forms. J. London Math. **85**, 455–471 (2012)
6. Y. Kitaoka, Dirichlet series in the theory of Siegel modular forms, Nagoya Math. J. **95**, 73–84 (1984)
7. W. Kohnen, New forms of half-integral weight, J. fur die reine und angew. Math. **333**, 32–72 (1982)
8. G. Shimura, On the periods of modular forms. Math. Ann. **229**, 211–221 (1977)
9. G. Shimura, The critical values of certain zeta functions associated with modular forms of half-integral weight. J. Math. Soc. Japan **33**, 649–672 (1981)

On Explicit Dimension Formulas for Spaces of Siegel Cusp Forms of Degree Two and Their Applications

Hidetaka Kitayama

1 Introduction

The purpose of this article is to survey explicit dimension formulas for spaces of Siegel cusp forms of degree two and their applications. Additionally, we will discuss some new results obtained in a joint work with Ibukiyama.

As is well known, dimension formulas for sufficiently large weight can be derived, at least in principle, from Selberg trace formula or Riemann-Roch theorem. As for Selberg trace formula, Hashimoto [8] and recently Wakatsuki [36] obtained arithmetic dimension formula for Siegel cusp forms of degree two for general arithmetic subgroups. However, in practice, it still requires elaborate and complicated works to derive explicit dimension formulas from their general formula. In Sect. 3, we will summarize some known results on Siegel cusp forms of degree two. We will treat not only the split $\mathbb{Q}$-form but also non-split $\mathbb{Q}$-forms. We will also review some facts on spaces of small weight proved by Ibukiyama [16].

In Sect. 4, we will explain our new project for generalizing known dimension formulas. Especially, we will present an explicit dimension formula for Siegel cusp forms with respect to paramodular groups of square-free level (Theorem 4.1), which was obtained in a joint work with Ibukiyama. This formula is a generalization of the formulas in [12, 14]. This study is motivated by our attempt to generalize Eichler-Jacquet-Langlands correspondence to the case of symplectic groups of higher degree. We will discuss a relation of dimensions between Siegel paramodular cusp forms and Siegel cusp forms with respect to certain discrete subgroups in non-split symplectic group. The details will be explained in Sect. 5.1.

H. Kitayama (✉)
Department of Mathematics, Faculty of Education, Wakayama University, Sakaedani 930, Wakayama-city, 640-8510, Japan
e-mail: hkitayam@center.wakayama-u.ac.jp

B. Heim et al. (eds.), *Automorphic Forms*, Springer Proceedings in Mathematics & Statistics 115, DOI 10.1007/978-3-319-11352-4_10

In another direction, explicit dimension formulas are also useful for studies of explicit construction of Siegel modular forms. We will review some known results in this direction in Sect. 5.2.

2 Preliminaries

2.1 Notations

Let $\mathbb{Z}$ and $\mathbb{Z}_p$ be the ring of integers and p-adic integers. Let $\mathbb{Q}$, $\mathbb{R}$, $\mathbb{C}$, and $\mathbb{Q}_p$ be the field of rational numbers, real numbers, complex numbers, and p-adic numbers, respectively. For a ring R, we denote the ring of matrices over R of degree n by $M(n; R)$. We denote the general linear group over R by $GL(n; R)$. We denote the identity matrix of degree n by I_n.

2.2 $\mathbb{Q}$-Forms of $Sp(2; \mathbb{R})$

Let $Sp(2; \mathbb{R})$ be the real symplectic group of degree two, that is,

$$Sp(2;\mathbb{R}) = \left\{ g \in GL(4,\mathbb{R}) \;\middle|\; g \begin{pmatrix} 0_2 & 1_2 \\ -1_2 & 0_2 \end{pmatrix} {}^t g = \begin{pmatrix} 0_2 & 1_2 \\ -1_2 & 0_2 \end{pmatrix} \right\}.$$

Let B be an indefinite quaternion algebra over $\mathbb{Q}$. We fix an isomorphism $B \otimes_{\mathbb{Q}} \mathbb{R} \simeq M(2; \mathbb{R})$ and we identify B with a subalgebra of $M(2; \mathbb{R})$. Let D be the discriminant of B, that is, the product of all prime numbers p for which $B \otimes_{\mathbb{Q}} \mathbb{Q}_p$ is a division algebra. Let W be a left free B-module of rank 2. Let f be a map on $W \times W$ to B defined by

$$f(x, y) = x_1 \overline{y_2} + x_2 \overline{y_1}, \quad x = (x_1, x_2), y = (y_1, y_2) \in W,$$

where $\bar{\ }$ is the canonical involution of B. Any non-degenerate quaternion hermitian form on W is equivalent to f (cf. [31]). We denote by G an algebraic group over $\mathbb{Q}$ defined as the unitary group with respect to the hermitian space (W, f). We denote the group of $\mathbb{Q}$-, $\mathbb{R}$-, and $\mathbb{Q}_p$-rational points of G by $G_{\mathbb{Q}}$, $G_{\mathbb{R}}$, and $G_{\mathbb{Q}_p}$, respectively. Then we have

$$\begin{aligned} G_{\mathbb{Q}} = U(2; B) &= \{ g \in GL(2; B) \mid f(xg, yg) = f(x, y) \text{ for } \forall x, y \in W \} \\ &= \left\{ g \in GL(2; B) \;\middle|\; g \begin{pmatrix} 0 & 1 \\ 1 & 0 \end{pmatrix} {}^t\overline{g} = \begin{pmatrix} 0 & 1 \\ 1 & 0 \end{pmatrix} \right\}, \end{aligned}$$

where ${}^t\overline{g} = \begin{pmatrix} \overline{a} & \overline{c} \\ \overline{b} & \overline{d} \end{pmatrix}$ for $g = \begin{pmatrix} a & b \\ c & d \end{pmatrix}$ and $G_{\mathbb{R}}$ is isomorphic to $Sp(2;\mathbb{R})$ by

$$\phi : G_{\mathbb{R}} \xrightarrow{\sim} Sp(2;\mathbb{R})$$

$$\phi(g) = \begin{pmatrix} a_1 & a_2 & b_2 & -b_1 \\ a_3 & a_4 & b_4 & -b_3 \\ c_3 & c_4 & d_4 & -d_3 \\ -c_1 & -c_2 & -d_2 & d_1 \end{pmatrix}, \quad g = \begin{pmatrix} A & B \\ C & D \end{pmatrix} \in G_{\mathbb{R}}$$

where $A = \begin{pmatrix} a_1 & a_2 \\ a_3 & a_4 \end{pmatrix}$, $B = \begin{pmatrix} b_1 & b_2 \\ b_3 & b_4 \end{pmatrix}$, $C = \begin{pmatrix} c_1 & c_2 \\ c_3 & c_4 \end{pmatrix}$, $D = \begin{pmatrix} d_1 & d_2 \\ d_3 & d_4 \end{pmatrix} \in B \otimes_{\mathbb{Q}} \mathbb{R}$. It is known that any $\mathbb{Q}$-form of $Sp(2;\mathbb{R})$ can be obtained as $U(2;B)$ by taking some indefinite quaternion algebra B (cf. [27]). If $B = M(2;\mathbb{Q})$, then $U(2;B)$ is isomorphic to $Sp(2;\mathbb{Q})$ by ϕ. On the other hand, if B is a division algebra, then $U(2;B)$ is a non-split $\mathbb{Q}$-form of $Sp(2;\mathbb{R})$.

2.3 Siegel Cusp Forms

Let $\mathfrak{H}_2$ be the Siegel upper half space of degree two, that is,

$$\mathfrak{H}_2 = \{Z \in M(2;\mathbb{C}) \mid {}^tZ = Z,\ \mathrm{Im}(Z) \text{ is positive definite}\}.$$

The group $Sp(2;\mathbb{R})$ acts on $\mathfrak{H}_2$ by

$$\gamma\langle Z\rangle := (AZ+B)(CZ+D)^{-1}$$

for any $\gamma = \begin{pmatrix} A & B \\ C & D \end{pmatrix} \in Sp(2;\mathbb{R})$ and $Z \in \mathfrak{H}_2$.

Let Γ be a discrete subgroup of $Sp(2;\mathbb{R})$ such that $\mathrm{vol}(\Gamma\backslash\mathfrak{H}_2) < \infty$. Let $\rho_{k,j} : GL(2;\mathbb{C}) \to GL(j+1;\mathbb{C})$ be the irreducible rational representation of the signature $(j+k,k)$ for $k, j \in \mathbb{Z}_{\geq 0}$, i.e. $\rho_{k,j} = \det^k \otimes \mathrm{Sym}_j$, where Sym_j is the symmetric j-tensor representation of $GL(2;\mathbb{C})$. We denote by $S_{k,j}(\Gamma)$ the space of Siegel cusp forms of weight $\rho_{k,j}$ with respect to Γ, that is, the space which consists of holomorphic functions $f : \mathfrak{H}_2 \to \mathbb{C}^{j+1}$ satisfying the following two conditions:

(i) $f(\gamma\langle Z\rangle) = \rho_{k,j}(CZ+D)f(Z)$, for $\forall\gamma = \begin{pmatrix} A & B \\ C & D \end{pmatrix} \in \Gamma, \forall Z \in \mathfrak{H}_2$,

(ii) $\left|\rho_{k,j}(\mathrm{Im}(Z)^{1/2})f(Z)\right|_{\mathbb{C}^{j+1}}$ is bounded on $\mathfrak{H}_2$,

where we define $|u|_{\mathbb{C}^{j+1}} = ({}^tu\overline{u})^{\frac{1}{2}}$ for $u \in \mathbb{C}^{j+1}$. It is known that $S_{k,j}(\Gamma)$ is a finite dimensional $\mathbb{C}$-vector space.

2.4 Split Cases

Let N be a positive integer. We define global analogues of the local parahoric subgroups as follows:

$$Sp(2;\mathbb{Z}) = Sp(2;\mathbb{R}) \cap M(4;\mathbb{Z}), \qquad \Gamma(N) = \{\gamma \in Sp(2;\mathbb{Z}) \mid \gamma \equiv I_4 \bmod N\},$$

$$\mathrm{K}(N) = Sp(2;\mathbb{Q}) \cap \begin{pmatrix} \mathbb{Z} & N\mathbb{Z} & \mathbb{Z} & \mathbb{Z} \\ \mathbb{Z} & \mathbb{Z} & \mathbb{Z} & N^{-1}\mathbb{Z} \\ \mathbb{Z} & N\mathbb{Z} & \mathbb{Z} & \mathbb{Z} \\ N\mathbb{Z} & N\mathbb{Z} & N\mathbb{Z} & \mathbb{Z} \end{pmatrix},$$

$$\Gamma_0'(N) = Sp(2;\mathbb{Z}) \cap \begin{pmatrix} \mathbb{Z} & N\mathbb{Z} & \mathbb{Z} & \mathbb{Z} \\ \mathbb{Z} & \mathbb{Z} & \mathbb{Z} & \mathbb{Z} \\ \mathbb{Z} & N\mathbb{Z} & \mathbb{Z} & \mathbb{Z} \\ N\mathbb{Z} & N\mathbb{Z} & N\mathbb{Z} & \mathbb{Z} \end{pmatrix},$$

$$\Gamma_0(N) = Sp(2;\mathbb{Z}) \cap \begin{pmatrix} \mathbb{Z} & \mathbb{Z} & \mathbb{Z} & \mathbb{Z} \\ \mathbb{Z} & \mathbb{Z} & \mathbb{Z} & \mathbb{Z} \\ N\mathbb{Z} & N\mathbb{Z} & \mathbb{Z} & \mathbb{Z} \\ N\mathbb{Z} & N\mathbb{Z} & \mathbb{Z} & \mathbb{Z} \end{pmatrix},$$

$$\mathrm{I}(N) = Sp(2;\mathbb{Z}) \cap \begin{pmatrix} \mathbb{Z} & N\mathbb{Z} & \mathbb{Z} & \mathbb{Z} \\ \mathbb{Z} & \mathbb{Z} & \mathbb{Z} & \mathbb{Z} \\ N\mathbb{Z} & N\mathbb{Z} & \mathbb{Z} & \mathbb{Z} \\ N\mathbb{Z} & N\mathbb{Z} & N\mathbb{Z} & \mathbb{Z} \end{pmatrix}.$$

2.5 Non-split Cases

Let B be an indefinite division quaternion algebra over $\mathbb{Q}$ with discriminant D. Let $\mathfrak{O}$ be the maximal order of B. Since B is indefinite, $\mathfrak{O}$ is unique up to $B^\times$-conjugation. We take a quaternion hermitian space (W, f). Let L be a left $\mathfrak{O}$-lattice of W, that is, L is a finitely generated $\mathbb{Z}$-module which satisfies $L \otimes_{\mathbb{Z}} \mathbb{Q} = W$ and $aL \subset L$ for any element $a \in \mathfrak{O}$. We define

$$U(2;B)_L := \{g \in U(2;B) \mid Lg = L\},$$

then $U(2;B)_L$ can be regarded as a discrete subgroup of $Sp(2;\mathbb{R})$ by the isomorphism ϕ, and it satisfies $\mathrm{vol}(U(2;B)_L \backslash \mathfrak{H}_2) < \infty$.

The two-sided $\mathfrak{O}$-ideal generated by elements of the form $f(x,y)$ $(x, y \in L)$ is called the norm of L. If L is maximal among left $\mathfrak{O}$-lattices with the same norm, then we call L a maximal lattice. For any maximal lattice L and any prime number p, it is known by [31] that

$$L \otimes_{\mathbb{Z}} \mathbb{Z}_p = \begin{cases} (\mathfrak{O}_p, \mathfrak{O}_p)g_p & \cdots \text{ if } p \nmid D \\ (\mathfrak{O}_p, \mathfrak{O}_p)g_p \text{ or } (\pi\mathfrak{O}_p, \mathfrak{O}_p)g_p & \cdots \text{ if } p \mid D \end{cases}$$

for some $g_p \in G_{\mathbb{Q}_p}$, where $\mathfrak{O}_p := \mathfrak{O} \otimes \mathbb{Z}_p$ and π is a prime element of $\mathfrak{O}_p$. We define $D_1, D_2 \in \mathbb{N}$ such that $D = D_1 D_2$ and

$$L \otimes_{\mathbb{Z}} \mathbb{Z}_p = \begin{cases} (\mathfrak{O}_p, \mathfrak{O}_p) g_p \text{ if } p \mid D_1, \\ (\pi\mathfrak{O}_p, \mathfrak{O}_p) g_p \text{ if } p \mid D_2, \end{cases}$$

for some $g_p \in G_{\mathbb{Q}_p}$. It is known that if two maximal lattices L_1 and L_2 correspond to the same pair (D_1, D_2), then L_1 and L_2 belong to the same class (i.e. $L_1 = L_2 g$ for some $g \in U(2; B)$) since B is indefinite; hence we have $U(2; B)_{L_1} = U(2; B)_{L_2}$. Therefore we can write

$$\Gamma^*_{D_1,D_2} := U(2; B)_L = U(2; B) \cap \begin{pmatrix} \mathfrak{O} & \mathfrak{A}^{-1} \\ \mathfrak{A} & \mathfrak{O} \end{pmatrix},$$

where $\mathfrak{A}$ is the two-sided ideal of $\mathfrak{O}$ determined by $\mathfrak{A}_p = \mathfrak{O}_p$ for $p \nmid D_2$ and $\mathfrak{A}_p = \pi\mathfrak{O}_p$ for $p \mid D_2$. For a natural number N, we define

$$\Gamma^*_{D_1,D_2}(N) = \left\{\gamma \in \Gamma^*_{D_1,D_2} \mid \gamma \equiv I_2 \bmod N\right\}.$$

We treat the spaces of Siegel cusp forms with respect to $\Gamma^*_{D_1,D_2}$ and $\Gamma^*_{D_1,D_2}(N)$. For studies of Siegel modular forms with respect to $\Gamma^*_{D_1,D_2}$, see [32] and [11].

3 Known Dimension Formulas

In this section, we summarize some known dimension formulas. It is well known that dimension formulas can be derived, in principle, from Selberg trace formula or Riemann-Roch theorem if weight is sufficiently large. Hashimoto [8] and recently Wakatsuki [36] derived an arithmetic dimension formula for general discrete subgroups of $Sp(2; \mathbb{R})$ from Selberg trace formula. On the other hand, Tsushima [33–35] used Riemann-Roch theorem.

3.1 Split Cases

First, we treat the split cases. An explicit dimension formula for $\Gamma(N)$ $(N \geq 3)$ for $k \geq 5, j = 0$ was obtained in Morita [26] and Christian [4] by Selberg trace formula and in Yamazaki [39] by Riemann-Roch theorem. As for $Sp(2; \mathbb{Z})$, the dimensions of the spaces $S_{k,0}(Sp(2; \mathbb{Z}))$ for any k can be calculated by the following generating functions by Igusa [20]:

$$\sum_{k=1}^{\infty} \dim S_{k,0}(Sp(2; \mathbb{Z})) t^k = \frac{1+t^{35}}{(1-t^4)(1-t^6)(1-t^{10})(1-t^{12})} - \frac{1}{(1-t^4)(1-t^6)}.$$

Tsushima [34] obtained explicit dimension formulas of vector-valued Siegel cusp forms for $Sp(2;\mathbb{Z})$ and $\Gamma(N)$ ($N \geq 2$) for $k \geq 4, j = 0$ or $k \geq 5, j > 0$ by Riemann-Roch theorem. Hashimoto [8], Hashimoto-Ibukiyama [10], and Ibukiyama [12] obtained explicit dimension formulas for $K(p)$, $\Gamma_0'(p)$ ($p \neq 2,3$), $\Gamma_0(p)$ ($p \neq 2$), and $I(p)$ ($p \neq 2,3$) for $k \geq 5, j = 0$, where p is a prime number, by Selberg trace formula. Ibukiyama [16, Theorem 2.6] proved that the above dimension formulas in [8, 10, 12] are valid also for $k = 4, j = 0$.

Very recently, Wakatsuki [37] obtained explicit multiplicity formulas for discrete series representations in the discrete spectrum of $L^2(\Gamma\backslash Sp(2;\mathbb{R}))$ for $\Gamma = Sp(2;\mathbb{Z})$, $\Gamma(N)$ ($N \geq 2$), $K(p)$, $\Gamma_0'(p)$, $\Gamma_0(p)$, and $I(p)$ (p: prime). It is known that $Sp(2;\mathbb{R})$ has four types of discrete series representations which are parametrized by Harish-Chandra parameters. Out of them, the multiplicity of the holomorphic discrete series representation of the Harish-Chandra parameter $(j+k-1, k-2)$ is equal to the dimension of the space of cusp forms of weight $\det^k \otimes \mathrm{Sym}_j$.

For readers' convenience, we collect some useful references in the following theorem.

Theorem 3.1. *The dimensions of the spaces $S_{k,j}(\Gamma)$ for $\Gamma = Sp(2;\mathbb{Z})$, $\Gamma(N)$, $K(p)$, $\Gamma_0'(p)$, $\Gamma_0(p)$, $I(p)$ for any positive integer N and any prime number p can be calculated by the following theorems respectively if $k \geq 4, j = 0$, or $k \geq 5, j > 0$:*

$S_{k,j}(Sp(2;\mathbb{Z}))$	[34, Theorem 4], [36, Theorem 7.1]
$S_{k,j}(\Gamma(2))$	[34, Theorem 3], [36, Theorem 7.2]
$S_{k,j}(\Gamma(N))$ ($N \geq 3$)	[34, Theorem 2], [36, Theorem 7.3]
$S_{k,j}(K(p))$	[12, Theorem 4] ($j = 0$), [14, Theorem 6.1]
$S_{k,j}(\Gamma_0'(p))$	[10, Theorem 3.3] ($j = 0$, $p \neq 2,3$), [37, Theorem A.1]
$S_{k,j}(\Gamma_0(p))$	[35, Corollary 4.12], [36, Theorem 7.4]
$S_{k,j}(\mathrm{I}(p))$	[10, Theorem 3.2] ($j = 0$, $p \neq 2,3$), [37, Theorem A.2]

As for the case where$(k, j) = (3,0)$, the following theorem is derived from Theorem 5.4 of Ibukiyama[16].

Theorem 3.2 (Ibukiyama [16]). *The dimensions of the spaces $S_{3,0}(\Gamma)$ for $\Gamma = K(p)$, $\Gamma_0'(p)$, $\Gamma_0(p)$, $I(p)$ for any prime number p can be calculated by substituting $k = 3, j = 0$ into the theorems in Theorem 3.1 and adding "+1."*

As for the case where $k = 1$, Ibukiyama proved the following vanishing theorem in [16, Theorem 6.1]. For $\Gamma = \Gamma_0(N)$ and $j = 0$, this theorem was proved by Ibukiyama and Skoruppa [19].

Theorem 3.3 (Ibukiyama [16]). *We have* $\dim S_{1,j}(\Gamma) = 0$ *for any j for $\Gamma = Sp(2;\mathbb{Z})$, $K(N)$, $\Gamma_0'(N)$, $\Gamma_0(N)$, $I(p)$, where N is any positive integer and p is any prime number.*

As for the case where $(k, j) = (2,0)$, Poor and Yuen determined spaces of small level for $\Gamma_0(p)$ in [28] and for $K(p)$ in [29].

3.2 Non-split Cases

Next, we treat the non-split cases. An explicit dimension formula for $\Gamma^*_{D,1}(N)$ ($N \geq 3$) for $k \geq 5$, $j = 0$ was obtained by Arakawa [2,3] by Selberg trace formula and by Yamaguchi [38] by Riemann-Roch theorem. Recently, Wakatsuki [36] generalize it to vector-valued case.

Theorem 3.4. *The dimensions of the spaces $S_{k,j}(\Gamma^*_{D,1}(N))$ for any j and N can be calculated by the following theorems if $k \geq 5$:*

$S_{k,j}(\Gamma^*_{D,1}(1))$	[9, Theorem 4.1] ($j = 0$), [36, Theorem 6.1]
$S_{k,j}(\Gamma^*_{D,1}(2))$	[36, Theorem 6.2]
$S_{k,j}(\Gamma^*_{D,1}(N))$ ($N \geq 3$)	[3, Theorem on page 126] ($j = 0$), [36, Theorem 6.3]

Recently, the author [22] obtained a dimension formula for $\Gamma^*_{D_1,D_2} = \Gamma^*_{D_1,D_2}(1)$ for any pair (D_1, D_2).

Theorem 3.5 (Kitayama). *The dimensions of the spaces $S_{k,j}(\Gamma^*_{D_1,D_2})$ for $k \geq 5$ and any j can be calculated by Theorem 3.1 in* [22].

4 Generalization to Square-Free Level

In this section, we explain our new project. As explained in the previous section, there are many results of explicit dimension formulas for Siegel cusp forms of degree two. However, many of them are restricted to the case of prime level, especially, for some fundamental groups such as $K(p)$, $\Gamma_0'(p)$, $\Gamma_0(p)$, and $I(p)$. The eventual goal of our project is to obtain dimension formulas for them of generalized level and make use of them to the purposes we will mention in Sect. 5. Until now, we have achieved a generalization of the dimension formula for paramodular groups to square-free level.

We use the following notations. We denote by $\omega(N)$ the number of prime divisors of N. For natural number m and n, we denote by $[a_0, \ldots, a_{m-1}; m]_n$ the function on n which takes the value a_i if $n \equiv i \bmod m$. We define the set $N(m;n) = \{p \mid N\ ;\ p \equiv m \bmod n\}$ and denote the number of elements of $N(m;n)$ by $\sharp N(m;n)$. We use the Legendre symbols:

$$\left(\frac{-1}{p}\right) = \begin{cases} 0 & \cdots \text{ if } p = 2 \\ 1 & \cdots \text{ if } p \equiv 1 \bmod 4 \\ -1 & \cdots \text{ if } p \equiv 3 \bmod 4 \end{cases}, \qquad \left(\frac{-3}{p}\right) = \begin{cases} 0 & \cdots \text{ if } p = 3 \\ 1 & \cdots \text{ if } p \equiv 1 \bmod 3 \\ -1 & \cdots \text{ if } p \equiv 2 \bmod 3 \end{cases}.$$

We define the following functions:

$$C_3(k,j) = [(k-2)(-1)^{j/2}, -(j+k-1), -(k-2)(-1)^{j/2}, j+k-1; 4]_k$$

$$C_4(k,j) = (j+k-1)\cdot[1,-1,0;3]_k + (k-2)\cdot[1,0,-1;3]_{j+k}$$

$$C_5(k,j) = (j+k-1)\cdot[-1,-1,0,1,1,0;6]_k + (k-2)\cdot[1,0,-1,-1,0,1;6]_{j+k}$$

$$C_8(k,j) = \begin{cases} [1,0,0,-1,-1,-1,-1,0,0,1,1,1;12]_k \cdots & \text{if } j \equiv 0 \bmod 12 \\ [-1,1,0,1,1,0,1,-1,0,-1,-1,0;12]_k \cdots & \text{if } j \equiv 2 \bmod 12 \\ [1,-1,0,0,-1,1,-1,1,0,0,1,-1;12]_k \cdots & \text{if } j \equiv 4 \bmod 12 \\ [-1,0,0,-1,1,-1,1,0,0,1,-1,1;12]_k \cdots & \text{if } j \equiv 6 \bmod 12 \\ [1,1,0,1,-1,0,-1,-1,0,-1,1,0;12]_k \cdots & \text{if } j \equiv 8 \bmod 12 \\ [-1,-1,0,0,1,1,1,1,0,0,-1,-1;12]_k \cdots & \text{if } j \equiv 10 \bmod 12 \end{cases}$$

$$C_9(k,j) = \begin{cases} [1,0,0,-1,0,0;6]_k \cdots & \text{if } j \equiv 0 \bmod 6 \\ [-1,1,0,1,-1,0;6]_k \cdots & \text{if } j \equiv 2 \bmod 6 \\ [0,-1,0,0,1,0;6]_k \cdots & \text{if } j \equiv 4 \bmod 6 \end{cases}$$

$$C_{10}(k,j) = \begin{cases} [1,0,0,-1,0;5]_k \cdots & \text{if } j \equiv 0 \bmod 10 \\ [-1,1,0,0,0;5]_k \cdots & \text{if } j \equiv 2 \bmod 10 \\ 0 \cdots & \text{if } j \equiv 4 \bmod 10 \\ [0,0,0,1,-1;5]_k \cdots & \text{if } j \equiv 6 \bmod 10 \\ [0,-1,0,0,1;5]_k \cdots & \text{if } j \equiv 8 \bmod 10 \end{cases}$$

$$C_{11}(k,j) = \begin{cases} [1,0,0,-1;4]_k \cdots & \text{if } j \equiv 0 \bmod 8 \\ [-1,1,0,0;4]_k \cdots & \text{if } j \equiv 2 \bmod 8 \\ [-1,0,0,1;4]_k \cdots & \text{if } j \equiv 4 \bmod 8 \\ [1,-1,0,0;4]_k \cdots & \text{if } j \equiv 6 \bmod 8 \end{cases}$$

Theorem 4.1 below is an explicit dimension formula for $S_{k,j}(K(N))$ with N square-free. If j is odd, then we have $\dim S_{k,j}(K(N)) = 0$ for any k since $K(N)$ contains $-I_4$.

Theorem 4.1 (Ibukiyama and Kitayama). *We assume N is a square-free positive integer. If $k \geq 5$ and j is an even non-negative integer, then we have*

$$\dim_{\mathbb{C}} S_{k,j}(K(N)) = \sum_{i=1}^{12} H_i(k,j,N) + \sum_{i=1}^{10} I_i(k,j,N).$$

If $k = 3, 4$ and $j = 0$, then we have

$$\dim_{\mathbb{C}} S_{k,0}(K(N)) = \sum_{i=1}^{12} H_i(k,0,N) + \sum_{i=1}^{10} I_i(k,0,N) + \begin{cases} 1 \cdots \textit{if } k = 3, \\ 0 \cdots \textit{if } k = 4. \end{cases}$$

Here $H_i(k, j, N)$ and $I_i(k, j, N)$ are explicit elementary functions on k, j, and N defined as follows:

$$H_1(k, j, N) = \prod_{p|N}(p^2+1)\cdot 2^{-7}3^{-3}5^{-1}\cdot(j+1)(k-2)(j+k-1)(j+2k-3)$$

$$H_2(k, j, N) = 2^{\omega(N)-8}3^{-2}\cdot(j+k-1)(k-2)(-1)^k\cdot\begin{cases} 11 \cdots \text{ if } 2 \mid N \\ 14 \cdots \text{ if } 2 \nmid N \end{cases}$$

$$H_3(k, j, N) = 2^{\omega(N)-6}3^{-1}\cdot C_3(k, j)\cdot\begin{cases} 5 \cdots \text{ if } 2 \mid N \\ 2 \cdots \text{ if } 2 \nmid N \end{cases}$$

$$H_4(k, j, N) = 2^{\omega(N)-3}3^{-3}\cdot C_4(k, j)\cdot\begin{cases} 5 \cdots \text{ if } 3 \mid N \\ 1 \cdots \text{ if } 3 \nmid N \end{cases}$$

$$H_5(k, j, N) = 2^{\omega(N)-3}3^{-2}\cdot C_5(k, j)$$

$$H_6(k, j, N) = H_{6,1}(N)\times(-1)^{j/2}(2k+j-3)+H_{6,2}(N)\times(-1)^{j/2+k}(j+1)$$

$$H_{6,1}(N) = \frac{1}{2^5\cdot 3}\prod_{p|N}(p+\left(\frac{-1}{p}\right)) + \frac{1}{2^7\cdot 3}\prod_{p|N}(p\left(\frac{-1}{p}\right)+1)$$

$$H_{6,2}(N) = \frac{1}{2^5\cdot 3}\prod_{p|N}(p+\left(\frac{-1}{p}\right)) - \frac{1}{2^7\cdot 3}\prod_{p|N}(p\left(\frac{-1}{p}\right)+1)$$

$$H_7(k, j, N) = H_{7,1}(N)\times[1,-1,0;3]_j(2k+j-3)+H_{7,2}(N)\times[0,1,-1;3]_{j+2k}(j+1)$$

$$H_{7,1}(N) = \frac{1}{2^3\cdot 3^2}\prod_{p|N}(p+\left(\frac{-3}{p}\right)) + \frac{1}{2^3\cdot 3^3}\prod_{p|N}(p\left(\frac{-3}{p}\right)+1)$$

$$H_{7,2}(N) = \frac{1}{2^3\cdot 3^2}\prod_{p|N}(p+\left(\frac{-3}{p}\right)) - \frac{1}{2^3\cdot 3^3}\prod_{p|N}(p\left(\frac{-3}{p}\right)+1)$$

$$H_8(k, j, N) = 2^{\omega(N)-2}3^{-1}\cdot C_8(k, j)$$

$$H_9(k, j, N) = 2^{\omega(N)-2}3^{-2}\cdot C_9(k, j)\cdot\begin{cases} 1 \cdots \text{ if } 2 \mid N \\ 4 \cdots \text{ if } 2 \nmid N \end{cases}$$

$$H_{10}(k, j, N) = 5^{-1}\cdot C_{10}(k, j)\cdot\begin{cases} 2^{\sharp N(\pm 1;5)} \cdots \text{ if } N(2;5)\cup N(3;5)=\emptyset \\ 0 \quad \cdots \text{ otherwise} \end{cases}$$

$$H_{11}(k, j, N) = 2^{-3}\cdot C_{11}(k, j)\cdot\begin{cases} 2^{\sharp N(\pm 1;8)} \cdots \text{ if } N(3;8)\cup N(5;8)=\emptyset \\ 0 \quad \cdots \text{ otherwise} \end{cases}$$

$$H_{12}(k, j, N) = H_{12,1}(N)\times[1,-1,0;3]_j(-1)^{j/2+k}+H_{12,2}(N)\times[0,-1,1;3]_{j+2k}(-1)^{j/2}$$

$$H_{12,1}(N) = \frac{1}{2^3\cdot 3}\prod_{p|N}(1+\left(\frac{3}{p}\right)) - \frac{1}{2^3\cdot 3}\prod_{p|N}(\left(\frac{-1}{p}\right)+\left(\frac{-3}{p}\right))$$

$$H_{12,2}(N) = \frac{1}{2^3 \cdot 3} \prod_{p|N} (1 + \left(\frac{3}{p}\right)) + \frac{1}{2^3 \cdot 3} \prod_{p|N} (\left(\frac{-1}{p}\right) + \left(\frac{-3}{p}\right))$$

$$I_1(k,j,N) = 2^{-3}3^{-1} \cdot (j+1) \cdot \prod_{p|N} (p+1)$$

$$I_2(k,j,N) = -2^{\omega(N)-4}3^{-1} \cdot (j+1)$$

$$I_3(k,j,N) = -2^{\omega(N)-5}3^{-2} \cdot N \cdot (j+1)(2k+j-3)$$

$$I_4(k,j,N) = 2^{\omega(N)-5} \cdot (-1)^k \cdot (4 - \left(\frac{-1}{N}\right))$$

$$I_5(k,j,N) = -2^{\omega(N)-4} \cdot 3^{-1} \cdot (-1)^k \cdot (2k+j-3)$$

$$I_6(k,j,N) = -2^{\omega(N)-3} \cdot [(-1)^{j/2}, -1, -(-1)^{j/2}, 1; 4]_k$$

$$I_7(k,j,N) = -2^{\omega(N)-2} \cdot 3^{-2} \cdot \begin{cases} [3,-3,0;3]_k + [3,0,-3;3]_{j+k} & \cdots \textit{ if } N \equiv 0 \bmod 3 \\ [5,-1,4;3]_k + [1,-4,-5;3]_{j+k} & \cdots \textit{ if } N \equiv 1 \bmod 3 \\ [1,-5,-4;3]_k + [5,4,-1;3]_{j+k} & \cdots \textit{ if } N \equiv 2 \bmod 3 \end{cases}$$

$$I_8(k,j,N) = -2^{\omega(N)-2} \cdot 3^{-1} \cdot ([-1,-1,0,1,1,0;6]_k + [1,0,-1,-1,0,1;6]_{j+k})$$

$$I_9(k,j,N) = -2^{-3} \cdot (-1)^{j/2} \cdot \prod_{p|N} (1 + \left(\frac{-1}{p}\right))$$

$$I_{10}(k,j,N) = -2^{-1}3^{-1} \cdot [1,-1,0;3]_j \cdot \prod_{p|N} (1 + \left(\frac{-3}{p}\right))$$

We will give a proof of Theorem 4.1 in the forthcoming joint paper with Ibukiyama.

5 Applications

5.1 Relations of Dimensions

One of the motivations for our studies of dimension formulas is to use them to investigate possible correspondences between Siegel cusp forms with respect to different discrete subgroups. Especially, our motivation for Theorem 4.1 is as follows. Eichler [6,7] proved the correspondence between automorphic forms on $SL(2;\mathbb{R})$ and on $SU(2)$, and Jacquet-Langlands [21] developed it from the viewpoint of representation theory. Langlands [25] gave a general conjecture for any reductive algebraic groups. On the other hand, Ihara [18] raised independently a problem of generalizing Eichler correspondence to symplectic groups of higher degree in 1960s. Hashimoto and Ibukiyama developed detailed studies by using explicit dimension formulas in [10,12] and [14]. They found some relations of dimensions of automorphic forms on $Sp(2;\mathbb{R})$ and its compact twist and described some detailed conjectures of correspondences. However, the conjectures were restricted to the case of prime level because explicit dimension formulas were not known except for prime level. We can prove the following theorem by using Theorem 3.5 and Theorem 4.1. We denote by $S_k^{new}(\Gamma_0^{(1)}(N))$ the spaces of new cusp forms of weight k with respect to $\Gamma_0^{(1)}(N) = \left\{ \begin{pmatrix} a & b \\ c & d \end{pmatrix} \in SL(2;\mathbb{Z}) \mid c \equiv 0 \bmod N \right\}$.

Theorem 5.1. *We assume N is a square-free positive integer and $\omega(N)$ is even. If $k \geq 5$ and j is an even non-negative integer, then we have*

$$\sum_{M|N} (-1)^{\omega(M)} 2^{\omega(M)} \dim S_{k,j}(K(N/M))$$

$$= \dim S_{k,j}(\Gamma_{1,N}^*) - \sum_{M|N,\omega(M)=odd} \dim S_{j+2}^{new}(\Gamma_0^{(1)}(M)) \times \dim S_{2k+j-2}^{new}(\Gamma_0^{(1)}(N/M)),$$

If $j = 0$, then we should replace $\dim S_{j+2}^{new}(\Gamma_0^{(1)}(M))$ *to* $\dim S_{j+2}^{new}(\Gamma_0^{(1)}(M)) + 1$.

This theorem makes us expect a correspondence between $S_{k,j}(K(N))$ and $S_{k,j}(\Gamma_{1,N}^*)$. Ibukiyama [12, 14] proved Theorem 5.1 already in the case where $\omega(N) = 1$. The case means a relation of dimensions between the space $S_{k,j}(K(p))$ and the space $\mathfrak{M}_{k+j-3,k-3}(U(p))$ of automorphic forms on the compact twist of $Sp(2;\mathbb{R})$. (See [12] and [14] for details.) He proposed a conjecture of correspondence between spaces of "newforms" of $S_{k,j}(K(p))$ and $\mathfrak{M}_{k+j-3,k-3}(U(p))$. Theorem 5.1 makes us expect that we can propose a generalized conjecture if we can find suitable definitions of newforms of $S_{k,j}(K(N))$ and $S_{k,j}(\Gamma_{1,N}^*)$. We will explain the details in the forthcoming joint paper with Ibukiyama.

5.2 *Explicit Structure of Siegel Modular Forms*

Another application of explicit dimension formulas is explicit construction of Siegel modular forms. The generating functions of dimensions imply the explicit structure. As illustrations of the graded rings of scalar-valued Siegel modular forms, see [1, 13, 15].

As for vector-valued Siegel modular forms with respect to $Sp(2;\mathbb{Z})$, explicit structure was obtained by Satoh [30] (k: even) if $j = 2$, by Ibukiyama [17] (k: odd) if $j = 2$, by Ibukiyama [17] (k: any) if $j = 4$, by Ibukiyama [17] (k: even) if $j = 6$, by Dorp [5] (k: odd) if $j = 6$, and by Kiyuna [24] (k: any) if $j = 8$.

Finally, we turn to the non-split cases. We can derive the generating function

$$\sum_{k=0}^{\infty} \dim M_{k,0}(\Gamma_{1,6}^*) t^k = \frac{(1+t^5)(1+t^{15})}{(1-t^2)(1-t^4)(1-t^5)(1-t^6)}$$

from Theorem 3.5 and ad hoc calculations for small weight. The author obtained explicit structure of the graded ring of scalar-valued Siegel modular forms with respect to $\Gamma_{1,6}^*$ in [23]. The following theorem is the only complete result of the graded rings for the non-split cases as far as the author knows.

Theorem 5.2 (Kitayama [23]). *The graded ring of Siegel modular forms with respect to $\Gamma_{1,6}^*$ is given explicitly by*

$$\bigoplus_{k=0}^{\infty} M_{k,0}(\Gamma_{1,6}^*) = \mathbb{C}[E_2, E_4, \chi_{5a}, E_6] \oplus \chi_{5b}\mathbb{C}[E_2, E_4, \chi_{5a}, E_6]$$
$$\oplus\, \chi_{15}\mathbb{C}[E_2, E_4, \chi_{5a}, E_6] \oplus \chi_{5b}\chi_{15}\mathbb{C}[E_2, E_4, \chi_{5a}, E_6],$$

where we denote by E_k ($k = 2, 4, 6$) the Eisenstein series which were defined in [11] *and denote by χ_{5a}, χ_{5b}, and χ_{15} the Siegel cusp forms of weight* 5, 5, *and* 15 *respectively, which are defined in Proposition 3.2 and Proposition 3.3 in* [23]. *The four modular forms E_2, E_4, χ_{5a}, and E_6 are algebraically independent over $\mathbb{C}$. The fundamental relations among the generators are given in* [23].

Judging from Theorem 5.1, there seem to exist lifted cusp forms in $S_{k,0}(\Gamma_{1,N}^*)$ from $S_2^{new}(\Gamma_0^{(1)}(M)) \times S_{2k-2}^{new}(\Gamma_0^{(1)}(N/M))$. It is an unknown phenomenon as far as the author knows. We can do numerical experiments on Euler factors of L-function and investigate such liftings by using Theorem 5.2. We will explain the details in the forthcoming joint paper with Ibukiyama.

References

1. H. Aoki, T. Ibukiyama, Simple graded rings of Siegel modular forms, differential operators and Borcherds products. Internat. J. Math. **16**, 249–279 (2005)
2. T. Arakawa, *Automorphic forms on quaternion unitary group of degree 2* (in Japanese). Master thesis. University of Tokyo (1975).
3. T. Arakawa, The dimension of the space of cusp forms on the Siegel upper half plane of degree two related to a quaternion unitary groups. J. Math. Soc. Japan **33**, 125–145
4. U. Christian, Berechnung des Ranges der Schar der Spitzenformen zur Modulgruppe zweiten Grades und Stufe $q > 2$. J. Reine Angew. Math. **277** (1975) 130–154; Zur Berechnung des Ranges der Schar der Spitzenformen zur Modulgruppe zweiten Grades und Stufe $q > 2$. J. Reine Angew. Math. **296** (1977) 108–118.
5. C.H. van Dorp, Generators for a module of vector-valued Siegel modular forms of degree 2. arXiv:1301.2910v1 [math.AG].
6. M. Eichler,Über die darstellbarkeit von Modulformen durch Theta Reihen. J. Reine Angew. Math. **195**, 159–171 (1956)
7. M. Eichler, Quadratische formen und modulformen. Acta arith. **4**, 217–239 (1958)
8. K. Hashimoto, The dimension of the spaces of cusp forms on Siegel upper half plane of degree two I. J. Fac. Sci. Univ. Tokyo sect. IA Math. **30**, 403–488 (1983)

9. K. Hashimoto, The dimension of the spaces of cusp forms on Siegel upper half plane of degree two II.The $\mathbb{Q}$-rank one case. Math. Ann. **266** 539–559 (1984)
10. K. Hashimoto, T. Ibukiyama, On relations of dimensions of automorphic forms of $Sp(2, \mathbb{R})$ and its compact twist $Sp(2)$ (II), Automorphic forms and number theory. Adv. Stud. Pure Math. **7**, 31–102 (1985)
11. Y. Hirai, On Eisenstein series on quaternion unitary groups of degree 2, J. Math. Soc. Japan **51**, 93–128 (1999)
12. T. Ibukiyama, On relations of dimensions of automorphic forms of $Sp(2, \mathbb{R})$ and its compact twist $Sp(2)$ (I), Automorphic forms and number theory, Adv. Stud. Pure Math. **7**, 7–30 (1985)
13. T. Ibukiyama, On Siegel modular varieties of level 3. Internat. J. Math. **2**, 17–35 (1991)
14. T. Ibukiyama, Paramodular forms and compact twist, in Automorphic Forms on GSp(4) (Proceedings of the 9th Autumn Workshop on Number Theory; M. Furusawa, ed.) (2007), 37–48.
15. T. Ibukiyama, F. Onodera, On the graded ring of Modular forms of the Siegel paramodular group of level 2. Abh. Math. Sem. Univ. Hamburg **67**, 297–305 (1997)
16. T. Ibukiyama, *Dimension Formulas of Siegel Modular Forms of Weight 3 and Supersingular Abelian Surfaces* (Proceedings of the 4-th Spring Conference), 39–60 (2007)
17. T. Ibukiyama, Vector valued Siegel modular forms of symmetric tensor weight of small degrees, Comment. Math. Univ. St. Pauli **61**, 51–75 (2012)
18. Y. Ihara, On certain arithmetical Dirichlet series. J. Math. Soc. Japan, **16**, 214–225 (1964)
19. T. Ibukiyama, N.-P. Skoruppa, A vanishing theorem for Siegel modular forms of weight one. Abh. Math. Sem. Univ. Hamburg **77**, 229–235 (2007)
20. J. Igusa, On Siegel modular forms of genus two II. Amer. J. Math. **86**, 392–412 (1964)
21. H. Jacquet, R.P. Langlands, *Automorphic Forms on GL(2)*, Lecture Notes in Mathematics, vol. 260 (Springer, NewYork, 1972).
22. H. Kitayama, An explicit dimension formula for Siegel cusp forms with respect to the non-split symplectic groups. J. Math. Soc. Japan **63**, 1263–1310 (2011)
23. H. Kitayama, On the graded ring of Siegel modular forms of degree two with respect to a non-split symplectic group. Internat. J. Math. **23** (2012).
24. T. Kiyuna, Vector-valued Siegel modular forms of weight $\det^k \otimes$ Sym(8), preprint (2012).
25. R.P. Langlands, *Problems in the Theory of Automorphic Forms*, Lecture Notes in Mathematics, vol. 170 (Springer, NewYork, 1970) pp. 18–61.
26. Y. Morita, An explicit formula for the dimension of spaces of Siegel modular forms of degree two. J. Fac. Sci. Univ. Tokyo Sect. IA Math. **21**, 167–248 (1974)
27. V. Platonov, A. Rapinchuk, *Algebraic Groups and Number Theory* (Academic Press, 1994)
28. C. Poor, D. Yuen, Dimensions of cusp forms for $\Gamma_0(p)$ in degree two and small weights. Abh. Math. Sem. Univ. Hamburg **77**, 185–222 (2007)
29. C. Poor and D. Yuen, Paramodular cusp forms. arXiv:0912.0049v1. (2009)
30. T. Satoh, On certain vector valued Siegel modular forms of degree two, Math. Ann. **274**, 335–352 (1986)
31. G. Shimura,Arithmetic of alternating forms and quaternion hermitian forms. J. Math. Soc. Japan **15**, 33–65 (1963)
32. T. Sugano, On holomorphic cusp forms on quaternion unitary groups of degree 2, J. Fac. Sci. Univ. Tokyo Sect. IA Math. **31**, 521–568 (1984)
33. R. Tsushima, On the spaces of Siegel cusp forms of degree two. Amer. J. Math. **65**, 843–885 (1982)
34. R. Tsushima, An explicit dimension formula for the spaces of generalized automorphic forms with respect to $Sp(2; \mathbb{Z})$. Proc. Japan Acad. Ser. A **59**, 139–142 (1983)
35. R. Tsushima, Dimension formula for the spaces of Siegel cusp forms and a certain exponential sum. Mem. Inst. Sci. Tech. Meiji Univ. **36**, 1–56 (1997) http://www.math.meiji.ac.jp/~tsushima.

36. S. Wakatsuki, Dimension formulas for spaces of vector-valued Siegel cusp forms of degree two. J. Number Theory **132**, 200–253 (2012)
37. S. Wakatsuki, Multiplicity formulas for discrete series representations in $L^2(\Gamma\backslash Sp(2;\mathbb{R}))$, J. Number Theory **133**, 3394–3425 (2013)
38. H. Yamaguchi, The parabolic contribution to the dimension of the space of cusp forms on Siegel space of degree two, unpublished (1976).
39. T. Yamazaki, On Siegel modular forms of degree two. Amer. J. Math. **98**, 39–53 (1976)

Borchers Lift on the Paramodular Group of Level 3

Judith Kreuzer

1 Introduction

This paper deals with Borcherds products on the paramodular group of level 3. After introducing the notation, we present in a more general setting how to construct Borcherds lifts. The approach used in this paper is based on work of V. Gritsenko and V. Nikulin (compare [8]). In section 3, we will go into more detail on the paramodular group of level 3. We will determine the characters and divisors on this group. Section 4 deals with weakly Jacobi forms of weight 0 and index 3. These functions are of special interest as they are used as input functions for the Borcherds lift which we want to consider. We will learn that it is possible to construct weakly Jacobi forms. Even more, we can influence properties of the Borcherds lift in the construction of weakly Jacobi forms. In Sect. 5, we will calculate the Borcherds lifts of the functions created in Section 4. Moreover, we will examine those Borcherds products and give statements on divisors and characters.

1.1 Notation

In this section, we give a short introduction to the notation and the most important terms.

The *Siegel upper half-space of degree 2* will be denoted by $\mathcal{H}$, i.e.

$$\mathcal{H} = \{X + iY;\ X, Y \in \mathrm{Sym}(2, \mathbb{R}), Y > 0\},$$

J. Kreuzer (✉)
Lehrstuhl A für Mathematik, RWTH Aachen, 52056 Aachen, Germany
e-mail: judith.kreuzer@matha.rwth-aachen.de

B. Heim et al. (eds.), *Automorphic Forms*, Springer Proceedings in Mathematics & Statistics 115, DOI 10.1007/978-3-319-11352-4_11

where $\mathrm{Sym}(2,\mathbb{R})$ denotes the symmetric 2×2-matrices over $\mathbb{R}$ and $Y > 0$ means that Y shall be positive definite. An element $Z \in \mathcal{H}$ will always be written as $Z = \left(\begin{smallmatrix} \tau & z \\ z & \omega \end{smallmatrix}\right)$.

$$\mathrm{Sp}(2;\mathbb{Z}) = \left\{M \in \mathbb{Z}^{4\times 4};\ M^t J M = J\right\}, \text{ where } J = \left(\begin{smallmatrix} 0 & -I_2 \\ I_2 & 0 \end{smallmatrix}\right),$$

is the *Siegel modular group of degree 2*. It is well known that $\mathrm{Sp}(2;\mathbb{Z})$ acts on the Siegel upper half-space. Let $M = \left(\begin{smallmatrix} A & B \\ C & D \end{smallmatrix}\right) \in \mathrm{Sp}(2;\mathbb{Z})$, $k \in \mathbb{Z}$, and $f : \mathcal{H} \to \mathbb{C}$. The *Petersson slash operator* is defined by

$$(f\,|_k\, M)(Z) = (\det(CZ + D))^{-k}\, f(M\langle Z\rangle) \quad \text{for } Z \in \mathcal{H},$$

where $M\langle Z\rangle = (AZ + B)(CZ + D)^{-1}$ denotes the action of the Siegel modular group on the Siegel upper half-space.

Furthermore, we consider the parabolic subgroup of $\mathrm{Sp}(2;\mathbb{Z})$:

$$\Gamma_\infty = \left\{ \begin{pmatrix} * & 0 & * & * \\ * & * & * & * \\ * & 0 & * & * \\ 0 & 0 & 0 & * \end{pmatrix} \in \mathrm{Sp}(2;\mathbb{Z}) \right\}.$$

It is known that $\Gamma_\infty \cong \mathrm{SL}_2(\mathbb{Z}) \ltimes \mathrm{H}(\mathbb{Z})$, where $\mathrm{H}(\mathbb{Z})$ denotes the Heisenberg group. Hence, we can write all characters on Γ_∞ in the form $\nu_\eta^a \times \nu_{\mathrm{H}}^b$, $a, b \in \mathbb{Z}$, where ν_η is the character on $\mathrm{SL}_2(\mathbb{Z})$ of order 24 and ν_{H} is the character on $\mathrm{H}(\mathbb{Z})$ of order 2.

The *paramodular group of level t* is defined by

$$\Gamma_t = \left\{ \begin{pmatrix} * & *t & * & * \\ * & * & * & */t \\ * & *t & * & * \\ *t & *t & *t & * \end{pmatrix} \in \mathrm{Sp}(2,\mathbb{Q});\ * \in \mathbb{Z} \right\},\ t \in \mathbb{N}.$$

There exists a normal extension of Γ_t of index 2 defined by

$$\Gamma_t^+ = \Gamma_t \cup \Gamma_t V_t, \quad \text{where } V_t = \frac{1}{\sqrt{t}} \begin{pmatrix} 0 & t & 0 & 0 \\ 1 & 0 & 0 & 0 \\ 0 & 0 & 0 & 1 \\ 0 & 0 & t & 0 \end{pmatrix}.$$

We will concentrate on the case $t = 3$ where the extension is even maximal discrete (cp. [8]).

Now we want to recall the definition of modular forms as the Borcherds products we consider in Sect. 2 belong to this class of functions.

Definition 1.1. *Let $k \in \mathbb{Z}$ and let $\nu : \mathrm{Sp}(2,\mathbb{Z}) \to \mathbb{C}$ be a character. A holomorphic function $f : \mathcal{H} \to \mathbb{C}$ is called a Siegel modular form of weight k with respect to* $\mathrm{Sp}(2,\mathbb{Z})$ *and ν if*

$$f(M\langle Z\rangle) = \nu(M)\det(CZ + D)^k f(Z) \text{ for all } M \in \mathrm{Sp}(2,\mathbb{Z}),\ Z \in \mathcal{H}.$$

The set of Siegel modular forms of weight k with respect to $\mathrm{Sp}(2,\mathbb{Z})$ *and ν will be denoted by* $M_k(\mathrm{Sp}(2,\mathbb{Z}),\nu)$.

Moreover, we want to give the definition of Jacobi forms as it can be found in [7]. In the definition below, $\mathbb{H}$ denotes the complex upper half-plane, namely

$$\mathbb{H} = \{\tau = x + iy \in \mathbb{C};\ x, y \in \mathbb{R}, y > 0\}.$$

Definition 1.2. *Let $k,\ t \in \mathbb{N}_0$. A holomorphic function $\phi : \mathbb{H} \times \mathbb{C} \to \mathbb{C}$ is called Jacobi form of weight k and index t if and only if its associated function*

$$\tilde{\phi} : \mathcal{H} \to \mathbb{C}, \quad \tilde{\phi}(Z) = \phi(\tau, z)e^{2\pi i t\omega}$$

satisfies

$$\left(\tilde{\phi} \mid_k M\right)(Z) = \tilde{\phi}(Z) \textit{ for all } M \in \Gamma_\infty \textit{ and } Z \in \mathcal{H}$$

and if ϕ has a Fourier expansion of the form

$$\phi(\tau, z) = \sum_{\substack{n \in \mathbb{N}_0, l \in \mathbb{Z} \\ 4nt - l^2 \geq 0}} c(n, l) \exp(2\pi i (n\tau + lz)).$$

The set of Jacobi forms of weight k and index t will be denoted by $J_{k,t}$.

We want to relax the conditions in the Fourier expansion. This will give us Jacobi forms which are still holomorphic on $\mathbb{H} \times \mathbb{C}$ but not necessarily holomorphic at infinity. So we allow $4nt - l^2 \geq -\alpha$ for an $\alpha \in \mathbb{N}$, and we allow $n \in \mathbb{Z}$. We will call ϕ a *weakly Jacobi form*, then. The set of weakly Jacobi forms is denoted by $J_{k,t}^{\text{weakly}}$.

2 Borcherds Lift

In this section, we introduce the Borcherds lift in a general setting. The following theorem describes the construction of Borcherds products as well as their most important properties.

From now on, we use the abbreviations

$$q = e^{2\pi i \tau},\ r = e^{2\pi i z},\ s = e^{2\pi i \omega}, \quad \tau, \omega \in \mathbb{H},\ z \in \mathbb{C},$$

and we use π_t for the natural projection, namely

$$\pi_t : \mathcal{H} \to \Gamma_t \backslash \mathcal{H},\ t \in \mathbb{N}.$$

Theorem 2.1 (Borcherds Lift). *Let* $\phi \in J_{0,t}^{weakly}$ *with Fourier expansion*

$$\phi(\tau, z) = \sum_{n,l \in \mathbb{Z}} c(n,l) e^{2\pi i (n\tau + lz)} \text{ with } c(n,l) \in \mathbb{Z}.$$

We set

$$A = \frac{1}{24} \sum_{l \in \mathbb{Z}} c(0,l), \; B = \frac{1}{2} \sum_{l>0} l c(0,l), \; C = \frac{1}{4} \sum_{l \in \mathbb{Z}} l^2 c(0,l)$$

and

$$\Lambda = \{(n,l,m) \in \mathbb{Z}^3;\, m > 0\} \cup \{(n,l,m) \in \mathbb{Z}^3;\, m = 0,\, n > 0\} \\ \cup \{(n,l,m) \in \mathbb{Z}^3;\, m = n = 0,\, l < 0\} \subset \mathbb{Z}^3.$$

Then

$$B_\phi(Z) = q^A r^B s^C \prod_{(n,l,m) \in \Lambda} (1 - q^n r^l s^{tm})^{c(nm,l)} \in M^{mero}_{\frac{1}{2}c(0,0)}(\Gamma_t^+, \nu_\eta^{24A} \times \nu_H^{2B}).$$

The divisors of B_ϕ *are given by a* $\mathbb{Z}$*-linear combination of Humbert surfaces*

$$H_d(b) = \pi_t(\{Z \in \mathcal{H};\, a\tau + bz + t\omega = 0\})$$

with discriminant $d = b^2 - 4ta$ *and multiplicity* $\sum_{n>0} c(n^2 a, nb)$.

Furthermore, $B_\phi(V_t\langle Z \rangle) = (-1)^D B_\phi(Z)$ *holds where* $D = \sum_{n<0, l \in \mathbb{Z}} \sigma_1(-n) c(n,l)$.

The proof of this theorem can be found in [8, Theorem 2.1]. According to [2, Theorem 13.3], it suffices to demand that $c(n,l)$ is integral for the principal part of ϕ only, i.e.

$$c(n,l) \in \mathbb{Z} \text{ for } 4tn - l^2 < 0.$$

The following property of Borcherds products is needed later in Sect. 5.

Remark 2.2. The Borcherds lift is multiplicative, i.e. for $\phi, \psi \in J_{0,t}^{weakly}$ it holds:

$$B_{\phi+\psi}(Z) = B_\phi(Z) B_\psi(Z)$$

3 The Paramodular Group Γ_3^+

This section deals with the paramodular group of level 3. We want to go into more details about characters on that group. This way we can determine which characters can appear as characters of a Borcherds lift. Moreover, we want to determine all possible divisors of a Borcherds lift on Γ_3^+.

We consider the following characters on Γ_3 given by their values on the generators of Γ_3:

$$\mu(M) = (v_\eta^8 \times 1_H)(M)$$
$$= \begin{cases} e^{2\pi i(s_1+s_4)/3}, & \text{if } M = \begin{pmatrix} 1 & 0 & s_1 & s_2 \\ 0 & 1 & s_2 & s_4/3 \\ 0 & 0 & 1 & 0 \\ 0 & 0 & 0 & 1 \end{pmatrix} \\ 1, & \text{if } M = \begin{pmatrix} U & 0 \\ 0 & (U^t)^{-1} \end{pmatrix}, U = \begin{pmatrix} u_1 & 3u_2 \\ u_3 & u_4 \end{pmatrix} \in \mathrm{GL}_2(\mathbb{Z}) \text{ or } M = \begin{pmatrix} 0 & 0 & -1 & 0 \\ 0 & 0 & 0 & -1/3 \\ 1 & 0 & 0 & 0 \\ 0 & 3 & 0 & 0 \end{pmatrix} \end{cases}$$

$$\kappa(M) = (v_\eta^{12} \times v_H)(M)$$
$$= \begin{cases} (-1)^{s_1+s_2+s_4}, & \text{if } M = \begin{pmatrix} 1 & 0 & s_1 & s_2 \\ 0 & 1 & s_2 & s_4/3 \\ 0 & 0 & 1 & 0 \\ 0 & 0 & 0 & 1 \end{pmatrix} \\ (-1)^{(1+u_1+u_4)(1+u_2+u_3)+u_1u_4}, & \text{if } M = \begin{pmatrix} U & 0 \\ 0 & (U^t)^{-1} \end{pmatrix}, U = \begin{pmatrix} u_1 & 3u_2 \\ u_3 & u_4 \end{pmatrix} \in \mathrm{GL}_2(\mathbb{Z}) \\ 1, & \text{if } M = \begin{pmatrix} 0 & 0 & -1 & 0 \\ 0 & 0 & 0 & -1/3 \\ 1 & 0 & 0 & 0 \\ 0 & 3 & 0 & 0 \end{pmatrix} \end{cases}$$

Furthermore, we consider the character χ on Γ_3^+:

$$\chi(M) = \begin{cases} -1, & \text{if } M = V_3 \\ 1, & \text{if } M \in \Gamma_3 \end{cases}$$

The characters μ and κ can be extended to Γ_3^+ by setting $\mu(V_3) = \kappa(V_3) = 1$. Due to [5, Lemma 2.5] and [6, §7] we have the following corollary.

Corollary 3.1. *The abelian characters of Γ_3^+ are generated by the characters μ of order 3, κ of order 2 and χ of order 2. All these 12 characters appear as characters of Borcherds lifts.*

As a next step, we want to have a closer look at rational quadratic divisors. We consider the lattice $L := \mathbb{Z}^2 \times \mathbb{Z}^2 \times \mathbb{Z}$ equipped with the quadratic form

$$q((l_1, l_2, l_3, l_4, \beta)) = l_1 l_2 + l_3 l_4 - 3\beta^2.$$

We use the bilinear form associated with q, i.e. $b(x, y) = q(x + y) - q(x) - q(y)$. Then L has signature $(2, 3)$ and the dual lattice is of the form $L' = \mathbb{Z}^2 \times \mathbb{Z}^2 \times \frac{1}{6}\mathbb{Z}$. For $\lambda = (l_1, l_2, l_3, l_4, \beta) \in L'$ with $q(\lambda) < 0$ we set

$$\lambda^{\perp} = \left\{ Z \in \mathcal{H};\ l_1 - 3l_2 \det(Z) + \text{trace}\left(\begin{pmatrix} l_3 & -3\beta \\ -3\beta & 3l_4 \end{pmatrix} Z \right) = 0 \right\}.$$

Then $\lambda^{\perp}$ is a rational quadratic divisor with discriminant $-12q(\lambda)$, if λ is primitive.

Corollary 3.2. *Let $L := \mathbb{Z}^2 \times \mathbb{Z}^2 \times \mathbb{Z}$ be the lattice equipped with the quadratic form*

$$q((l_1, l_2, l_3, l_4, \beta)) = l_1 l_2 + l_3 l_4 - 3\beta^2.$$

Γ_3^+ acts on the set of rational quadratic divisors of fixed discriminant. Up to this action there are 4 rational quadratic divisors $\lambda_j^{\perp}$ with discriminant $j \leq 12$:

$$\lambda_1^{\perp} = \{\left(\begin{smallmatrix} \tau & z \\ z & \omega \end{smallmatrix}\right);\ z = 0\}$$
$$\lambda_4^{\perp} = \{\left(\begin{smallmatrix} \tau & z \\ z & \omega \end{smallmatrix}\right);\ z = \tfrac{1}{2}\}$$
$$\lambda_9^{\perp} = \{\left(\begin{smallmatrix} \tau & z \\ z & \omega \end{smallmatrix}\right);\ z = \tfrac{1}{3}\}$$
$$\lambda_{12}^{\perp} = \{\left(\begin{smallmatrix} \tau & z \\ z & \omega \end{smallmatrix}\right);\ \tau = 3\omega\}$$

4 Weakly Jacobi Forms of Weight 0 and Index 3

In this section, we will concentrate on weakly Jacobi forms of weight 0 and index 3 as these functions form the input set for the Borcherds lifts which we want to consider. We will construct weakly Jacobi forms such that their Borcherds lifts have certain properties. This is possible because of the following property of $J_{0,3}^{\text{weakly}}$ (cp. [5]):

Proposition 4.1. *Given a principal part $\{c(n,l);\ 12n - l^2 < 0\}$ there exists a weakly Jacobi form*

$$\phi(\tau, z) = \sum_{n,l \in \mathbb{Z}} c(n,l) q^n r^l \in J_{0,3}^{\text{weakly}}.$$

Moreover, ϕ is uniquely determined by its principal part.

Now, we can construct weakly Jacobi forms of weight 0 and index 3 by fixing the principal part only.

Definition 4.2. *Let $(d \bmod 12) \in \{0, 1, 4, 9\}$. We set*

$$c_{\phi_d}(N) = \begin{cases} 1, & \text{if } N < 0, N = -d \\ 0, & \text{if } N < 0, N \neq -d \end{cases}$$

where $N = 12n - l^2$, $n, l \in \mathbb{Z}$.

Similarly to [7, Theorem 2.2], the Fourier coefficients of a Jacobi form of weight 0 and index 3 depend only on the discriminant $12n - l^2$ and not on the choice of (n, l). Hence, the construction above determines uniquely a Jacobi form $\phi_d \in J_{0,3}^{\text{weakly}}$. In the next section, we will construct the Borcherds lifts of these Jacobi forms.

5 Borcherds Lifts on Γ_3^+

Knowing the principal part of the input function for a Borcherds lift gives us almost every information we need to characterize the lift. The only information missing is the constant term of the Jacobi form and therefore the weight of the lift.

According to [3, Theorem 3.1], there is a condition of the obstruction space involving the principal part and the constant term of the input function. This condition helps us to determine directly the constant term of the functions ϕ_d which were created in Sect. 4. In our case the obstruction space is given by vector-valued modular forms of weight $\frac{5}{2}$ and character ρ_L' where ρ_L' is the dual of the Weil representation with respect to $(L'/L, q \bmod \mathbb{Z})$. Moreover, it is known that this space can be generated by the Eisenstein series $E_{5/2}$. For a realization of this Eisenstein series compare [1, Theorem 4.8].

Hence, we can calculate the constant coefficient of ϕ_d using the equation

$$\sum_{\mu=0}^{5} \sum_{\substack{n \in q((0,0,0,0,\frac{\mu}{6}))+\mathbb{Z} \\ n \leq 0}} \tilde{c}_{\phi_d,\mu}(12n) c_{E_{5/2}}(\tfrac{\mu}{6}, -n) = 0$$

where

$$\tilde{c}_{\phi_d,\mu}(N) = \begin{cases} 1, & \text{if } N = -d = 12n - l^2 \text{ and } l \equiv \mu \pmod 6 \\ 0, & \text{otherwise} \end{cases}$$

and $c_{E_{5/2}}(\frac{\mu}{6}, n)$ stands for the Fourier coefficients of the Eisenstein series.

From the equation above we can see that the Fourier coefficients of the Eisenstein series $E_{5/2}$ determine directly the constant term of the input function of the Borcherds lift and therewith also the weight of the Borcherds lift. For that reason, we are interested in a formula for the coefficients which is easy to handle and which provides us with boundaries or approximations on the weight of a Borcherds lift. Having a closer look at the formula given in [1, Theorem 4.8] we get the following.

Conjecture 5.1. *Let* $(d \bmod 12) \in \{0, 1, 4, 9\}$. *If* $\nu_3(d) \not\equiv 0 \pmod 2$, *then the following identity for the coefficients of the Eisenstein series* $E_{5/2}$ *holds:*

$$c_{E_{5/2}}(\frac{\mu}{6}, \frac{d}{12}) = 12H(2, d)$$

where $H(2, d)$ *denotes the Cohen numbers (cp. [4, §2]).*

Finally, we determine the Borcherds lift of ϕ_d for $(d \bmod 12) \in \{0, 1, 4, 9\}$. According to Theorem 2.1, knowing the principal part and the constant term of ϕ_d is sufficient to get the most important properties of the Borcherds lift:

input	weight	divisor	character
ϕ_1	1	$\lambda_1^\perp$	$\chi\mu^2\kappa$
ϕ_4	7	$\lambda_1^\perp + \lambda_4^\perp$	$\chi\mu^2$
$\phi_4 - \phi_1$	6	$\lambda_4^\perp$	κ
ϕ_9	17	$\lambda_1^\perp + \lambda_9^\perp$	$\chi\kappa$
$\phi_9 - \phi_1$	16	$\lambda_9^\perp$	μ
ϕ_{12}	12	$\lambda_{12}^\perp$	χ

The Borcherds lift of $\phi_4 - \phi_1$ and $\phi_9 - \phi_1$ can be determined easily by using the multiplicativity of the Borcherds product.

We can see from the table above that there exists a Borcherds product for each rational quadratic divisor on Γ_3^+. Moreover, there is a link between divisors and characters.

Theorem 5.2.

(i) Let $F \in M_{2k+1}(\Gamma_3^+, \nu)$, $k \in \mathbb{Z}$, where ν is an arbitrary character of Γ_3^+. Then

$$F(Z) = 0 \text{ for all } Z \in \lambda_1^\perp.$$

(ii) Let $F \in M_{2k}(\Gamma_3^+, \nu)$, $k \in \mathbb{Z}$, $\nu \in \{\kappa(\chi\mu)^j; j \in \mathbb{Z}\}$. Then

$$F(Z) = 0 \text{ for all } Z \in \lambda_4^\perp.$$

(iii) Let $F \in M_{2k}(\Gamma_3^+, \nu)$, $k \in \mathbb{Z}$, $\nu \in \{\chi(\kappa\mu)^j; j \in \mathbb{Z}\}$. Then

$$F(Z) = 0 \text{ for all } Z \in \lambda_{12}^\perp.$$

Proof.

(i) Let $M = \begin{pmatrix} U & 0 \\ 0 & (U^t)^{-1} \end{pmatrix}$ with $U = \begin{pmatrix} 1 & 0 \\ 0 & -1 \end{pmatrix}$ and let $Z \in \lambda_1^\perp$, i.e. $Z = \begin{pmatrix} \tau & 0 \\ 0 & \omega \end{pmatrix}$. Then

$$M\langle Z\rangle = Z \text{ and } \nu(M) = 1$$

(since $\mu(M) = \kappa(M) = \chi(M) = 1$). Hence, for $Z \in \lambda_1^\perp$

$$(F|_{2k+1} M)(Z) = \left(\det\begin{pmatrix} 1 & 0 \\ 0 & -1 \end{pmatrix}^{-t}\right)^{-2k-1} F(M\langle Z\rangle) \underset{\substack{F \text{ modular} \\ \text{form}}}{=} \nu(M)F(Z)$$

holds if and only if

$$-F(Z) = F(Z)$$

and the claim follows.

(ii) Use analogous arguments for

$$M = \begin{pmatrix} U & 0 \\ 0 & (U^t)^{-1} \end{pmatrix} \begin{pmatrix} E & S \\ 0 & E \end{pmatrix} \quad \text{where } U = \begin{pmatrix} 1 & 0 \\ 0 & -1 \end{pmatrix} \text{ and } S = \begin{pmatrix} 0 & -1 \\ -1 & 0 \end{pmatrix}.$$

(iii) Use analogous arguments for $M = V_3$.

□

Remark 5.3. There are no similar results for $\lambda_9^{\perp}$ as there does not exist a matrix $M \in \Gamma_3^+$ with $M\langle Z\rangle = Z$ for all $Z \in \lambda_9^{\perp}$ and $\mu(M) = -1$.

By Theorem 5.2, the knowledge of the weight and character of an arbitrary given Borcherds lift induces knowledge on some of the divisors. Thus dividing by B_{ϕ_1}, $B_{\phi_4-\phi_1}$, or $B_{\phi_{12}}$ gives us the possibility to control divisors or characters of that given Borcherds lift. This way it is possible to reduce a Borcherds lift to a modular form with trivial character or without the divisors $\lambda_1^{\perp}$, $\lambda_4^{\perp}$ and $\lambda_{12}^{\perp}$. Because of the multiplicativity of Borcherds lifts, the resulting function is a Borcherds lift as well. Due to the remark above we do not get any control on functions having divisor $\lambda_9^{\perp}$.

In the last part of this paper, we will use Theorem 5.2 and the Borcherds lifts constructed above to determine

$$M_{2\mathbb{Z}}(\Gamma_3^+, \nu) := \bigoplus_{k\in 2\mathbb{Z}} M_k(\Gamma_3^+, \nu)$$

for every abelian character ν on Γ_3^+ (cp. Corollary 3.1). According to [5, Lemma 5.3], we have

$$M_{2\mathbb{Z}}(\Gamma_3^+, 1) := \bigoplus_{k\in 2\mathbb{Z}} M_k(\Gamma_3^+, 1) = \mathbb{C}[B_{\phi_1}^6, B_{\phi_1}^4 f_4, B_{\phi_1}^2 f_4^2, E_4, E_6, E_{12}] =: \mathcal{R}, \quad (*)$$

where $f_4 \in M_4(\Gamma_3^+, \mu)$ and $E_k \in M_k(\Gamma_3^+, 1)$, $k \in \{4, 6, 12\}$, are Eisenstein series. Furthermore, $B_{\phi_1}^6$, E_4, E_6 and E_{12} are algebraically independent. The generators of $\mathcal{R}$ satisfy the relations

$$\begin{aligned}
\left(B_{\phi_1}^4 f_4\right)^2 &= B_{\phi_1}^6 \cdot B_{\phi_1}^2 f_4^2, \\
\left(B_{\phi_1}^2 f_4^2\right)^2 &= B_{\phi_1}^4 f_4 \left(c\, B_{\phi_1}^4\, f_4\, E_4 + p(B_{\phi_1}^6, E_4, E_6, E_{12})\right), \\
B_{\phi_1}^4 f_4\, B_{\phi_1}^2 f_4^2 &= B_{\phi_1}^6 \left(c\, B_{\phi_1}^4\, f_4\, E_4 + p(B_{\phi_1}^6, E_4, E_6, E_{12})\right)
\end{aligned}$$

where $c \in \mathbb{C}$ and $p \in \mathbb{C}[X_1, X_2, X_3, X_4]$ is a polynomial (see [5, Section 5] for details). We apply this in the following:

Theorem 5.4. *One has*

(i) $M_{2\mathbb{Z}}(\Gamma_3^+, \kappa) = B_{\phi_4-\phi_1}\mathcal{R}$,
(ii) $M_{2\mathbb{Z}}(\Gamma_3^+, \chi) = B_{\phi_{12}}\mathcal{R}$,
(iii) $M_{2\mathbb{Z}}(\Gamma_3^+, \mu) = B_{\phi_1}^2\mathcal{R} \oplus f_4\mathcal{R}$,
(iv) $M_{2\mathbb{Z}}(\Gamma_3^+, \mu^2) = B_{\phi_1}^4\mathcal{R} \oplus B_{\phi_1}^2 f_4\mathcal{R} \oplus f_4^2\mathcal{R}$,
(v) $M_{2\mathbb{Z}}(\Gamma_3^+, \mu\kappa) = B_{\phi_1}^2 B_{\phi_4-\phi_1}\mathcal{R} \oplus B_{\phi_4-\phi_1} f_4\mathcal{R}$,
(vi) $M_{2\mathbb{Z}}(\Gamma_3^+, \mu^2\kappa) = B_{\phi_1}^4 B_{\phi_4-\phi_1}\mathcal{R} \oplus B_{\phi_1}^2 B_{\phi_4-\phi_1} f_4\mathcal{R} \oplus B_{\phi_4-\phi_1} f_4^2\mathcal{R}$,
(vii) $M_{2\mathbb{Z}}(\Gamma_3^+, \chi\kappa) = B_{\phi_4-\phi_1} B_{\phi_{12}}\mathcal{R}$,
(viii) $M_{2\mathbb{Z}}(\Gamma_3^+, \chi\mu) = B_{\phi_1}^2 B_{\phi_{12}}\mathcal{R} \oplus B_{\phi_{12}} f_4\mathcal{R}$,
(ix) $M_{2\mathbb{Z}}(\Gamma_3^+, \chi\mu^2) = B_{\phi_1}^4 B_{\phi_{12}}\mathcal{R} \oplus B_{\phi_1}^2 B_{\phi_{12}} f_4\mathcal{R} \oplus B_{\phi_{12}} f_4^2\mathcal{R}$,
(x) $M_{2\mathbb{Z}}(\Gamma_3^+, \chi\mu\kappa) = B_{\phi_1}^2 B_{\phi_4-\phi_1} B_{\phi_{12}}\mathcal{R} \oplus B_{\phi_4-\phi_1} B_{\phi_{12}} f_4\mathcal{R}$,
(xi) $M_{2\mathbb{Z}}(\Gamma_3^+, \chi\mu^2\kappa) = B_{\phi_1}^4 B_{\phi_4-\phi_1} B_{\phi_{12}}\mathcal{R} \oplus B_{\phi_1}^2 B_{\phi_4-\phi_1} B_{\phi_{12}} f_4\mathcal{R}$
$\oplus B_{\phi_4-\phi_1} B_{\phi_{12}} f_4^2\mathcal{R}$.

Proof. Let $f \in M_k(\Gamma_3^+, \kappa)$, k even. Because of Theorem 5.2(ii) $f(Z) = 0$ for all $Z \in \lambda_4^\perp$. As the divisor of $B_{\phi_4-\phi_1}$ is $\lambda_4^\perp$, the quotient $f/B_{\phi_4-\phi_1}$ is holomorphic and a modular form of weight $k-6$. Because $k-6$ is even, we get

$$\frac{f}{B_{\phi_4-\phi_1}} \in M_{k-6}(\Gamma_3^+, 1).$$

Hence, the claim of *(i)* follows by using $(*)$.

Let $f \in M_k(\Gamma_3^+, \chi)$, k even. Using Theorem 5.2(iii) and similar arguments as above, we receive

$$\frac{f}{B_{\phi_{12}}} \in M_{k-12}(\Gamma_3^+, 1).$$

and, therefore, the claim *(ii)*.

Let $f \in M_k(\Gamma_3^+, \mu)$, k even. Using the same arguments as in the proof of [5, Theorem 5.2], there exists a polynomial $P \in \mathbb{C}[X_1, X_2, X_3]$ such that

$$f - f_4\, P(E_4, E_6, E_{12}) = 0 \text{ on } \lambda_1^\perp$$

(of order at least 2). Hence,

$$\tilde{f} := (f - f_4\, P(E_4, E_6, E_{12}))/B_{\phi_1}^2 \in M_{k-2}(\Gamma_3^+, 1).$$

Now *(iii)* follows.

Let $f \in M_k(\Gamma_3^+, \mu^2)$, k even. As above, there exists a polynomial $P \in \mathbb{C}[X_1, X_2, X_3]$ such that

$$f - f_4^2\, P(E_4, E_6, E_{12}) = 0 \text{ on } \lambda_1^\perp$$

(of order at least 2). Hence,

$$\tilde{f} := (f - f_4^2\, P(E_4, E_6, E_{12}))/B_{\phi_1}^2 \in M_{k-2}(\Gamma_3^+, \mu).$$

Then the claim of *(iv)* follows by using *(iii)*.

To prove *(v)* to *(xi)*, use Theorem 5.2 in order to start the reduction process. Then, apply *(i)* to *(iv)*. □

Acknowledgement We would like to thank the referee for valuable suggestions.

References

1. J.H. Bruinier, M. Kuss, Eisenstein series attached to lattices and modular forms on orthogonal groups. Manuscripta Math. **106**, 443–459 (2001)
2. R.E. Borcherds. Automorphic forms with singularities on Grassmannians. Invent. Math. **132**, 491–562 (1998)
3. R.E. Borcherds. The Gross-Kohnen-Zagier theorem in higher dimensions. Duke Math. J. **97**, 219–233 (1999)
4. H. Cohen. Sums involving the values at negative integers of L-functions of quadratic characters. Math. Ann. **217**, 271–285 (1975)
5. T. Dern. Paramodular forms of degree 2 and level 3. Comment. Math. Univ. St. Paul. **51**, 157–194 (2002)
6. T. Dern, A. Marschner. Characters of paramodular groups and some extensions. Comm. Algebra **30**, 4589–4604 (2002)
7. M. Eichler, D. Zagier. *The Theory of Jacobi Forms*, Progress in mathematics, vol. 55 (Birkhäuser Boston Inc., Boston, MA, 1985)
8. V.A. Gritsenko, V.V. Nikulin, Automorphic forms and Lorentzian Kac-Moody algebras. Part II. Internat. J. Math. **9**, 201–275 (1998)

The Restriction of Octonionic Theta Constants to the Quaternionic Half-Space

C. Dieckmann and A. Krieg

1 Introduction

Freitag and Salvati Manni [4] studied restrictions of octonionic theta constants to the quaternionic half-space. They proved that the vector space, which is spanned by all restricted octonionic theta constants, coincides with the vector space spanned by quaternionic theta series of first respectively second kind. In this paper we will show the same statement, where the proof makes use of a completely different idea.

2 Quaternions and Octonions

We use the same notation as in [5] to describe quaternions and we use the same notation as in [1, 2] respectively [4] to describe octonions. The most important properties and definitions are mentioned in the next two subsections. More details are given in the literature mentioned above.

2.1 *Quaternions*

We define the skew-field of *quaternions* as

$$\mathbb{H} := \mathbb{R} + \mathbb{R}i_1 + \mathbb{R}i_2 + \mathbb{R}i_3$$

C. Dieckmann • A. Krieg (✉)
Lehrstuhl A für Mathematik, RWTH Aachen, Aachen, Germany
e-mail: cornelia.dieckmann@matha.rwth-aachen.de; krieg@matha.rwth-aachen.de

B. Heim et al. (eds.), *Automorphic Forms*, Springer Proceedings in Mathematics & Statistics 115, DOI 10.1007/978-3-319-11352-4_12

with defining relations

$$i_1^2 = i_2^2 = i_3^2 = -1,\ i_1 i_2 = i_3 \text{ and } i_2 i_1 = -i_3.$$

Every $z \in \mathbb{H}$ has a unique representation

$$z = z_0 + z_1 i_1 + z_2 i_2 + z_3 i_3 \text{ with } z_0, z_1, z_2, z_3 \in \mathbb{R}.$$

In this notation we call z_0 the *real part* of z. The *conjugation*

$$\tau_{\mathbb{H}} : \mathbb{H} \to \mathbb{H},\ z \mapsto \overline{z} := z_0 - z_1 i_1 - z_2 i_2 - z_3 i_3$$

defines an involution on $\mathbb{H}$ and the map

$$N_{\mathbb{H}} : \mathbb{H} \to \mathbb{R},\ z \mapsto N_{\mathbb{H}}(z) := z_0^2 + z_1^2 + z_2^2 + z_3^2 = z\overline{z} = \overline{z}z$$

satisfies $N_{\mathbb{H}}(ab) = N_{\mathbb{H}}(a)N_{\mathbb{H}}(b)$ for all $a, b \in \mathbb{H}$. We call $N_{\mathbb{H}}(z)$ the *norm* of z. In addition to that, for $0 \neq z \in \mathbb{H}$ the inverse is given by

$$z^{-1} = \frac{1}{N_{\mathbb{H}}(z)}\overline{z}.$$

First we consider

$$\mathcal{O}_{\mathbb{H}} := \mathbb{Z} + \mathbb{Z}i_1 + \mathbb{Z}i_2 + \mathbb{Z}\omega,\ \omega := \frac{1}{2}(1 + i_1 + i_2 + i_3),$$

which is a maximal order of $\mathbb{H}$ and which we call the *Hurwitz quaternions*. The unit group $\mathcal{O}_{\mathbb{H}}^{\times}$ in $\mathcal{O}_{\mathbb{H}}$ is a non-commutative group of order 24. It consists of all elements $z \in \mathcal{O}_{\mathbb{H}}$ with $N_{\mathbb{H}}(z) = 1$. One has

$$\mathcal{O}_{\mathbb{H}}^{\times} = \{\pm 1, \pm i_1, \pm i_2, \pm i_3, \tfrac{1}{2}(\pm 1 \pm i_1 \pm i_2 \pm i_3)\}.$$

Finally we consider the set of even Hurwitz quaternions $\wp$ defined by

$$\wp := \{z \in \mathcal{O}_{\mathbb{H}},\ N_{\mathbb{H}}(z) \in 2\mathbb{Z}\},$$

which is a two-sided prime ideal in $\mathcal{O}_{\mathbb{H}}$ with $\mathcal{O}_{\mathbb{H}}/\wp \cong \mathbb{F}_4$. We have

$$\wp = (1 + i_1)\,\mathcal{O}_{\mathbb{H}} = \mathcal{O}_{\mathbb{H}}\,(1 + i_1).$$

2.2 Octonions

We define the set of *octonions*, also called *Cayley numbers*, as

$$\mathcal{C} = \mathbb{H} + j\mathbb{H} \text{ with } j^2 = -1$$

and the following addition and multiplication rules

$$(\alpha + j\beta) + (\gamma + j\delta) = (\alpha + \gamma) + j(\beta + \delta),$$
$$(\alpha + j\beta)(\gamma + j\delta) = \left(\alpha\gamma - \delta\overline{\beta}\right) + j\left(\overline{\alpha}\delta + \gamma\beta\right)$$

for $\alpha, \beta, \gamma, \delta \in \mathbb{H}$.
The octonions form an 8-dimensional normed division algebra over $\mathbb{R}$. Similarly to quaternions every $z \in \mathcal{C}$ has a unique representation

$$z = \alpha + j\beta \text{ with } \alpha, \beta \in \mathbb{H}.$$

Here, the *real part* of z is defined as the real part of α. The *conjugation*

$$\tau_{\mathcal{C}} : \mathcal{C} \to \mathcal{C},\ z = \alpha + j\beta \mapsto \overline{z} := \overline{\alpha} - j\beta$$

defines an involution on $\mathcal{C}$ and the map

$$N_{\mathcal{C}} : \mathcal{C} \to \mathbb{R},\ z = \alpha + j\beta \mapsto N_{\mathcal{C}}(z) := N_{\mathbb{H}}(\alpha) + N_{\mathbb{H}}(\beta) = z\overline{z} = \overline{z}z$$

satisfies $N_{\mathcal{C}}(ab) = N_{\mathcal{C}}(a)N_{\mathcal{C}}(b)$ for all $a, b \in \mathcal{C}$. We call $N_{\mathcal{C}}(z)$ the *norm* of z. In addition to that, for $0 \neq z \in \mathcal{C}$ the inverse of z is given by

$$z^{-1} = \frac{1}{N_{\mathcal{C}}(z)}\overline{z}.$$

Just as in [4, 19.1] we consider the maximal order $\mathcal{O}_{\mathcal{C}}$ given by

$$\mathcal{O}_{\mathcal{C}} = \bigcup_{k=0}^{3} (\alpha_k + \mathcal{O}_{\mathbb{H}} + j\mathcal{O}_{\mathbb{H}})$$

of integral octonions with

$$\alpha_0 = 0,\ \ \alpha_1 = \tfrac{1+i_1}{2} + j\tfrac{1+i_3}{2},\ \ \alpha_2 = \tfrac{1+i_3}{2} + j\tfrac{1+i_1}{2},\ \ \alpha_3 = \tfrac{1+i_2}{2} + j\tfrac{1+i_2}{2}.$$

Another special set we are looking at is the set of units $\mathcal{O}_{\mathcal{C}}^{\times}$, which consists of all elements $z \in \mathcal{O}_{\mathcal{C}}$ with $N_{\mathcal{C}}(z) = 1$. There are exactly 240 units.

3 Theta Series

In this section we introduce four different kinds of theta series. We begin with the octonionic and quaternionic theta constants and at the end of this section we consider quaternionic theta series of first and second kind. In every subsection we also state the most important properties of the theta series.

3.1 *Octonionic Theta Constants*

In this section we use the same notation as in [1] respectively [4]. We denote the *octonionic half-space* of degree 2 by

$$\mathcal{H}(\mathcal{C}) = \{Z = X + iY \in \mathrm{Mat}(2, \mathcal{C}) \otimes_{\mathbb{R}} \mathbb{C};\ Z = \overline{Z}^{tr} := \overline{X}^{tr} + i\overline{Y}^{tr}, Y > 0\}$$

and define the octonionic theta constant of characteristic p by

$$\vartheta_p(Z) = \sum_{h \in p + 2\mathcal{O}_{\mathcal{C}}^2} \exp(\pi i (N(h_1)z_1 + N(h_2)z_2 + 2\mathrm{Re}(\overline{h_1} z h_2))),$$

where $p \in \mathcal{O}_{\mathcal{C}}^2$ and $Z = \begin{pmatrix} z_1 & z \\ \overline{z} & z_2 \end{pmatrix} \in \mathcal{H}(\mathcal{C})$ (cf [1,4]). A simple verification yields

Lemma 1. *For arbitrary $p \in \mathcal{O}_{\mathcal{C}}^2$ the following properties hold:*

a) $\vartheta_{p+q}(Z) = \vartheta_p(Z)$ *for all* $q \in 2\mathcal{O}_{\mathcal{C}}^2$,
b) $\vartheta_p(Z[U]) = \vartheta_{Up}(Z)$ *for all* $U \in \{\begin{pmatrix} 1 & u \\ 0 & 1 \end{pmatrix}, \begin{pmatrix} 1 & 0 \\ u & 1 \end{pmatrix}, u \in \mathcal{O}_{\mathcal{C}}, \pm\begin{pmatrix} 0 & 1 \\ -1 & 0 \end{pmatrix}\} \cup \{\begin{pmatrix} \varepsilon & 0 \\ 0 & \overline{\varepsilon} \end{pmatrix}, \varepsilon \in \mathcal{O}_{\mathcal{C}}^{\times}, \begin{pmatrix} n & t \\ \overline{t} & m \end{pmatrix}, n, m \in \mathbb{N}, t \in \mathcal{O}_{\mathcal{C}}, nm - N(t) = \pm 1\}$,
where $Z[U]$ *is defined by* $\overline{U}^{tr} Z U$.

Since the octonions are non-associative one has to check that the expression $Z[U] = \overline{U}^{tr}(ZU) = (\overline{U}^{tr} Z)U$ is well-defined for the special U's mentioned above.

3.2 *Quaternionic Theta Constants*

We use the same notation as in [3] except that we write $2k$ for the weight (as in [5]) instead of k (as in [3]). Therefore let

$$\mathcal{H}(\mathbb{H}) = \{Z = X + iY \in \mathrm{Mat}(2, \mathbb{H}) \otimes_{\mathbb{R}} \mathbb{C};\ Z = \overline{Z}^{tr} = \overline{X}^{tr} + i\overline{Y}^{tr}, Y > 0\}.$$

denote the *quaternionic half-space* of degree 2 (cf. [3, Sect.3]). It is clear that $\mathcal{H}(\mathbb{H})$ is a submanifold of $\mathcal{H}(\mathcal{C})$.

More generally than in [3, Sect.10] we define the theta constants of characteristic $a \in \mathcal{O}_{\mathbb{H}}^2$ by

$$\Theta_a(Z) := \sum_{g \in \frac{1}{2}a + \wp^2} \exp(\pi i Z[g]) \text{ for } Z \in \mathcal{H}(\mathbb{H})$$

where $Z[g] := \overline{g}^{tr} Z g$.

Lemma 2. *For every $a \in \mathcal{O}_{\mathbb{H}}$ and $Z \in \mathcal{H}(\mathbb{H})$ the following properties hold:*

(a) $\Theta_{a+2b}(Z) = \Theta_a(Z)$ *for all* $b \in \wp^2$,
(b) $\Theta_{a\varepsilon}(Z) = \Theta_a(Z)$ *for all* $\varepsilon \in \mathcal{O}_{\mathbb{H}}^{\times}$,
(c) $\Theta_a(Z[U]) = \Theta_{Ua}(Z)$ *for all* $U \in \mathrm{GL}(2, \mathcal{O}_{\mathbb{H}})$,
(d) $\Theta_a(Z[\rho E_2]) = \Theta_{\rho a \overline{\rho}}(Z)$ *for* $E_2 = \begin{pmatrix} 1 & 0 \\ 0 & 1 \end{pmatrix}$, $\rho = \frac{1+i_1}{\sqrt{2}}$, $\rho \mathcal{O}_{\mathbb{H}} \overline{\rho} = \mathcal{O}_{\mathbb{H}}$.

In [6] the behaviour of quaternionic theta constants under the transpose

$$\mathcal{H}(\mathbb{H}) \to \mathcal{H}(\mathbb{H}), \; Z \mapsto Z^{tr}$$

was studied. We quote the result as

Theorem 1. *Given* $a = \begin{pmatrix} \alpha \\ \beta \end{pmatrix} \in \mathcal{O}_{\mathbb{H}}^2$ *one has*

$$\Theta_a(Z^{tr}) = \Theta_{a^*}(Z),$$

where

$$a^* = \begin{cases} \varepsilon \overline{a} \varepsilon^{-1}, & \text{if } \alpha \text{ or } \beta \equiv \varepsilon \bmod 2\mathcal{O}_{\mathbb{H}}, \; \varepsilon \in \mathcal{O}_{\mathbb{H}}^{\times}, \\ \eta \overline{a} \eta^{-1}, & \text{if } a \in \wp^2 \text{ and } \alpha \text{ or } \beta \equiv \eta \bmod 2\wp, \; \eta \in \wp, \; N(\eta) = 2, \\ \overline{a}, & \text{if } a \in 2\mathcal{O}_{\mathbb{H}}^2. \end{cases}$$

3.3 *Quaternionic Theta Series of Second Kind*

We now consider another type of theta series which occurs in [4] when considering the restriction of octonionic theta series to the quaternionic half-space $\mathcal{H}(\mathbb{H})$.
Let S denote the matrix $\begin{pmatrix} 2 & 1+i_1 \\ 1-i_1 & 2 \end{pmatrix}$. For $A \in \mathrm{Mat}(2, \mathcal{O}_{\mathbb{H}})$ the quaternionic theta series of second kind of characteristic A is defined by

$$\Theta_A(Z; S) := \sum_{G \in \frac{1}{2}A + \mathrm{Mat}(2, \mathcal{O}_{\mathbb{H}})} \exp(2\pi i \; \tau(S \cdot Z[G])), \; Z \in \mathcal{H}(\mathbb{H}),$$

where

$$\tau : \mathrm{Mat}(2,\mathbb{H}) \to \mathbb{R},\ M \mapsto \tau(M) := \tfrac{1}{2}\mathrm{trace}(M + \overline{M}^{tr}).$$

Lemma 3. *For every* $A \in \mathrm{Mat}(2,\mathcal{O}_{\mathbb{H}})$ *and* $Z \in \mathcal{H}(\mathbb{H})$ *the following properties hold:*

(a) $\Theta_{A+2B}(Z;S) = \Theta_A(Z;S)$ *for all* $B \in \mathrm{Mat}(2,\mathcal{O}_{\mathbb{H}})$,
(b) $\Theta_A(Z[U];S) = \Theta_{UA}(Z;S)$ *for all* $U \in \mathrm{GL}(2,\mathcal{O}_{\mathbb{H}})$,
(c) $\Theta_A(Z[\rho E_2];S) = \Theta_{\rho A\overline{\rho}}(Z;S)$ *for* $\rho = \frac{1+i_1}{\sqrt{2}}$,
(d) $\Theta_{A\overline{V}^{tr}}(Z;S) = \Theta_A(Z;S)$ *for all* $V \in \mathrm{Aut}(S)$.

The relation

$$\Theta_A(Z;S) = \sum_{u:\wp^2\backslash\mathcal{O}_{\mathbb{H}}^2} \Theta_{2u+a(1+i_1)+b}(2Z)\Theta_{2u+b}(2Z) \text{ for } A = \begin{pmatrix} a & b \end{pmatrix}. \tag{1}$$

is the main tool to calculate the behaviour under the transpose (cf. [6]).

Theorem 2. *Given* $A \in \mathrm{Mat}(2,\mathcal{O}_{\mathbb{H}})$ *there exists a matrix* $A^* \in \mathrm{Mat}(2,\mathcal{O}_{\mathbb{H}})$ *depending only on* A *such that*

$$\Theta_A(Z^{tr};S) = \Theta_{A^*}(Z;S),\ Z \in \mathcal{H}(\mathbb{H}).$$

Remark 1. The matrix $A^* \in \mathrm{Mat}(2,\mathcal{O}_{\mathbb{H}})$ can be specified explicitly (cf. [6]).

3.4 Quaternionic Theta Series of First Kind

Freitag and Salvati Manni [4] moreover considered the quaternionic theta series of first kind for $P, Q \in \mathrm{Mat}(2,\mathcal{O}_{\mathbb{H}})$ and $S = \begin{pmatrix} 2 & 1+i_1 \\ 1-i_1 & 2 \end{pmatrix}$ given by

$$\Theta^*_{P,Q}(Z;S) = \sum_{G\in S^{-1}P+\mathrm{Mat}(2,\mathcal{O}_{\mathbb{H}})} \exp(\pi i\ \tau(S[G]\cdot Z + 2\overline{G}^{tr}Q)),\ Z \in \mathcal{H}(\mathbb{H}).$$

Lemma 4. *For all* $Z \in \mathcal{H}(\mathbb{H})$ *one has*

(a) $\Theta^*_{P+A,Q+B}(Z;S) = \exp(\pi i\tau(\overline{P}^{tr}S^{-1}B)) \cdot \Theta^*_{P,Q}(Z;S)$ *for all* $A, B \in \mathrm{Mat}(2,\wp)$,
(b) $\Theta^*_{P,Q}(Z[U];S) = \Theta^*_{P\overline{U}^{tr},QU^{-1}}(Z;S)$ *for all* $U \in \mathrm{GL}(2,\mathcal{O}_{\mathbb{H}})$,
(c) $\Theta^*_{P,Q}(Z[\rho E_2];S) = \Theta^*_{\rho P\overline{\rho},\rho Q\overline{\rho}}(Z;S)$ *for* $\rho = \frac{1+i_1}{\sqrt{2}}$,
(d) $\Theta^*_{\overline{V}^{-tr}P,\overline{V}^{-tr}Q}(Z;S) = \Theta^*_{P,Q}(Z;S)$ *for all* $V \in \mathrm{Aut}(S)$.

4 Restricted Octonionic Theta Constants

In this section we consider restrictions of octonionic theta constants to the quaternionic half-space. The restriction can be computed to be (cf. [4, Sect.19])

$$\vartheta_p(Z) = \sum_{u:\wp^2\backslash\mathcal{O}_{\mathbb{H}}^2} \Theta_{2u+\alpha(1-i_2)}(2Z)\Theta_{2\overline{u}+\beta(1-i_2)}(2Z^{tr}) \tag{2}$$

for all $Z \in \mathcal{H}(\mathbb{H}) \subseteq \mathcal{H}(\mathcal{C})$ and $p = \alpha + j\beta \in \mathcal{O}_{\mathcal{C}}^2$.
In particular the case $p = 0$ leads to (cf. [3,5])

$$\begin{aligned}\vartheta_0(Z) &= \sum_{u:\wp^2\backslash\mathcal{O}_{\mathbb{H}}^2} \Theta_{2u}(2Z)\Theta_{2\overline{u}}(2Z^{tr}) = \sum_{u:\wp^2\backslash\mathcal{O}_{\mathbb{H}}^2} \Theta_{2u}(2Z)^2 \\ &= \Theta_0(Z;S), \quad Z \in \mathcal{H}(\mathbb{H}).\end{aligned}$$

The relation between restricted octonionic theta constants and quaternionic theta constants leads to the following.

Lemma 5. *Let $p = \alpha + j\beta \in \mathcal{O}_{\mathcal{C}}^2$. Then the following properties hold for all $Z \in \mathcal{H}(\mathbb{H})$:*

(a) $\vartheta_{p+q}(Z) = \vartheta_p(Z)$ for all $q \in 2\mathcal{O}_{\mathcal{C}}^2$,
(b) $\vartheta_p(Z[U]) = \vartheta_{p^*}(Z)$ for a certain $p^* \in \mathcal{O}_{\mathcal{C}}^2$ depending on $U \in \mathrm{GL}(2,\mathcal{O}_{\mathbb{H}})$ and $p \in \mathcal{O}_{\mathcal{C}}^2$,
(c) $\vartheta_p(Z[\eta E_2]) = \vartheta_{p'}(Z)$, with $p' = \eta\alpha\overline{\eta} + j(\eta\beta\overline{\eta})$, $\eta = \frac{1-i_2}{\sqrt{2}}$.
(d) $\vartheta_p(Z^{tr}) = \vartheta_{jp}(Z)$.

The main statement of this paper is due to Freitag and Salvati Manni (cf. [4, 21.4]) with a completely different proof.

Theorem 3. *Let $S = \begin{pmatrix} 2 & 1+i_1 \\ 1-i_1 & 2 \end{pmatrix}$. Then the three vector spaces*

$$\begin{aligned}\mathcal{A}_1 &= \langle \Theta_A(\cdot;S);\ A \in \mathrm{Mat}(2,\mathcal{O}_{\mathbb{H}})\rangle, \\ \mathcal{A}_2 &= \langle \Theta^*_{P,Q}(\cdot;S);\ P,\ Q \in \mathrm{Mat}(2,\mathcal{O}_{\mathbb{H}})\rangle, \\ \mathcal{A}_3 &= \langle \vartheta_{p|\mathcal{H}(\mathbb{H})};\ p \in \mathcal{O}_{\mathcal{C}}^2\rangle\end{aligned}$$

coincide.

Proof. The first two vector spaces coincide as shown in [6, Corollary 1.7]. Unfortunately there is no explicit formula between restricted octonionic theta constants and quaternionic theta series of first or second kind such as in [6, Corollary 1.7]. Consequently one needs to show that $\mathcal{A}_1 \subseteq \mathcal{A}_3$ and $\mathcal{A}_3 \subseteq \mathcal{A}_1$. Therefore one uses the properties from Lemma 3 and Lemma 5 to restrict oneself to special representatives. For these finitely many representatives one has to verify the result by a direct calculation. The calculation is always done by using the representation of restricted octonionic theta constants respectively quaternionic theta series of second kind as quaternionic theta constants (cf. (2) for the restricted octonionic theta constants and (1) for the quaternionic theta series of second kind) and moreover the knowledge of the explicit behaviour of quaternionic theta constants and quaternionic theta series of second kind under the transpose (cf. Theorem 1 and Theorem 2). For the reader's convenience we include some of the calculations. In order to show that $\mathcal{A}_1 \subseteq \mathcal{A}_3$ it suffices to consider

$$\Theta_A(Z;S) \text{ for } \begin{cases} A = \begin{pmatrix} a & b \end{pmatrix} \text{ with } b = \begin{pmatrix} 0 \\ 1 \end{pmatrix}, \text{ if } A \notin \mathrm{Mat}(2,\wp), \\ A = \begin{pmatrix} 0 & 0 \\ 0 & 1-i_2 \end{pmatrix} \text{ and } \begin{pmatrix} 1-i_1 & 0 \\ 0 & 1-i_2 \end{pmatrix}, \text{ if } A \in \mathrm{Mat}(2,\wp), \\ A = \begin{pmatrix} 0 & 0 \\ 0 & 0 \end{pmatrix}, \text{ if } A \in \mathrm{Mat}(2, 2\mathcal{O}_{\mathbb{H}}). \end{cases}$$

We will now calculate Θ_A in the first case. Let $A = \begin{pmatrix} a & b \end{pmatrix}$ with $b = \begin{pmatrix} 0 \\ 1 \end{pmatrix}$. Then equation (1) yields

$$\Theta_A(Z;S) = \sum_{u:\wp^2 \backslash \mathcal{O}_{\mathbb{H}}^2} \Theta_{2u+a(1+i_1)+b}(2Z)\Theta_{2u+b}(2Z).$$

Due to $(2u+d)_{2,1} = 2u_2 + 1 \equiv 1 \bmod 2\mathcal{O}_{\mathbb{H}}$ and by using Theorem 1 we have

$$\Theta_A(Z;S) = \sum_{u:\wp^2 \backslash \mathcal{O}_{\mathbb{H}}^2} \Theta_{2u+a(1+i_1)+b}(2Z)\Theta_{2\overline{u}+\overline{b}}(2Z^{tr}).$$

With $p := [a(1+i_1)+b]\,(1-i_2)^{-1} + j\left[\overline{b}(1-i_2)^{-1}\right]$ and equation (2) we finally have

$$\Theta_A(Z;S) = \vartheta_p(Z).$$

Furthermore one can show

$$\Theta_A(Z;S) = \begin{cases} \vartheta_{\begin{pmatrix} 0 \\ 0 \end{pmatrix}}(Z), \text{ if } A = \begin{pmatrix} 0 & 0 \\ 0 & 0 \end{pmatrix}, \\ \vartheta_{\begin{pmatrix} 0 \\ 1+i_2 \end{pmatrix}}(Z), \text{ if } A = \begin{pmatrix} 0 & 0 \\ 0 & 1-i_2 \end{pmatrix}, \\ \vartheta_{\begin{pmatrix} 0 \\ 1+j \end{pmatrix}}(Z) + \vartheta_{\begin{pmatrix} 1+i_2 \\ 1+j \end{pmatrix}}(Z) - \vartheta_{\begin{pmatrix} 0 \\ 1+i_2 \end{pmatrix}}(Z), \text{ if } A = \begin{pmatrix} 1-i_1 & 0 \\ 0 & 1-i_2 \end{pmatrix}. \end{cases}$$

The proof of $\mathcal{A}_3 \subseteq \mathcal{A}_1$ is done by a case analysis. If $p = r + js \notin \mathcal{O}_{\mathbb{H}}^2 + j\mathcal{O}_{\mathbb{H}}^2$ and when we define $a := r(1-i_2)$ and $b := s(1-i_2)$, then one can show that $b \notin \wp^2$ because of the representation of the maximal order $\mathcal{O}_{\mathcal{C}}$. Therefore $b_1 \equiv \varepsilon \bmod 2\mathcal{O}_{\mathbb{H}}$ or $b_2 \equiv \varepsilon \bmod 2\mathcal{O}_{\mathbb{H}}$ for an $\varepsilon \in \mathcal{O}_{\mathbb{H}}^{\times}$. Using (2), Theorem 1, and (1) we obtain for $Z \in \mathcal{H}(\mathbb{H})$

$$\vartheta_p(Z) = \Theta_A(Z;S) \text{ for } A = \left((a - \varepsilon\overline{b}\overline{\varepsilon})(1+i_1)^{-1},\ \ \varepsilon\overline{b}\overline{\varepsilon}\right).$$

If $p = r + js \in \mathcal{O}_{\mathbb{H}}^2 + j\mathcal{O}_{\mathbb{H}}^2$ and $r \in \wp^2$ one gets for $Z \in \mathcal{H}(\mathbb{H})$

$$\vartheta_p(Z) = \Theta_A(Z^{tr};S) = \Theta_{A^*}(Z;S) \text{ for } A = \left((\overline{a} - b)(1+i_1)^{-1},\ \ b\right)$$

and A^* as in Theorem 2. For $p = r + js \in \mathcal{O}_{\mathbb{H}}^2 + j\mathcal{O}_{\mathbb{H}}^2$ and $s \in \wp^2$ we obtain

$$\vartheta_p(Z)_{|\mathcal{H}(\mathbb{H})} = \Theta_A(Z;S) \text{ for } A = \left((a - \overline{b})(1+i_1)^{-1},\ \ \overline{b}\right).$$

If $p = r + js \in \mathcal{O}_{\mathbb{H}}^2 + j\mathcal{O}_{\mathbb{H}}^2$ and $r, s \notin \wp^2$ one finally has to calculate a direct representation for the following restricted octonionic theta constants:

$$\vartheta_p(Z), \text{ where } p = \begin{cases} \binom{\varepsilon}{0} + j\binom{0}{1}, & \varepsilon \in \mathcal{O}_{\mathbb{H}}^{\times}, \\ \binom{0}{\varepsilon} + j\binom{0}{1}, & \varepsilon \in \{1, \omega, \overline{\omega}\}, \\ \binom{\alpha(1+i_2)}{r_2} + j\binom{0}{1}, & \alpha, r_2 \in \{1, \omega, \overline{\omega}\}. \end{cases}$$

Here, for example, $\vartheta_p(Z)$, $Z \in \mathcal{H}(\mathbb{H})$, with $p = \binom{1+i_2}{1} + j\binom{0}{1}$ is given by

$$\begin{aligned}\vartheta_{\binom{1+i_2}{1}+j\binom{0}{1}}(Z)_{|\mathcal{H}(\mathbb{H})} &= \Theta_{\begin{pmatrix}1-i_1 & 0\\ 0 & 1-i_2\end{pmatrix}}(Z;S) + \Theta_{\begin{pmatrix}0 & 0\\ 0 & 1-i_2\end{pmatrix}}(Z;S) \\ &\quad - \Theta_{\begin{pmatrix}0 & 0\\ \omega & 1-i_2\end{pmatrix}}(Z;S) - \Theta_{\begin{pmatrix}1-i_2 & 0\\ \omega & 1-i_2\end{pmatrix}}(Z;S) \\ &\quad + \frac{1}{2}\left[\Theta_{\begin{pmatrix}-i_1-i_3 & 0\\ 0 & 1-i_2\end{pmatrix}}(Z;S) + \Theta_{\begin{pmatrix}i_1+i_3 & 0\\ 0 & 1-i_2\end{pmatrix}}(Z;S)\right].\end{aligned}$$

For the other representatives one can find similar expressions. □

Acknowledgement The authors would like to thank the referee for valuable suggestions.

References

1. W.L. Baily, Introductory lectures on automorphic forms, Iwanami Shoten and Princeton (University Press, Tokyo-Princeton 1973)
2. H.-D. Ebbinghaus et al., Numbers (Springer, New York/Berlin/Heidelberg 1990)
3. E. Freitag, C.F. Hermann, Some modular varieties of low dimension. Adv. Math. **152**, 203–287 (2000)
4. E. Freitag, R. *Salvati Manni, Modular Forms for the Even Unimodular Lattice of Signature (2,10)* (Heidelberg, 2005) Preprint
5. A. Krieg, *Modular Forms on Half-Spaces of Quaternions*, Lect. Notes Math. vol. 1143 (Springer, Berlin/Heidelberg/New York 1985)
6. A. Krieg, Theta series over the Hurwitz quaternions. Int. J. Number Theor. **6**, 25–36 (2010)

On the Proportion of Quadratic Twists for Non-vanishing and Vanishing Central Values of L-Functions Attached to Primitive Forms

Kenji Makiyama

1 Introduction

Let $S_{2k}^{\text{new}}(\Gamma_0(N))$ be the space of newforms of weight $2k$ on $\Gamma_0(N)$. Let k and N be positive integers and $f \in S_{2k}^{\text{new}}(\Gamma_0(N))$ a primitive form. Let r be either 0 or 1. Then a conjecture of Goldfeld [4] states that

$$\sharp\{|D| \le X \mid \text{ord}_{s=k} L(s, f \otimes \left(\frac{D}{\cdot}\right)) = r\} \gg X$$

(cf. Ono [10, Conjecture 9.10]), i.e., there exists a positive constant c such that for sufficiently large $X > 0$, we have

$$\sharp\{|D| \le X \mid \text{ord}_{s=k} L(s, f \otimes \left(\frac{D}{\cdot}\right)) = r\} \ge cX,$$

where D stands for fundamental discriminants, that is, discriminants of quadratic fields and $\text{ord}_{s=k} L(s, f \otimes (\frac{D}{\cdot}))$ is the order of the L-function $L(s, f \otimes (\frac{D}{\cdot}))$ attached to f at $s = k$.

Note that $L(s, f \otimes (\frac{D}{\cdot}))$ has a functional equation relating s with $2k - s$. Thus, $L(k, f \otimes (\frac{D}{\cdot}))$ is called the central value.

Remark 1.1 (cf. Vatsal [15]). The conjecture is true for $k = 1, N = 19$ and any primitive form f corresponding to an elliptic curve E over $\mathbb{Q}$. Here the word "corresponding" means that the L-functions attached to f and E coincide. Note that any primitive form with integral Fourier coefficients at the cusp ∞ always

K. Makiyama (✉)
Department of Mathematics, Kyoto Sangyo University, Kyoto 603-8555, Japan
e-mail: kenji_m@cc.kyoto-su.ac.jp

B. Heim et al. (eds.), *Automorphic Forms*, Springer Proceedings in Mathematics & Statistics 115, DOI 10.1007/978-3-319-11352-4_13

corresponds to an elliptic curve over $\mathbb{Q}$ such that their L-functions coincide via the Eichler-Shimura theory and that for any given elliptic curve over $\mathbb{Q}$, we can find a primitive form by the modularity theorem which is called the Shimura-Taniyama conjecture and proved by Breuil, Conrad, Diamond and Taylor [1].

2 Non-vanishing

2.1 Known Results on Non-vanishing

We shall recall some known results on non-vanishing and note that $\mathrm{ord}_{s=k} L(s, f \otimes (\frac{D}{\cdot})) = 0$ means $L(k, f \otimes (\frac{D}{\cdot})) \neq 0$. We put

$$\mathcal{N}_{k,f}(X) := \sharp\{|D| \leq X \mid L(k, f \otimes \left(\frac{D}{\cdot}\right)) \neq 0\}.$$

Currently, the best estimate has been given by Ono and Skinner [11] who showed that

$$\mathcal{N}_{k,f}(X) \gg \frac{X}{\log X}.$$

Galois representations attached to modular forms and a theorem of Waldspurger play important roles in the proof of this result.

By Waldspurger [16], under certain conditions, the central values $L(k, f \otimes (\frac{D}{\cdot}))$ are proportional to the squares of Fourier coefficients of modular forms of weight $k + 1/2$ corresponding to f under the Shimura correspondence.

James [6] gave the first example which shows that $\mathcal{N}_{k,f}(X) \gg X$ for some k, N and f. He reduced the problem to estimating the proportion of class numbers of imaginary quadratic fields which are not divisible by 3. This can be estimated by the work of Davenport and Heilbronn [2] which has been refined by Nakagawa and Horie [9]. His method to estimate was used by Vatsal [14] and Kohnen [7] in the process of proving their results on non-vanishing.

Vatsal has showed that for any semistable elliptic curve over $\mathbb{Q}$ with a rational point of order 3 and good reduction at 3, any primitive form f corresponding to such an elliptic curve satisfies $\mathcal{N}_{1,f}(X) \gg X$ (cf. [14, Theorem 0.3]).

Suppose that k is even. If $f \in S_{2k}^{\mathrm{new}}(\Gamma_0(N))$ is a primitive form, then $L(k, f \otimes (\frac{D}{\cdot})) = 0$ for $D < 0$, since the sign of the functional equation of $L(k, f \otimes (\frac{D}{\cdot}))$ is -1 in this case. Thus, our problem has been reduced to the estimation for $D > 0$ in this case.

For $\varepsilon > 0$, let $X \gg_\varepsilon 0$ mean that X is sufficiently large depending on ε. For $X > 0$, we denote by $F^+(X)$ the set of all positive fundamental discriminants which are less than X. For a positive integer N, let $\mathcal{N}_{k,N}^+(X)$ be the cardinality of elements D in $F^+(X)$ such that there exists a primitive form f in $S_{2k}^{\mathrm{new}}(\Gamma_0(N))$ satisfying $L(k, f \otimes (\frac{D}{\cdot})) \neq 0$.

Kohnen [7] proved that if $\varepsilon > 0$ and $X \gg_\varepsilon 0$, then for any even integer $k \geq 6$,

$$\mathcal{N}_{k,1}^{+}(X) \geq \left(\frac{9}{16\pi^2} - \varepsilon\right) X.$$

We put

$$d_{k,N} := \dim_{\mathbb{C}}(S_{2k}^{\mathrm{new}}(\Gamma_0(N))).$$

For a primitive form $f \in S_{2k}^{\mathrm{new}}(\Gamma_0(N))$, we put

$$\mathcal{N}_{k,f}^{+}(X) := \sharp\{D \in F^+(X) \mid L(k, f \otimes \left(\frac{D}{\cdot}\right)) \neq 0\}.$$

Note that $S_{2k}^{\mathrm{new}}(\Gamma_0(1)) = S_{2k}(\mathrm{SL}_2(\mathbb{Z}))$. The above Kohnen's result immediately implies that there exists a primitive form $f \in S_{2k}(\mathrm{SL}_2(\mathbb{Z}))$ such that

$$\mathcal{N}_{k,f}^{+}(X) \geq \left(\frac{1}{d_{k,1}} \cdot \frac{9}{16\pi^2} - \varepsilon\right) X.$$

Moreover, Kohnen [7] obtained that for arbitrary primitive form f in $S_{2k}(\mathrm{SL}_2(\mathbb{Z}))$,

$$\mathcal{N}_{k,f}^{+}(X) \geq \left(\frac{1}{d_{k,1}} \cdot \frac{9}{16\pi^2} - \varepsilon\right) X$$

as a corollary of the above result under the assumption that a conjecture of Maeda holds for all k (cf. [5, Conjecture 1.2]). Note that this conjecture has been verified for $6 \leq k \leq 2000$ (cf. [3, Theorem 1]).

2.2 Main Results on Non-vanishing

For an even integer $k \geq 6$, we put $q := e^{2\pi\sqrt{-1}z}$ for $z \in \mathbb{C}$ with $\mathrm{Im}(z) > 0$ and define

$$\delta_k(z) := \frac{1}{4\pi\sqrt{-1}}\left((\frac{k}{2} - 1)G_{k-2}(4z)\frac{d}{dz}\theta(z) - \frac{d}{dz}G_{k-2}(4z)\theta(z)\right),$$

where

$$G_k(z) := \frac{1}{2}\zeta(1-k) + \sum_{n\geq 1}\sigma_{k-1}(n)q^n$$

with

$$\sigma_{k-1}(n) := \sum_{0<d\,|\,n} d^{k-1}$$

and

$$\theta(z) := \sum_{n\in\mathbb{Z}} q^{n^2} = 1 + 2\sum_{n\geq 1} q^{n^2}.$$

Note that δ_k is a cusp form of weight $k + 1/2$ on $\Gamma_0(4)$ which has appeared in the paper of Kohnen and Zagier [8]. Let Δ_k be the primitive form in $S_{2k}(\mathrm{SL}_2(\mathbb{Z}))$ corresponding to δ_k under the Shimura correspondence.

One of the main results on non-vanishing is the following.

Theorem 2.1. *Suppose that* $k = 6, 8, 10$. *Let* p *be a prime number with* $p \equiv 1 \pmod 4$. *If the* p*-th Fourier coefficient of* Δ_k *is not zero, then the assertion that* $\mathcal{N}^+_{k,p}(X) \gg X$ *holds. More precisely, if* $\varepsilon > 0$ *and* $X \gg_\varepsilon 0$, *then*

$$\mathcal{N}^+_{k,p}(X) \geq \left(\frac{9p}{32\pi^2(p+1)} - \varepsilon\right) X.$$

As Kohnen obtained the result on $\mathcal{N}^+_{k,f}(X)$ from the result on $\mathcal{N}^+_{k,1}(X)$, we obtain the following.

Corollary 2.2. *Let the assumptions be the same as in Theorem 2.1. Then there exists a primitive form* $f \in S^{\mathrm{new}}_{2k}(\Gamma_0(p))$ *such that the assertion that* $\mathcal{N}^+_{k,f}(X) \gg X$ *holds. More precisely, if* $\varepsilon > 0$ *and* $X \gg_\varepsilon 0$, *then*

$$\mathcal{N}^+_{k,f}(X) \geq \left(\frac{1}{d_{k,p}} \cdot \frac{9p}{32\pi^2(p+1)} - \varepsilon\right) X.$$

Comparing Theorem 2.1 with Kohnen's result above, we can improve Theorem 2.1 at the point where the level N can be taken as some prime number. In return for the improvement, weight k must be either $6, 8$ or 10.

Let us see an outline of the proof of Theorem 2.1. We can determine the sign $\epsilon \in \{\pm 1\}$, letting the set

$$\{D \in F^+(X) \| \; D \equiv -1 \pmod 3,\; 3 \nmid h(-3D),\; \left(\frac{D}{p}\right) = \epsilon\}$$

be included in the set giving the cardinality $\mathcal{N}^+_{k,f}(X)$, where $h(-3D)$ is the class number of the quadratic field with the discriminant $-3D$. And the determined subset can be estimated by [9, Proposition 2] which is a result giving the estimate for indivisibility of class numbers of quadratic fields. Then, indivisibility of class numbers relates to the non-vanishing of Fourier coefficients of the half-integral weight cusp form δ_k and it relates to the non-vanishing of the central L-values of primitive forms by the Kohnen-Zagier formula.

3 Vanishing

3.1 A Known Result on Vanishing

We shall recall a known result on vanishing. Let $\varepsilon > 0$. The current best estimate seems to be due to Perelli and Pomykala [12], who showed that for $k = 1$, any N and any f corresponding to an elliptic curve over $\mathbb{Q}$,

$$\sharp\{|D| \leq X \mid \operatorname{ord}_{s=1} L(s, f \otimes \left(\frac{D}{\cdot}\right)) = 1\} \gg X^{1-\varepsilon}.$$

3.2 The Main Result on Vanishing

We denote by $F_p(X)$ the set of fundamental discriminants D with $(-1)^k D > 0$ and $(D, p) = 1$ for a prime number p and $X > 0$. For a primitive form $f \in S_{2k}^{\text{new}}(\Gamma_0(N))$ and any prime divisor p of N, we put

$$\mathcal{V}_{k,f}(X) := \sharp\{D \in F_p(X) \mid L(k, f \otimes \left(\frac{D}{\cdot}\right)) = 0\}.$$

Note that

$$\mathcal{V}_{k,f}(X) = \sharp\{D \in F_p(X) \mid \operatorname{ord}_{s=k} L(s, f \otimes \left(\frac{D}{\cdot}\right)) \geq 1\}.$$

For an integer N and a prime number p, we denote by $\operatorname{ord}_p(N)$ the non-negative integer satisfying $p^{\operatorname{ord}_p(N)} \mid N$ and $(p, Np^{-\operatorname{ord}_p(N)}) = 1$.

Theorem 3.1. *Let $k \geq 2$ be an integer and N a positive odd integer such that $\operatorname{ord}_p(N) \in 2\mathbb{Z} + 1$ holds for any prime divisor p of N. For any primitive form $f \in S_{2k}^{\text{new}}(\Gamma_0(N))$, if $\varepsilon > 0$ and $X \gg_\varepsilon 0$, then*

$$\mathcal{V}_{k,f}(X) \geq \left(\frac{3p}{2\pi^2(p+1)} - \varepsilon\right) X.$$

In particular,

$$\mathcal{V}_{k,f}(X) \gg X.$$

Comparing Theorem 3.1 with Perelli-Pomykala's result, we can improve Theorem 3.1 at the point where the estimate is given by X and k can be taken as any weight which is greater than 1. In return for the improvement, the estimate is given only for the part where the order is greater than or equal to 1, that is, the order is not just 1, and the level N need to satisfy the technical assumption.

Let us see an outline of the proof of Theorem 3.1. Sakata [13] has showed that for a fixed prime divisor p of N, if a fundamental discriminant D with $(-1)^k D > 0$ and $p \nmid D$ satisfies that the eigen value of f for the Atkin-Lehner involution at p and $\left(\frac{D}{p}\right)$ do not coincide, then the central L-value of f is zero. Namely, thanks to the work of Sakata, it is sufficient to estimate some classes modulo p consisting of fundamental discriminants and it is possible by using the result in the paper [9].

Acknowledgements The author is very grateful to Professor Atsushi Yamagami for variable guidance and kind help. He is also grateful to Professor Atsushi Murase and Professor Bernhard Heim for valuable comments on improving the manuscript and supports for my participation in the international workshop on mathematics held at Oman in February 2012.

References

1. C. Breuil, B. Conrad, F. Diamond, R. Taylor, On the modularity of elliptic curves over $\mathbb{Q}$, or wild 3-adic exercises. J. Amer. Math. Soc. **14**(4), 843–939 (2001)
2. H. Davenport, H. Heilbronn, On the density of discriminants of cubic fields II. Proc. Roy. Soc. Lond. A **322**, 405–420 (2001)
3. D.W. Farmer, K. James, The irreducibility of some level 1 Hecke polynomials. Math. Comp. **71**(239), 1263–1270 (2002)
4. D. Goldfeld, Conjectures on elliptic curves over quadratic fields. Number Theory, Carbondale, Springer Lect. Notes. **751**, 108–118 (1979)
5. H. Hida, Y. Maeda, Non-abelian base change for totally real fields. Olga Taussky-Todd: in memoriam. Pacific J. Math. Special Issue, 189–217 (1997)
6. K. James, L-series with non-zero central critical value. J. AMS **11**(3), 635–641 (1998)
7. W. Kohnen, On the proportion of quadratic character twists of L-functions attached to cusp forms not vanishing at the central point. J. Rein. Angew. Math. **508**, 179–187 (1999)
8. W. Kohnen, D. Zagier, Values of L-series of modular forms at the center of the critical strip. Invent. Math. **64**, 175–198 (1981)
9. J. Nakagawa, K. Horie, Elliptic curves with no rational points. Proc. AMS **104**(1), 20–24 (1988)
10. K. Ono, The web of modularity: arithmetic of the coefficients of modular forms and q-series, CBMS Regional Conference Series in Mathematics, vol. 102 (American Mathematical Society, 2004)
11. K. Ono, C. Skinner, Nonvanishing of quadratic twists of modular L-function. Invent. Math. **134**, 651–660 (1998)
12. A. Perelli, J. Pomykala, Averages of twisted L-functions. Acta Arith. **80**, 149–163 (1997)
13. H. Sakata, On the Kohnen-Zagier formula in the general case of '4× general odd' level. Nagoya Math. J. **190**, 63–85 (2008)
14. V. Vatsal, Canonical periods and congruence formulae. Duke Math. J. **98**, 397–419 (1999)
15. V. Vatsal, Rank-one twists of a certain elliptic curve. Math. Ann. **311**, 791–794 (1998)
16. J.-L. Waldspurger, Sur les coefficients de Fourier des formes modulaires de poids demi-entier. J. Math. Pures et Appl. **60**, 375–484 (1997)

Bessel Periods of Theta Lifts to $GSp(1,1)$ and Central Values of Some L-Functions of Convolution Type

Hiro-aki Narita*

1 Introduction

This article overviews our results presented in the international workshop on mathematics held at GUtech and Sultan Qaboos University in Oman. The main result is an explicit relation between a Bessel period of some theta lift to the indefinite symplectic group $GSp(1,1)$ and the central value of an L-function of convolution type for the lift (cf. Theorem 3.2).

In our recent work [10] such convolution-type L-functions of the theta lifts to $GSp(1,1)$ are proved to coincide with those of some paramodular theta lifts to the symplectic group $GSp(2)$ of degree two, for which we note that the two theta lifts just mentioned satisfy expected properties of the local and global Jacquet-Langlands-Shimizu correspondence. The Bessel periods above then turn out to be also related to the central values of the L-functions of convolution type for the paramodular theta lifts.

As an application our formula yields the existence of the theta lifts to $GSp(1,1)$ with strictly positive central values of such convolution-type L-functions (cf. Theorem 4.4). More precisely, to discuss such existence property, we specify a definite quaternion algebra over $\mathbb{Q}$ defining $GSp(1,1)$. For this we should remark that the strict positivity of the central L-values is essentially reduced to the following problem on special values of hypergeometric functions:

*Partially supported by Grand-in-Aid for Young Scientists (B) 21740025, JSPS.

H. Narita (✉)
Department of Mathematics, Faculty of Science
Kumamoto University, Kurokami, Kumamoto 860-8555, Japan
e-mail: narita@sci.kumamoto-u.ac.jp

B. Heim et al. (eds.), *Automorphic Forms*, Springer Proceedings
in Mathematics & Statistics 115, DOI 10.1007/978-3-319-11352-4_14

$$\begin{pmatrix} \frac{\kappa_1+\kappa_2}{2} \\ \frac{\kappa_2-\kappa_1}{2} \end{pmatrix} {}_2F_1(-\frac{\kappa_2-\kappa_1}{2}, -\frac{\kappa_2-\kappa_1}{2}; \kappa_1+1; -1) \neq (-1)^{\kappa_1/4-1} 2^{-\frac{\kappa_2-4}{2}} ?$$

Here $(\kappa_1, \kappa_2) \in (4\mathbb{Z}_{>0})^{\oplus 2}$. Recently, Akihito Ebisu has partially resolved this problem and the author has found further examples of special values of hypergeometric functions related to the strict positivity of the central L-values. We will report the progress elsewhere.

2 Basic Notations

2.1 *Algebraic Groups*

Let B be a definite quaternion algebra over the field $\mathbb{Q}$ of rational numbers with discriminant d_B and $B \ni x \mapsto \bar{x} \in B$ denote the main involution. Then $\mathrm{tr}(x) := x + \bar{x}$ and $n(x) := x\bar{x}$ are the reduced trace and the reduced norm of $x \in B$, respectively. We put $B^- := \{x \in B \mid \mathrm{tr}(x) = 0\}$, the set of pure quaternions.

By $G = GSp(1,1)$ and $Sp(1,1)$ we denote two $\mathbb{Q}$-algebraic groups defined by the groups

$$G(\mathbb{Q}) = GSp(1,1)(\mathbb{Q}) := \left\{ g \in M_2(B) \;\middle|\; {}^t\bar{g} \begin{pmatrix} 0 & 1 \\ 1 & 0 \end{pmatrix} g = \nu(g) \begin{pmatrix} 0 & 1 \\ 1 & 0 \end{pmatrix},\ \nu(g) \in \mathbb{Q}^\times \right\},$$

$$Sp(1,1)(\mathbb{Q}) := \{g \in GSp(1,1)(\mathbb{Q}) \mid \nu(g) = 1\}$$

of $\mathbb{Q}$-rational points, respectively. Here $\bar{g} = \begin{pmatrix} \bar{a} & \bar{b} \\ \bar{c} & \bar{d} \end{pmatrix}$ for $g = \begin{pmatrix} a & b \\ c & d \end{pmatrix} \in G(\mathbb{Q})$. As is well known, these two algebraic groups are inner $\mathbb{Q}$-forms of the usual symplectic groups $GSp(2)$ and $Sp(2)$ of degree two defined by

$$GSp(2)(\mathbb{Q}) := \left\{ g \in GL_4(\mathbb{Q}) \;\middle|\; {}^t g \begin{pmatrix} 0_2 & 1_2 \\ -1_2 & 0_2 \end{pmatrix} g = \mu(g) \begin{pmatrix} 0_2 & 1_2 \\ -1_2 & 0_2 \end{pmatrix},\ \mu(g) \in \mathbb{Q}^\times \right\},$$

$$Sp(2)(\mathbb{Q}) := \{g \in GSp(2)(\mathbb{Q}) \mid \mu(g) = 1\}$$

respectively.

For an algebraic group $\mathcal{G}$ we denote the center of $\mathcal{G}$ by $Z_{\mathcal{G}}$.

2.2 *Groups Over Local Fields*

In what follows, let $\mathbb{A}_F$ be the ring of adeles for a number field F and we will often deal with $\mathbb{A}_{\mathbb{Q}}$. Let $\mathbb{A}_f$ be the ring of finite adeles of $\mathbb{A}_{\mathbb{Q}}$. For each place $v \leq \infty$ of $\mathbb{Q}$, denote the v-adic completion of B at v by B_v.

Let G^1_∞ be the group $Sp(1,1)(\mathbb{R})$ of real points for $Sp(1,1)$ and denote its Lie algebra by $\mathfrak{g}$. The group

$$K_\infty := \{\begin{pmatrix} a & b \\ b & a \end{pmatrix} \in G^1_\infty \mid a \pm b \in B^1_\infty\}$$

forms a maximal compact subgroup of $Sp(1,1)(\mathbb{R})$, where $B^1_\infty := \{x \in B_\infty \mid n(x) = 1\}$. This group is isomorphic to $B^1_\infty \times B^1_\infty \simeq SU(2) \times SU(2)$, where $SU(2)$ denotes the special unitary group of degree two.

Let us fix a maximal order $\mathfrak{O}$ of B and $\mathfrak{O}_p$ denote the p-adic completion of $\mathfrak{O}$ at a finite prime p. The ring $\mathfrak{O}_p$ has the prime ideal $\mathfrak{P}_p$. For each fixed divisor D of d_B we introduce a maximal open compact subgroup

$$K_f(D) := \prod_{p<\infty} K_p$$

of $G(\mathbb{A}_f)$, where $K_p := \{g \in G(\mathbb{Q}_p) \mid gL_p = L_p\}$ with

$$L_p := \begin{cases} \mathfrak{O}_p \oplus \mathfrak{O}_p & (p \nmid d_B \text{ or } p|D), \\ \mathfrak{O}_p \oplus \mathfrak{P}_p^{-1} & (p|\frac{d_B}{D}). \end{cases}$$

2.3 Automorphic Forms

From now on we fix a divisor D of d_B. For a positive integer l we let (σ_l, V_l) be the l-th symmetric tensor representation of $M_2(\mathbb{C})$ on the $l+1$-dimensional representation space V_l. The restriction of σ_l to $B^1_\infty \simeq SU(2)$ is the irreducible representation of $B^1_\infty \simeq SU(2)$ with highest weight l.

For a positive integer κ we let $S_\kappa(D)$ be the space of elliptic cusp forms of weight κ with level D (cf. [7, Sect. 3.1]). For a non-negative integer κ' we let $\mathcal{A}_{\kappa'}$ be the space of automorphic forms of weight $\sigma_{\kappa'}$ with respect to $\prod_{p<\infty} \mathfrak{O}_p^\times$ (cf. [7, Sect. 3.2]), where $\mathfrak{O}_p^\times$ denotes the unit group of $\mathfrak{O}_p$.

In addition to these, we need the space of cusp forms on $G(\mathbb{A}_\mathbb{Q})$ as follows:

Definition 2.1. For $\lambda = (\lambda_1, \lambda_2) \in \{(m,n) \in \mathbb{Z}^2_{>0} \mid m < n\}$ let π_λ be the discrete series representation of G^1_∞ parametrized by λ and $\Lambda = (\Lambda_1, \Lambda_2) := (\lambda_1 - 1, \lambda_2)$ the highest weight of the minimal K_∞-type $\tau_\Lambda := \sigma_{\Lambda_1} \boxtimes \sigma_{\Lambda_2}$ of π_λ. We then introduce the space $\mathcal{S}_{\tau_\Lambda}(D)$ of $V_{\Lambda_1} \boxtimes V_{\Lambda_2}$-valued cusp forms F on $G(\mathbb{A})$ satisfying the following:

1. $F(z\gamma g k_f k_\infty) = \tau_\Lambda(k_\infty)^{-1} F(g)$ for $(z, \gamma, g, k_f, k_\infty) \in Z_G(\mathbb{A}) \times G(\mathbb{Q}) \times G(\mathbb{A}) \times K_f(D) \times K_\infty$.

2. For any fixed $g \in G(\mathbb{A})$, the constant term $\int_{B^-\backslash B^-(\mathbb{A})} F(\begin{pmatrix} 1 & x \\ 0 & 1 \end{pmatrix} g)dx$ of the Fourier expansion of F is zero, where see Sect. 3.1 for the Fourier expansion.
3. For each fixed $g_f \in G(\mathbb{A}_f)$, the right translations of the coefficients of $F|_{G^1_\infty}(g_f *)$ by G^1_∞ generate π_λ as a $(\mathfrak{g}, K_\infty)$-module (if F is non-zero).

2.4 Theta Lifts to G

For $v = p < \infty$ let $\mathbb{V}_p$ be the space of locally constant compactly supported functions on $B_p^2 \times \mathbb{Q}_p^\times$. Let $\mathcal{S}(\mathbb{H}^2)$ stand for the space of Schwartz functions on $\mathbb{H}^2$. We introduce the space $\mathbb{V}_\infty$ of smooth functions φ on $\mathbb{H}^2 \times \mathbb{R}^\times$ such that, for each fixed $t \in \mathbb{R}^\times$, $\mathbb{H}^2 \ni X \mapsto \varphi(X,t)$ belongs to $\mathcal{S}(\mathbb{H}^2) \otimes \mathrm{End}(V_{\frac{\kappa_1+\kappa_2}{2}} \boxtimes V_{\frac{\kappa_2-\kappa_1}{2}})$ for $(\kappa_1, \kappa_2) \in (2\mathbb{Z}_{\geq 0})^2$ with $\kappa_1 \leq \kappa_2$. We let $\varphi_{0,p} \in \mathbb{V}_p$ be the characteristic function of $L_p \times \mathbb{Z}_p^\times$. For $(\kappa_1, \kappa_2) \in (2\mathbb{Z}_{\geq 0})^2$ with $\kappa_1 \leq \kappa_2$ we define $\varphi_{0,\infty} \in \mathbb{V}_\infty$ by

$$\varphi_{0,\infty}(X,t) := \begin{cases} t^{\frac{\kappa_2+3}{2}} \sigma_{\frac{\kappa_1+\kappa_2}{2}}(X_1 + X_2) \boxtimes \sigma_{\frac{\kappa_2-\kappa_1}{2}}(X_1 - X_2) \exp(-2\pi t^t \bar{X} X) & (t > 0), \\ 0 & (t < 0). \end{cases}$$

We then put $\varphi_0 := \prod_{v \leq \infty} \varphi_{0,v}$.

Following [6, Sect. 3] we introduce a metaplectic representation $r = \otimes'_{v \leq \infty} r_v$ of $G(\mathbb{A}_\mathbb{Q}) \times GL_2(\mathbb{A}_\mathbb{Q}) \times B^\times(\mathbb{A}_\mathbb{Q})$ on the restricted tensor product $\mathbb{V} = \otimes'_{v \leq \infty} \mathbb{V}_v$ with respect to $\{\varphi_{0,p}\}_{p<\infty}$. It is associated with the standard additive character ψ of $\mathbb{A}$. With $\varphi_0 \in \mathbb{V}$ we define the $\mathrm{End}(V_{\frac{\kappa_1+\kappa_2}{2}} \boxtimes V_{\frac{\kappa_2-\kappa_1}{2}})$-valued theta function $\theta_{\kappa_1,\kappa_2}(g,h,h')$ by

$$\sum_{(X,t) \in B^2 \times \mathbb{Q}^\times} r(g,h,h')\varphi_0(X,t).$$

For $(\kappa_1, \kappa_2) \in (2\mathbb{Z}_{>0})^2$ with $\kappa_1 \leq \kappa_2$, we consider the theta lift

$$S_{\kappa_1}(D) \times \mathcal{A}_{\kappa_2} \ni (f, f') \mapsto \mathcal{L}(f, f')(g)$$

with

$$\mathcal{L}(f, f')(g) := \int_{\mathbb{R}_+^2 (H \times H')(\mathbb{Q}) \backslash (H \times H')(\mathbb{A})} \overline{f(h)} \theta_{\kappa_1,\kappa_2}(g,h,h') f'(h') dh dh'.$$

Theorem 2.2. *Let $(\kappa_1, \kappa_2) \in (2\mathbb{Z}_{>0})^2$ satisfy $1 < \kappa_1 < \kappa_2 + 2$. We have*

$$\mathcal{L}(f, f') \in \mathcal{S}_{\tau_\Lambda}(D)$$

with $\Lambda = (\frac{\kappa_2-\kappa_1}{2}, \frac{\kappa_1+\kappa_2}{2})$.

For this theorem see [9].

2.5 Automorphic L-Functions

For a Hecke eigenform $f \in S_{\kappa}(D)$ (respectively a Hecke eigenform $f' \in \mathcal{A}_{\kappa'}$) we denote by $\pi(f)$ (respectively $\pi(f')$) the irreducible automorphic representation of $GL_2(\mathbb{A}_{\mathbb{Q}})$ (respectively $B^{\times}(\mathbb{A}_{\mathbb{Q}})$) generated by f (respectively f'). Later we will need the adjoint L-functions $L(\pi, \mathrm{Ad}, s)$ for $\pi = \pi(f)$ and the Jacquet-Langlands-Shimizu lift $\mathrm{JL}(\pi(f'))$ (cf. [3, 14]) of $\pi(f')$ to state the main result, Theorem 3.2.

For a Hecke eigenform $F \in \mathcal{S}_{\tau_\Lambda}(D)$ with $\Lambda = (\frac{\kappa_2-\kappa_1}{2}, \frac{\kappa_2+\kappa_1}{2})$, we introduce the global spinor L-function and some global convolution-type L-function.

We first define the global spinor L-function for $\mathcal{S}_{\tau_\Lambda}(D)$, modifying the definition of [8, Sect. 2.6] at the archimedean place.

In [6, Sect. 5.1] we introduced three Hecke operator $\mathcal{T}_p^i$ with $0 \le i \le 2$ for $p \nmid d_B$. Let Λ_p^i be the Hecke eigenvalue of $\mathcal{T}_p^i$ for F with $0 \le i \le 2$. For $p \nmid d_B$ we put

$$Q_{F,p}(t) := 1 - p^{-\frac{3}{2}}\Lambda_p^1 t + p^{-2}(\Lambda_p^2 + p^2 + 1)t^2 - p^{-\frac{3}{2}}\Lambda_p^1 t^3 + t^4.$$

For this we note that $Q_{F,p}(p^{-s})^{-1}$ coincides with the local spinor L-function for an unramified principal series representation of the group $GSp(2)(\mathbb{Q}_p)$. On the other hand, in [6, Sect. 5.2], we introduced two Hecke operators $\mathcal{T}_p^i$ with $0 \le i \le 1$ for $p|d_B$. Let Λ'^i_p be the Hecke eigenvalue of $\mathcal{T}_p^i$ for F with $0 \le i \le 1$. For $p|d_B$ we put

$$Q_{F,p}(t) := \begin{cases} (1 - p^{-\frac{3}{2}}(\Lambda'^1_p - (p-1)\Lambda'^0_p)t + t^2)(1 - \Lambda'^0_p p^{-\frac{1}{2}}t) & (p|\frac{d_B}{D}), \\ (1 + \Lambda'^0_p p^{-\frac{1}{2}}t)(1 - \Lambda'^0_p p^{-\frac{1}{2}}t) & (p|D). \end{cases}$$

The first one is due to Sugano [15, (3–4)]. The first factor of the second one comes from the numerator of the formal Hecke series.

We define the global spinor L-function $L(F, \mathrm{spin}, s)$ of a Hecke eigenform F in $\mathcal{S}_{\tau_\Lambda}(D)$ by

$$L(F, \mathrm{spin}, s) := \prod_{v \le \infty} L_v(F, \mathrm{spin}, s),$$

where

$$L_v(F, \mathrm{spin}, s) := \begin{cases} Q_{F,p}(p^{-s})^{-1} & (v = p < \infty), \\ \Gamma_{\mathbb{C}}(s + \frac{\kappa_1-1}{2})\Gamma_{\mathbb{C}}(s + \frac{\kappa_2+1}{2}) & (v = \infty) \end{cases}$$

with $\Gamma_{\mathbb{C}}(s) := 2(2\pi)^{-s}\Gamma(s)$.

For $\xi \in B^- \setminus \{0\}$ we let $E_\xi := \mathbb{Q}(\xi)$, which is isomorphic to an imaginary quadratic field. Given a Hecke character χ of $\mathbb{A}_{E_\xi}^{\times}/E_\xi^{\times}\mathbb{R}_+^{\times}$, we next introduce the L-function

$$L(F,\chi^{-1},s) := \prod_{v\le\infty} L_v(F,\chi^{-1},s)$$

of convolution type for a Hecke eigenform $F \in \mathcal{S}_{\tau_\Lambda}(D)$. Here the local factors $L_v(F,\chi^{-1},s)$ are as follows:

$$L_v(F,\chi^{-1},s) :=$$

$$\begin{cases} Q_{F,p}(\alpha_p^\chi p^{-s})^{-1} Q_{F,p}(\beta_p^\chi p^{-s})^{-1} & (\chi \text{ is unramified at } v = p < \infty),\\ 1 & (\chi \text{ is ramified at } v = p < \infty),\\ \Gamma_{\mathbb{C}}(s+\kappa_1-\frac{1}{2})\Gamma_{\mathbb{C}}(s+\frac{1}{2})\Gamma_{\mathbb{C}}(s+\frac{\kappa_1+\kappa_2+1}{2})\Gamma_{\mathbb{C}}(s+\frac{\kappa_2-\kappa_1+1}{2}) & (v=\infty), \end{cases}$$

where, when χ is unramified at $v = p < \infty$,

$$(\alpha_v^\chi, \beta_v^\chi) := \begin{cases} (\chi_p(\varpi_{p,1})^{-1}, \chi_p(\varpi_{p,2})^{-1}) & (v = p\text{: split}),\\ (\chi_p(p)^{-1}, -\chi_p(p)^{-1}) = (1,-1) & (v = p\text{: inert}),\\ (\chi_p(\varpi_p)^{-1}, 0) & (v = p\text{: ramified}) \end{cases}$$

with two distinct prime elements $\varpi_{p,1}$ and $\varpi_{p,2}$ (respectively a prime element ϖ_p) for a split prime p (respectively a ramified prime p).

3 Explicit Formula for the Bessel Periods in Terms of Central L-Values (Main Theorem)

3.1 *Fourier Expansion of the Theta Lifts to* $GSp(1,1)$

We review the Fourier expansion of $\mathcal{L}(f,f')$ on $GSp(1,1)(\mathbb{A}_{\mathbb{Q}})$ following [7, Sect. 1.3]. We recall that we have put $B^- := \{x \in B \mid \mathrm{tr}(x) = 0\}$ (cf. Sect. 2.1). We write the expansion as

$$\mathcal{L}(f,f')(g) = \sum_{\xi\in B^-\setminus\{0\}} \mathcal{L}(f,f')_\xi(g),$$

where

$$\mathcal{L}(f,f')_\xi(g) := \int_{B^-\backslash B^-(\mathbb{A}_{\mathbb{Q}})} \mathcal{L}(f,f')(\begin{pmatrix} 1 & x\\ 0 & 1\end{pmatrix} g)\psi(-\mathrm{tr}(\xi x))dx$$

with the standard additive character ψ on $\mathbb{Q}\backslash\mathbb{A}_{\mathbb{Q}}$. Here we normalize the measure dx so that the volume of $B^-\backslash B^-(\mathbb{A}_{\mathbb{Q}})$ is one. For $\xi \in B^- \setminus \{0\}$ we let X_ξ be the set of unitary characters on $\mathbb{A}_{\mathbb{Q}}^\times E_\xi^\times\backslash\mathbb{A}_{E_\xi}^\times$, namely Hecke characters. The Fourier expansion is then refined as follows:

$$\mathcal{L}(f,f')(g) = \sum_{\xi\in B^-\setminus\{0\}} \sum_{\chi\in X_\xi} \mathcal{L}(f,f')_\xi^\chi(g),$$

with

$$\mathcal{L}(f,f')_\xi^\chi(g) := \mathrm{vol}(\mathbb{R}_+^\times \mathbb{A}_\mathbb{Q}^\times \backslash \mathbb{A}_{E_\xi}^\times)^{-1} \int_{\mathbb{R}_+^\times E_\xi^\times \backslash \mathbb{A}_{E_\xi}^\times} \mathcal{L}(f,f')_\xi(s1_2\cdot g)\chi(s)^{-1}ds.$$

The Fourier coefficient $\mathcal{L}(f,f')_\xi^\chi$ of $\mathcal{L}(f,f')$ is called a *Bessel period.*

3.2 Assumption and Notation for the Main Theorem (1)

To deduce our formula for $\mathcal{L}(f,f')_\xi^\chi$, we let $(f,f') \in S_{\kappa_1}(D) \times \mathcal{A}_{\kappa_2}$ and assume the following two conditions:

(1) The two forms f and f' are Hecke eigenforms and have the same eigenvalue for the "Atkin-Lehner involution." Namely, with the eigenvalue ϵ_p of f (respectively ϵ'_p of f'), $\epsilon_p = \epsilon'_p$ holds at $p|D$ for (f,f'). Otherwise $\mathcal{L}(f,f') \equiv 0$ (cf. [6, Remark 5.2 (ii)]).
(2) We assume that $\xi \in B^- \setminus \{0\}$ is primitive.

Namely, for each finite prime p, we let

$$\mathfrak{a}_p := \begin{cases} \mathfrak{O}_p & (p \nmid d_B \text{ or } p|D) \\ \mathfrak{P}_p & (p|\frac{d_B}{D}) \end{cases}, \quad (\mathfrak{a}_p^-)^* := \{z \in B_p^- \mid \mathrm{tr}(\bar{z}w) \in \mathbb{Z}_p, \text{ for any } w \in \mathfrak{a}_p \cap B_p^-\}$$

and assume that

$$\xi \in (\mathfrak{a}_p^-)^* \setminus p(\mathfrak{a}_p^-)^*.$$

For the second assumption we note that, in general, a Fourier coefficient F_ξ of an automorphic form F on $G(\mathbb{A}_\mathbb{Q})$ indexed by $\xi \in B^-$ satisfies

$$F_\xi(g) = F_\xi(\begin{pmatrix} t & 0 \\ 0 & 1 \end{pmatrix} g) = F_{t\xi}(g) \quad (t \in \mathbb{Q}^\times).$$

We then see that the problem determining F_ξ is reduced to the case where ξ is primitive.

3.3 Assumption and Notation for the Main Theorem (2)

For each $\xi \in B^- \setminus \{0\}$, d_ξ denotes the discriminant of E_ξ. Let $h(E_\xi)$ (respectively $w(E_\xi)$) be the class number of E_ξ (respectively the number of roots of unity in E_ξ). Put $r = \frac{2\sqrt{n(\xi)}}{\sqrt{d_\xi}} \in \mathbb{Q}^\times$. For a Hecke character $\chi = \prod_{v \le \infty} \chi_v$ of $\mathbb{R}_+^\times E_\xi^\times \backslash \mathbb{A}_{E_\xi}^\times$, we let $w_\infty(\chi) \in \mathbb{Z}$ be such that

$$\chi_\infty(u) = (\delta_\xi(u)/|\delta_\xi(u)|)^{w_\infty(\chi)} \ (u \in E_\infty),$$

where δ_ξ is the isomorphism between E_∞ and $\mathbb{C}$ defined by

$$\delta_\xi(x + y\xi) := x + \sqrt{-n(\xi)}y \quad (x,\ y \in \mathbb{R}).$$

For each prime $v = p < \infty$, we let $p^{i_p(\chi)}$ be the conductor of χ at p and put $A(\chi) := \prod_{p<\infty} p^{i_p(\chi)}$. We introduce

$$\mu_p := \frac{\mathrm{ord}_p(2\xi)^2 - \mathrm{ord}_p(d_\xi)}{2},$$

which coincides with $\mathrm{ord}_p(r)$.

3.4 Main Theorem

We need further notations to state our formula for $\mathcal{L}(f, f')_\xi^\chi$.

For D we put $\delta(D) := \sharp\{p < \infty \mid p|D\}$. We let

$$e_p(E_\xi) = \begin{cases} -1 & (p \text{ is inert in } E_\xi), \\ 0 & (p \text{ ramifies in } E_\xi), \\ 1 & (p \text{ splits in } E_\xi). \end{cases}$$

Let $(*, *)_{\kappa_2}$ (respectively $|| * ||$) denote a fixed unitary inner product of V_{κ_2} (respectively the norm of V_{κ_2} induced by this inner product). For $(f, f') \in S_{\kappa_1}(D) \times \mathcal{A}_{\kappa_2}$ we denote by $\langle f, f\rangle$ (respectively $\langle f', f'\rangle$) the norm of f (respectively f') (cf. [8, Sects. 2.3 and 2.4]). We define $f'_\infty,\ f'_{\infty,\kappa_1} \in V_{\kappa_2}$ as in [8, Sect. 2.4], where note that $w_\infty(\chi)$ is $-\kappa_1$ while $w_\infty(\chi) = -\kappa$ in [8, Sect. 2.4].

We now introduce

$$S_1 = \{p|D \mid p \text{ is inert in } E_\xi\},\ S_2^+(f, \chi) = \big\{p|D \mid p \text{ is ramified in } E_\xi \text{ and } \chi_p(\varpi_p) = -\epsilon_p\big\}$$

with a prime element ϖ_p of $E_{\xi,p}$, where recall that ϵ_p denotes the eigenvalue of the Atkin-Lehner involution of f at $p|D$ (cf. Sect. 2.2 (1)).

By the same reasoning as in [7, Theorem 5.1.1] we see the following:

Proposition 3.1. *For $\xi \in B^- \setminus \{0\}$, $\mathcal{L}(f,f')^{\chi}_{\xi} \equiv 0$ unless $i_p(\chi) = 0$ for any $p|d_B$ and $w_\infty(\chi) = -\kappa_1$.*

For $\eta_\infty \in \mathbb{R}_+^\times$, we introduce

$$W_\xi^{\kappa_1,\kappa_2}(\eta_\infty) := 2^{\kappa_1} n(\xi)^{\frac{\kappa_2+2}{4}} (-\sqrt{-1})^{\kappa_2-\kappa_1} \eta_\infty^{\frac{\kappa_2}{2}+2} \exp(-4\pi\eta_\infty\sqrt{n(\xi)})$$
$$\sum_{i=0}^{\frac{\kappa_2-\kappa_1}{2}} \sum_{j=0}^{\frac{\kappa_2-\kappa_1}{2}-i} \binom{\frac{\kappa_2-\kappa_1}{2}}{i}\binom{\frac{\kappa_2-\kappa_1}{2}-i}{j} \frac{2^{(\kappa_2-\kappa_1)-2(i+j)}\Gamma(i+\frac{1}{2})\Gamma(j+\frac{1}{2})}{(2\pi\eta_\infty\sqrt{n(\xi)})^{i+j+1}}$$

and let $g_0 = (g_{0,p})_{p<\infty} \in G_{\mathbb{A}_f}$ be given by

$$g_{0,p} := \begin{cases} \mathrm{diag}(p^{i_p(\chi)-\mu_p}, p^{2(i_p(\chi)-\mu_p)}, 1, p^{i_p(\chi)-\mu_p}) & (p \nmid d_B), \\ 1_2 & (p|d_B). \end{cases}$$

Then we are now able to state our formula for $\mathcal{L}(f,f')^{\chi}_{\xi}$ (cf. [10]).

Theorem 3.2. *Let (f,f') be Hecke eigenforms with the same signature of the Atkin-Lehner involutions (cf. (1) in Sect. 3.2) and f be primitive. Furthermore let $\xi \in B^- \setminus \{0\}$ be primitive (cf. (2) in Sect. 3.2). Suppose that χ satisfies $i_p(\chi) = 0$ for any $p|d_B$ and $w_\infty(\chi) = -\kappa_1$ (cf. Proposition 3.1). We have*

$$\frac{||\mathcal{L}(f,f')^{\chi}_{\xi}(g_0)||^2}{\langle f,f\rangle\langle f',f'\rangle} = C(f,f',\xi,\chi)L(\mathcal{L}(f,f'),\chi^{-1},\frac{1}{2})$$

where, if $\pi(f')_p|_{E_{\xi,p}^\times} = \chi_p$ for $p|d_B$ ramified in E_ξ and $S_1 = S_2^+(f,\chi) = \emptyset$,

$$C(f,f',\xi,\chi) = \frac{2^{|\delta(D)|-4}(\kappa_2+1)\binom{\kappa_2}{\frac{\kappa_1+\kappa_2}{2}}|d_\xi|^{\frac{3}{2}}w(E_\xi)^2}{h(E_\xi)^2A(\chi)^4D^{\frac{3}{2}}L(\pi(f),\mathrm{Ad},1)L(\mathrm{JL}(\pi(f')),\mathrm{Ad},1)}$$
$$\prod_{p\nmid d_B} p^{4\mu_p}(1-\delta(i_p(\chi)>0)e_p(E_\xi)p^{-1})^{-2}\prod_{p|d_B} r_p p^{-1}\prod_{p|D}(p+1)^{-2}\cdot W_\xi^{\kappa_1,\kappa_2}(1)^2$$
$$\cdot\frac{(f'_{\infty,\kappa_1},f'_{\infty,\kappa_1})_{\kappa_2}}{(f'_\infty,f'_\infty)_{\kappa_2}},$$

and $C(f,f',\xi,\chi) = 0$ otherwise.

For this theorem we explain the steps of the proof.

1. We begin with showing that
$$\mathcal{L}(f, f')^{\chi}_{\xi}(g_0) = \text{(explicit constant)} \cdot \overline{P(f,\chi)} P(f',\chi),$$
where $P(f,\chi)$ (respectively $P(f',\chi)$) denotes the toral integrals of f (respectively f') with respect to χ (cf. [16]). The method of the proof is due to [7], which deals with the case of $\kappa_1 = \kappa_2$.
2. According to Waldspurger [16], the square norms of the two toral integrals are proportional to the central values of the base change lift L-functions with χ^{-1}-twist for $\pi(f)$ or $\mathrm{JL}(\pi(f'))$. We can determine the constant of proportionality explicitly. For $\pi(f)$ we can do it by modifying the result of [5]. For $\mathrm{JL}(\pi(f'))$ we evaluate the constant explicitly with the help of an explicit formula for zonal spherical functions on $GL_2(\mathbb{Q}_p)$ by Macdonald (cf. [4, Chap.V, Sect. 3, (3.4)]). This leads to the explicit formula for the constant $C(f, f', \xi, \chi)$ of proportionality.
3. The last step is to verify that the L-function $L(\mathcal{L}(f, f'), \chi^{-1}, s)$ decomposes into the product of the two base change lift L-functions of $\pi(f)$ and $\mathrm{JL}(\pi(f'))$ with χ^{-1}-twist. This is done by a careful comparison of each local factor of these global L-functions.

Remark 3.3. According to Sugano [15, Theorem 2-1], the Bessel period $\mathcal{L}(f, f')^{\chi}_{\xi}$ is determined by the evaluation at $g_{0,f}\begin{pmatrix} \sqrt{\eta_\infty} & 0 \\ 0 & \sqrt{\eta_\infty}^{-1} \end{pmatrix}$ with $\eta_\infty \in \mathbb{R}_+^\times$.

For $(\kappa_1, \kappa_2) \in (2\mathbb{Z}_{>0})^{\oplus 2}$ with $\kappa_1 \neq \kappa_2$, Okazaki [12] constructs non-vanishing theta lifts from non-zero primitive forms $(f_1, f_2) \in S_{\kappa_1}(N_1) \times S_{\kappa_2}(N_2)$ to generic cusp forms $\mathcal{L}'(f_1, f_2)$ on $GSp(2)(\mathbb{A}_\mathbb{Q})$ with the following properties:

1. It is a paramodular form of level $N_1 N_2$ (for the definition see [13]).
2. At the archimedean place, it lies in the minimal K'_∞-type of the discrete series representation with Harish-Chandra parameter $(\frac{\kappa_1+\kappa_2}{2} - 1, -\frac{|\kappa_1-\kappa_2|}{2})$, where K'_∞ denotes a maximal compact subgroup of $Sp(2)(\mathbb{R})$.

By means of Novodvorsky's zeta integral [11] we can define the spinor L-function $L(F, \mathrm{spin}, s)$ for a generic cusp form F on $GSp(2)(\mathbb{A}_\mathbb{Q})$. For a Hecke eigenform $f' \in \mathcal{A}_{\kappa_2}$ we let $\mathrm{JL}(f')$ be the primitive form corresponding to f' via Eichler or Jacquet-Langlands-Shimizu correspondence (cf. [1–3, 14]). As in the case of $GSp(1,1)$, we can consider $L(\mathcal{L}'(f, \mathrm{JL}(f')), \mathrm{spin}, s)$ and can thus also consider the convolution-type L-function $L(\mathcal{L}(f, \mathrm{JL}(f')), \chi^{-1}, s)$ with a Hecke character χ. We can show that the correspondence $\mathcal{L}(f, f') \mapsto \mathcal{L}'(f, \mathrm{JL}(f'))$ satisfies the expected properties of the Jacquet-Langlands-Shimizu correspondence for $GSp(2)$ and $GSp(1,1)$. As a consequence, we have
$$L(\mathcal{L}(f, f'), \chi^{-1}, s) = L(\mathcal{L}'(f, \mathrm{JL}(f')), \chi^{-1}, s).$$

We thereby reformulate Theorem 3.2 as follows (cf. [10]):

Corollary 3.4. *Under the assumption as in Theorem 3.2 we have*

$$\frac{||\mathcal{L}(f,f')^{\chi}_{\xi}(g_0)||^2}{\langle f,f\rangle\langle f',f'\rangle} = C(f,f',\xi,\chi)L(\mathcal{L}(f,f'),\chi^{-1},\frac{1}{2}) = C(f,f',\xi,\chi)L(\mathcal{L}'(f,\mathrm{JL}(f')),\chi^{-1},\frac{1}{2}).$$

4 Non-vanishing Theta Lifts to $GSp(1,1)$ and Strict Positivity of Central L-Values

Following the approach of [7, Sect. 14] we discuss the existence of non-vanishing Bessel periods $\mathcal{L}(f,f')^{\chi}_{\xi}$, which leads to the existence of non-vanishing $\mathcal{L}(f,f')$'s satisfying the strict positivity of the central L-values for the convolution-type L-functions. For this we specify the definite quaternion algebra B and its maximal order $\mathfrak{O}$ as follows:

$$B := \mathbb{Q} + \mathbb{Q}i + \mathbb{Q}j + \mathbb{Q}k, \quad i^2 = j^2 = -1,\ ij = -ji = k,$$

$$\mathfrak{O} := \mathbb{Z}\frac{1+i+j+k}{2} + \mathbb{Z}i + \mathbb{Z}j + \mathbb{Z}k.$$

As a primitive $\xi \in B^-$ we take $\xi = i/2$. We note that $d_B = 2$ for this B, thus $D = 1$ or 2. For this section we assume that the Hecke character χ is unramified at every finite place p.

Our result on the existence of non-vanishing $\mathcal{L}(f,f')$'s is stated as follows (cf. [10]):

Theorem 4.1. *When $D = 1$ let both of $\kappa_1 \geq 12$ and $\kappa_2 \geq 8$ be divisible by 4, and when $D = 2$ let both of $\kappa_1 \geq 8$ and $\kappa_2 \geq 8$ be divisible by 8. Suppose further that*

$$1 + 2^{-\frac{\kappa_2-4}{2}}(-1)^{\kappa_1/4}\begin{pmatrix} \frac{\kappa_1+\kappa_2}{2} \\ \frac{\kappa_2-\kappa_1}{2} \end{pmatrix} {}_2F_1(-\frac{\kappa_2-\kappa_1}{2}, -\frac{\kappa_2-\kappa_1}{2}; \kappa_1+1; -1) \neq 0,$$

where ${}_2F_1(\alpha,\beta;\gamma;x)$ denotes the hypergeometric function with parameters α,β,γ (cf. [17, Chap.XIV]). There are Hecke eigenforms $(f,f') \in S_{\kappa_1}(D) \times \mathcal{A}_{\kappa_2}$ such that $\mathcal{L}(f,f')^{\chi}_{\xi} \not\equiv 0$, thus $\mathcal{L}(f,f') \not\equiv 0$.

This theorem follows from the two lemmas as below.

Lemma 4.2. *For $i = 1$ or 2 let $\kappa_i \in \mathbb{Z}_{>0}$ be divisible by 4 and satisfy*

$$1 + 2^{-\frac{\kappa_2-4}{2}}(-1)^{\kappa_1/4}\begin{pmatrix} \frac{\kappa_1+\kappa_2}{2} \\ \frac{\kappa_2-\kappa_1}{2} \end{pmatrix} {}_2F_1(-\frac{\kappa_2-\kappa_1}{2}, -\frac{\kappa_2-\kappa_1}{2}; \kappa_1+1; -1) \neq 0.$$

Suppose further that $\kappa_2 \geq 8$ *and* κ_2 *is divisible by* 8. *Then there is a Hecke eigenform* $f' \in \mathcal{A}_{\kappa_2}$ *such that* $(f'(1), v^*_{\kappa_2, \frac{\kappa_1+\kappa_2}{2}})_{\kappa_2} \neq 0$ *and* $\epsilon'_2 = 1$, *where, for* $0 \leq i \leq \kappa_2$, $v_{\kappa_2,i}$ *denotes a weight vector of* V_{κ_2} *with weight* $2i - \kappa_2$ *and* $v^*_{\kappa_2,i} \in V^*_{\kappa_2}$ *its dual vector.*

Lemma 4.3. *Let* $\kappa \geq 12$ *(respectively* $\kappa \geq 8$*) be divisible by* 4 *(respectively* 8*) when* $D = 1$ *(respectively* $D = 2$*). There is a Hecke eigenform* $f \in S_\kappa(D)$ *such that* $f(\gamma_0) \neq 0$ *when* $D = 1$ *(respectively a primitive form* $f \in S_\kappa(D)$ *such that* $f(\gamma_0) \neq 0$ *and* $\epsilon_2 = 1$ *when* $D = 2$*). For the notation* $\gamma_0 \in GL_2(\mathbb{A}_\mathbb{Q})$ *see [7, Sect. 5.2].*

The two conditions

$$(f'(1), v^*_{\kappa_2, \frac{\kappa_1+\kappa_2}{2}})_{\kappa_2} \neq 0, \quad f(\gamma_0) \neq 0$$

are equivalent to non-vanishing of the two toral integrals $P(f', \chi)$ and $P(f, \chi)$, respectively, where recall that these two toral integrals appear in the explicit formula for $\mathcal{L}(f, f')^\chi_\xi$ (cf. Sect. 3.4). We prove the first lemma (respectively the second lemma) in [10] (respectively [7, Proposition 14.4.1]).

As a result of Theorem 3.2 and Theorem 4.1 we have the following theorem (cf. [10]):

Theorem 4.4. *Under the same assumption as in Theorem 4.1 we have*

$$L(\mathcal{L}(f, f'), \chi^{-1}, \frac{1}{2}) = L(\mathcal{L}'(f, \mathrm{JL}(f')), \chi^{-1}, \frac{1}{2}) > 0.$$

Acknowledgements The author is very grateful to Berhard Heim for his invitation to the conference. His thank is also due to the referee for helpful comments.

References

1. M. Eichler, Über die Darstellbarkeit von Modulformen durch Thetareihen. J. Reine Angew. Math. **195**, 156–171 (1955)
2. M. Eichler, Quadratische Formen und Modulfunktionen. Acta Arith. **4**, 217–239 (1958)
3. H. Jacquet, R.P.Langlands, Automorphic Forms on $GL(2)$, *Lecture Notes in Math*, vol. 114 (Springer, NewYork 1970)
4. I. Macdonald, *Symmetric Functions and Hall polynomials* (Oxford University Press, 1979)
5. A. Murase, *CM-values and Central L-values of Elliptic Modular Forms (II)*, vol. 30 (Max-Planck-Institut für Mathematik Preprint Series, 2008)
6. A. Murase, H. Narita, Commutation relations of Hecke operators for Arakawa lifting. Tohoku Math. J. **60**, 227–251 (2008)
7. A. Murase, H. Narita, Fourier expansion of Arakawa lifting I: An explicit formula and examples of non-vanishing lifts. Israel J. Math. **187**, 317–369 (2012)
8. A. Murase, H. Narita, Fourier expansion of Arakawa lifting II: Relation with central L-values, preprint.
9. H. Narita, Jacquet-Langlands-Shimizu correspondence for theta lifts to $GSp(2)$ and its inner forms I: An explicit functorial correspondence, with an appendix by Ralf Schmidt, preprint.

10. H. Narita, Jacquet-Langlands-Shimizu correspondence for theta lifts to $GSp(2)$ and its inner forms II: An explicit formula for Bessel periods and non-vanishing of theta lifts, preprint.
11. M. Novodvorsky, Automorphic L-functions for symplectic group $GSp(4)$, Proc. Symp. Pure Math. **33**(2), 87–95 (1979)
12. T. Okazaki, Paramodular forms on $GSp_2(\mathbb{A})$, preprint.
13. B. Roberts, R. Schmidt, Local new forms for $GSp(4)$, *Lecture Notes in Math*, vol. 1918 (Springer, NewYork, 2007)
14. H. Shimizu, Theta series and automorphic forms on GL_2. J. Math. Soc. Japan **24**, 638–683 (1972)
15. T. Sugano, On holomorphic cusp forms on quaternion unitary groups of degree 2. J. Fac. Sci. Univ. Tokyo **31**, 521–568 (1985)
16. J. Waldspurger, Sur les valeurs de certaines fonctions L automorphes en leur centre de symetrie. Compos. Math. **54**, 173–242 (1985)
17. E.T. Whittaker, G.N. Watson, *A Course of Modern Analysis* (Cambridge university press, Cambridge 2000)

Intersections of Two Walls of the Gottschling Fundamental Domain of the Siegel Modular Group of Genus Two

Takayuki Oda

1 Introduction

If one has a chance to see the classical books of Fricke-Klein on modular functions of one variable, which were written one hundred years ago, one can find many beautiful pictures of fundamental domains bounded by geodesic lines with respect to the invariant hyperbolic metric in the complex upper half plane $\mathfrak{H}_1$.

Those pictures are suppressed in the modern books which are mainly written for different purposes, but we can find the disguised figure of these pictures in the theory of modular symbols of the elliptic modular curves that is developed after 1970.

After preliminary investigation of Hilbert modular forms by the students of David Hilbert (recall the famous Erich Hecke), the extension of the theory of automorphic forms to higher dimensional cases was developed by Carl L. Siegel for the case of the Siegel upper half space of degree n and the Siegel modular group of degree n.

In his paper "Symplectic Geometry" (1943) [6] on the Siegel modular varieties and related modular varieties, Siegel gave a construction of the fundamental domains for this higher dimensional cases, extending the method of the proof in the one-dimensional case. And in a follow-up paper now forgotten by people, a student of Siegel, Gottschling, found an explicit description of the fundamental domain obtained by Siegel, when the degree n is two. This was done about a half century ago.

T. Oda (✉)
The University of Tokyo, Tokyo, Japan
e-mail: takayuki@ms.u-tokyo.ac.jp

B. Heim et al. (eds.), *Automorphic Forms*, Springer Proceedings
in Mathematics & Statistics 115, DOI 10.1007/978-3-319-11352-4_15

Because my doctoral thesis was about "Periods of Hilbert modular surfaces" (1982), I have been interested in "periods" of automorphic forms of many variables. Among others I have been interested in establishing some fundamental materials to develop the theory of automorphic forms as harmonic differential forms, i.e., Whittaker functions which appear in the Fourier expansion of automorphic forms and other spherical functions. We had some results and applications in the last two decades.

But the periods of algebraic varieties are coupling of differential forms and cycles (at least, topological chains) by integration on the varieties in question.

For the de Rham cohomology groups of the arithmetic quotients $\Gamma\backslash\mathfrak{H}_2$ of the Siegel upper half space of genus two with respect to a lattice Γ in $G = Sp(2, \mathbf{R})$, we have a special case of the Matsushima isomorphism:

$$H^i(\Gamma\backslash\mathfrak{H}_2, \mathbf{C}) = \oplus_{\pi\in\hat{G},\chi_\pi=\chi_1}\{H^i(\mathfrak{g}, K; M_{\pi,K}) \oplus \mathrm{Hom}_G(M_{\pi,K}, \mathcal{A}(\Gamma\backslash G))\}.$$

Here $\mathcal{A}(\Gamma\backslash G)$ is the space of automorphic forms on G with respect to Γ in the sense of Harish-Chandra (see [4] for example).

This theorem is the generalization of the Eichler-Shimura isomorphism theorem on elliptic modular curves for higher dimensional case, and the meaning of this theorem is that the cohomology classes of a modular variety is a sum of harmonic differential forms with respect to the elliptic system of differential operators including the Casimir equation. The representations π which contribute to the $(\mathfrak{g}, K)$-cohomology groups $H^i(\mathfrak{g}, K; M_{\pi,K}) \neq \{0\}$ nontrivially play the role of the template of the type of harmonicity which is a generalization of the notion of weight and holomorphy.

As we notice soon, we know little about cycles on modular varieties, except for elliptic modular curves, for which we already have the theory of "modular symbols."

Even in the higher dimensional case, there is something called *modular symbols* which are chains obtained from some (real analytic) embedding $Y \to X$ from one arithmetic quotient Y to the other X. But we can obtain only very limited type of cycles (or chains) (*cf.* [3] the Springer Lecture Notes in Math., edited by Labesse and Schwermer, long time ago).

But these cycles and chains, though they give interesting cycles (we can compute the Green currents, say), seem to be very limited ones. Therefore it seems better to start something new on cycles on modular varieties from a more fundamental view point, even if the target that we aim at might be far.

Now we propose

Problem (A hope or an illusion). Find something dual to the harmonic differential forms on $V_\Gamma := \Gamma\backslash\mathfrak{H}_2$, i.e., which should be called *harmonic* or *rigid* chains or cycles, which are plenty enough to generate the homology groups of V_Γ.

Here I want to start from examples motivated by a previous paper [2]. In [2], heuristically speaking, we want to draw a "picture" of a fundamental domain $\mathcal{F}_2$ of the Siegel modular group $Sp(2, \mathbb{Z})$ of genus two in the Siegel upper half space $\mathfrak{H}_2$ of genus 2.

In this project [2], based on the classical result of Gottschling, we enumerate all the 0-cells in $\mathcal{F}_2$ obtained as the intersections of 28 5-dimensional walls in the boundary $\partial\mathcal{F}_2$. We could do that by some tools of computer algebra, more specifically speaking by some package of Risa/Asir.

In this paper, we discuss some examples of 4-cells which are intersections of two 5-dimensional walls. Our results here are only case studies. But as far as we know, there exists almost no result on the geometry of the fundamental domain of $Sp(2, \mathbf{Z})$.

Here is the plan of the article. In Sect. 2, we recall the case of elliptic modular group. In the next section, Sect. 3, we review the case of the Siegel modular group of genus 2. Among others we recall the results in a former paper, which is a rather recent result. An outline of the method proof is also given. In Sect. 4, we enumerate the cells of dimension 4 with small number of 0-cells. A case study of local structure of an intersection of two walls is given, by infinitesimal linearization of the problem. In Sect. 5, some simple differential geometric facts are recalled, and a case study of 4-dimensional cell is given.

The author thanks Takahiro Hayata, who is a coauthor of the previous paper and has been working together with the author for several years. He generously permits to include some unpublished computation in Sect. 6.

2 The Case of Elliptic Modular Group

Recall the elliptic modular group

$$SL(2,\mathbb{Z}) := \left\{ \begin{pmatrix} a & b \\ c & d \end{pmatrix} \mid a,\ b,\ c,\ d \text{ rational integers, with } ad - bc = 1. \right\}$$

and the complex upper half plane $\mathfrak{H}_1 := \{\tau \in \mathbb{C} |\ Im(\tau) > 0\}$.

Then $\gamma = \begin{pmatrix} a & b \\ c & d \end{pmatrix} \in SL(2,\mathbb{Z})$ acts on $\mathfrak{H}_1$ via the linear fractional transformation:

$$\gamma(\tau) = (a\tau + b)/(c\tau + d).$$

Recall the notion of fundamental domain.

Definition. It is a subset S in the whole space $\mathfrak{H}_1$ which is a complete representative system of the quotient space $SL(2,\mathbb{Z})\backslash\mathfrak{H}_1$. That is, for any point z of $\mathfrak{H}_1$, we can find a point z' in S uniquely such that $z = \gamma(z')$ for some element γ in $SL(2,\mathbb{Z})$.

Normally we require more conditions for S. Say, it should be connected, the boundary ∂S is "piecewise" smooth, etc.

Then we can find a fundamental domain $\mathcal{F}_1$ in $\mathfrak{H}_1$ with respect to $SL(2,\mathbb{Z})$, i.e., a connected subset of $\mathfrak{H}_1$, such that

(1) For any τ there is an element $\gamma \in SL(2, \mathbb{Z})$ and $\tau_0 \in \mathcal{F}_1$ such that $\tau = \gamma(\tau_0)$.
Moreover such γ and τ_0 is almost uniquely determined, i.e.:

(2) For any interior points τ_0 and τ_0', there is no element $\gamma' \in SL(2, \mathbf{Z})$ so that $\tau_0' = \gamma'(\tau_0)$, except for the unity $\gamma' = e$.

In this case as the closure of a fundamental domain, we can choose the simple domain

$$\bar{\mathcal{F}}_1 := \{\tau = x + \sqrt{-1}y \mid y > 0 \text{ and } |x| \leq \frac{1}{2},\ x^2 + y^2 \geq 1\}.$$

3 The Case of the Siegel Modular Group of Genus 2

3.1 The Siegel Upper Half Space and the Siegel Modular Group of Genus 2

Now we set

$$\Gamma_2 := \{\gamma = \begin{pmatrix} A & B \\ C & D \end{pmatrix} \mid \text{with } A,\ B,\ C,\ D \in M_2(\mathbb{Z}), \text{ and } {}^t\gamma J_2 \gamma = J_2\}.$$

Here $J_2 = \begin{pmatrix} 0_2 & -1_2 \\ 1_2 & 0_2 \end{pmatrix}$.

Moreover, we consider the Siegel upper half space

$$\mathfrak{H}_2 := \{Z := \begin{pmatrix} z_{11} & z_{12} \\ z_{12} & z_{22} \end{pmatrix} \in M_2(\mathbb{C}) \mid Z = {}^tZ, \text{ and } \mathrm{Im}(Z) \text{ positive definite}\}.$$

which is the generalization of the complex upper half space. It is of complex dimension 3, i.e., the real dimension is 6.

The action of $\gamma = \begin{pmatrix} A & B \\ C & D \end{pmatrix}$ on Z is also an analogue of the linear fractional transformation:

$$\gamma(Z) = (AZ + B)(CZ + D)^{-1} = {}^t(CZ + D)^{-1} \cdot {}^t(AZ + B).$$

From now on we utilize the involutive diffeomorphism

$$Z^\theta = -\bar{Z}.$$

For a subset $S \subset \mathfrak{H}_2$, we denote by S^θ the image of S by θ.

3.2 A Fundamental Domain in the Case of Genus 2

About 50 years ago, Gottschling [1] found an explicit construction of a fundamental domain $\mathcal{F}_2$ in $\mathfrak{H}_2$ with respect to $\Gamma_2 = Sp(2,\mathbb{Z}) = Sp_4(\mathbb{Z})$.

The solution was to find 28 inequalities of 3 different types ($W_0 \sim W_{27}$). Here we list up them by the following rule. Firstly we write "W_i" with the number i. After this we denote the label $[Label]$ to help our memory. Lastly the corresponding inequality:

$$f_i(x_{11}, x_{22}, x_{12}, y_{11}, y_{22}, y_{12}) \geq 0$$

is given.

(1) Firstly, there are 6 inequalities :

$$W_{22} = [X1] : \frac{1}{2} - x_{11} \geq 0; \qquad W_{23} = [X1^{\theta}] : x_{11} + \frac{1}{2} \geq 0;$$

$$W_{24} = [X2] : \frac{1}{2} - x_{22} \geq 0; \qquad W_{25} = [X2^{\theta}] : x_{22} + \frac{1}{2} \geq 0;$$

$$W_{26} = [X3] : \frac{1}{2} - x_{12} \geq 0; \qquad W_{27} = [X3^{\theta}] : x_{12} + \frac{1}{2} \geq 0.$$

These are necessary to consider the fundamental domain with respect to the unipotent radical of the Siegel parabolic subgroup in G. We have $f_{22}(Z) = \frac{1}{2} - x_{11}$, etc.

(2) There are 3 inequalities of "Minkowski" type with respect to the Levi part $GL(2,\mathbb{Z})$ of the Siegel parabolic subgroup:

$$W_{19} = [Y1] : y_{22} - y_{11} \geq 0; \quad W_{20} = [Y2] : y_{11} - 2y_{12} \geq 0;$$
$$W_{21} = [Y3] : y_{12} \geq 0.$$

We have $f_{19}(Z) = y_{22} - y_{11}$, etc.

(3) There are yet another 19 equalities of type

$$W_i = [Label] : |\det(CZ + D)| - 1 \geq 0; \quad (i = 0 \sim 18)$$

with the following 19 pairs (C, D):

Let E_{ij} be the matrix units of $M_2(\mathbb{C})$ with $i,\ j \in \{1,\ 2\}$, and let $J = E_{12} + E_{21}$.

(3a) For C with rank$C = 1$, we have 4 cases:

$$W_0 = [\mathbf{1}] : (C, D) = (E_{11}, E_{22}); \quad W_1 = [\mathbf{2}] : (C, D) = (E_{22}, E_{11});$$

$$W_2 = [R] : (C, D) = (E_{11} - E_{12}, E_{11} + E_{22} + E_{21});$$

$$W_3 = [R^{\theta}] : (C, D) = (E_{11} - E_{12}, -E_{11} - E_{22} - E_{21}).$$

(3b) The remaining 15 inequalities are given with $C = 1_2$ and D are given by the following list. We use the *format*

$$W_i = [Label] : D :$$

with $4 \leq i \leq 18$.

$W_4 = [\mathbf{O}] : D = 0_2;$
$W_5 = [E1] : D = E_{11};$ $W_6 = [E1^\theta] : D = -E_{11};$
$W_7 = [E2] : D = E_{22};$ $W_8 = [E2^\theta] : D = -E_{22};$
$W_9 = [I+] : D = 1_2;$ $W_{10} = [I+^\theta] : D = -1_2;$
$W_{11} = [I-] : D = E_{11} - E_{22};$ $W_{12} = [I-^\theta] : D = -E_{11} + E_{22};$
$W_{13} = [J0] : D = E_{12} + E_{21};$ $W_{14} = [J0^\theta] : D = -E_{12} - E_{21};$
$W_{15} = [J1] : D = J + E_{11};$ $W_{16} = [J1^\theta] : D = -J - E_{11};$
$W_{17} = [J2] : D = J + E_{22};$ $W_{18} = [J2^\theta] : D = -J - E_{22}.$

3.3 What Is Our Problem?

By an abuse of symbol, we denote by W_i the corresponding wall of 5 dimensions of $\mathcal{F}_2$ given by $W_i := \{Z \in \mathcal{F}_2 \mid f_i(Z) = 0\}$ which is a subset of the boundary $\partial\mathcal{F}_2$ of $\mathcal{F}_2$ and we have $\partial\mathcal{F}_2 = \cup_i W_i$.

In a previous paper [2], we had a partial result for the following problem.

Problem. Find the stratification of the fundamental domain $\mathcal{F}_2$ by the cells of smaller dimensions obtained from the various intersections of the 28 walls W_i $(i = 0, \cdots, 27)$.

Say, we expect that the intersections $W_{i_1} \cap W_{i_2} \cap \cdots \cap W_{i_k}$ of several different walls generically define a cell or a union of cells of codimension k. But actually situation seems to be quite complicated.

Our strategy is to find zero-dimensional cells firstly.

3.4 Search of 0-Cells and the Main Result

Firstly, a candidate of the set of 0-cells in the above sense is enumerated as follows.

We consider 28 real algebraic sets of dimension 5, which are zero point set of the equation $f_i(x_{11}, x_{22}, x_{12}, y_{11}, y_{22}, y_{12}) = 0$. Here $z_{11} = x_{11} + \sqrt{-1}y_{22}$, etc.

For each subset λ of 28 walls W_i, we consider the ideals I_λ of $R := \mathbb{R}[x_{ij}, y_{ij}]$ generated by the defining equations

$$f_i \in R \ (W_i \in \lambda).$$

There are 2^{28} such ideals.

We find the ideal I_λ such that R/I_λ is zero dimensional (primary decomposition of Emmy Noether). The key point is that we can compute effectively! By Z_0 we denote the set of points defined by these zero-dimensional ideals. Next we have to consider $Z_0 \cap \overline{\mathcal{F}_2}$.

Our main result in the previous paper [2] is that the cardinality of the set $Z_0 \cap \overline{\mathcal{F}_2}$ is 180.

3.5 *Partial List of* 0*-Cells*

We define 40 points e_k $(k = 0, \sim, 39)$ and the actions of involutive operators

$$\theta_i : z_{ii} \to -\bar{z}_{ii} \ (i = 1,2) \quad \theta_3 : z_{12} \to -\bar{z}_{12},$$

$$\theta = \theta_1 \cdot \theta_2 \cdot \theta_3, \quad \text{and } \sigma_0 : Z \to \begin{pmatrix} z_{22} & z_{12} \\ z_{12} & z_{11} \end{pmatrix}.$$

Then we have 180 points, which are the intersections of the 0-dimensional intersections of 28 walls inside $\mathfrak{H}_2$ and the closure of $\mathcal{F}_2$.

Here is a partial list of points e_i :

$$e_0 := \frac{1}{2} 1_2 + i\omega_0 1_2 \text{ with } \omega_0 = \frac{\sqrt{3}}{2}$$

$$e_1 := \frac{1}{3}\begin{pmatrix} 1 & 1 \\ 1 & 1 \end{pmatrix} + i\omega_1 \begin{pmatrix} 1 & \frac{1}{2} \\ \frac{1}{2} & 1 \end{pmatrix} \text{ with } \omega_1 = \frac{2\sqrt{2}}{3}$$

$$e_2 := \begin{pmatrix} \omega_6 & \frac{1}{2} \\ \frac{1}{2} & \omega_6 \end{pmatrix} + i \begin{pmatrix} 1 & \frac{1}{2} \\ \frac{1}{2} & 1 \end{pmatrix} \text{ with } \omega_6 = 0.1539\cdots$$

$$e_3 := \frac{1}{2}\begin{pmatrix} 1 & 1 \\ 1 & 1 \end{pmatrix} + i\omega_0 1_2$$

$$e_5 := \frac{1}{2} 1_2 + i \begin{pmatrix} 1 & \frac{1}{2} \\ \frac{1}{2} & 1 \end{pmatrix}$$

$$e_7 := \begin{pmatrix} \frac{1}{2} & \omega_0 - 3/4 \\ \omega_0 - 3/4 & \frac{1}{2} \end{pmatrix} + i \begin{pmatrix} \omega_0 & \frac{1}{2}\omega_0 \\ \frac{1}{2}\omega_0 & (3-\omega_0)/2 \end{pmatrix}$$

$$e_8 := \frac{1}{2}\begin{pmatrix} 1 & 1 \\ 1 & 1 \end{pmatrix} + i\omega_3 \begin{pmatrix} 1 & \frac{1}{2} \\ \frac{1}{2} & 1 \end{pmatrix} \text{ with } \omega_3 = \frac{1}{3}\sqrt{2\sqrt{37}-2} = 1.0627\cdots$$

$$e_9 := \frac{1}{2}\begin{pmatrix} 1 & 1 \\ 1 & 1 \end{pmatrix} + i \begin{pmatrix} \omega_0 & \omega_0/2 \\ \omega_0/2 & \omega_4 \end{pmatrix}$$

$$\text{with } \omega_4 = \sqrt{(3\sqrt{3045}+521)/2^9} = 1.1579\cdots$$

$$e_{11} := \frac{1}{2}1_2 + i\begin{pmatrix}\omega_0 & \omega_0/2\\ \omega_0/2 & \omega_5/16\end{pmatrix} \text{ with } \omega_5 = 3\sqrt{(\sqrt{285}+49)/2}$$

$$e_{18} := \frac{1}{2}\begin{pmatrix}1 & 1\\ 1 & 1\end{pmatrix} + i\begin{pmatrix}\omega_0 & \omega_{21}\\ \omega_{21} & \omega_0\end{pmatrix} \text{ with } \omega_{21} = 0.1723559\cdots$$

$$e_{25} := \begin{pmatrix}\frac{1}{2} & (1-\omega_{18})/4\\ (1-\omega_{18})/4 & \frac{1}{2}\end{pmatrix} + i\omega_{19}\begin{pmatrix}1 & \frac{1}{2}\\ \frac{1}{2} & 1\end{pmatrix}$$

$$\text{with } \omega_{18} = 0.5285\cdots \text{ and } \omega_{19} = 0.9861\cdots$$

Then we have

$P_0 = e_0 : [\mathbf{1}, \mathbf{2}, O, E1^\theta, E2^\theta, I+^\theta, Y1, Y3, X1, X2]$
$P_1 = e_0^{\theta_1} : [\mathbf{1}, \mathbf{2}, O, E1^\theta, E2, I-^\theta, Y1, Y3, X1, X2^\theta]$
$P_2 = e_0^{\theta_2} : [\mathbf{1}, \mathbf{2}, O, E1, E2^\theta, I-, Y1, Y3, X1^\theta, X2]$
$P_3 = e_0^{\theta_1\theta_2} : [\mathbf{1}, \mathbf{2}, O, E1, E2, I+, Y1, Y3, X1^\theta, X2^\theta]$
$P_4 = e_1^{\theta_3} : [\mathbf{1}, \mathbf{2}, R^\theta, E1^\theta, E2^\theta, I+^\theta, J0, Y1, Y2]$
$P_5 = e_1^{\theta_1\theta_2} : [\mathbf{1}, \mathbf{2}, R, E1, E2, I+, J0^\theta, Y1, Y2]$
$\vdots$
$P_* = e_{**} : [W_*, \cdots, \cdots, W_{***}]$
$\vdots$
$P_{172} = e_{39}^{\theta_2} : [\mathbf{2}, O, E2, J1^\theta, Y1, X1]$
$P_{173} = e_{39}^{\theta_3} : [\mathbf{2}, E1^\theta, I+^\theta, J0, Y1, X1]$
$P_{174} = e_{39}^{\theta_3\theta_2\sigma_0} : [\mathbf{1}, O, E1^\theta, J_2, Y1, X2^\theta]$
$P_{175} = e_{39}^{\theta_3\sigma_0} : [\mathbf{1}, E2^\theta, I+^\theta, J0, Y1, X2]$
$P_{176} = e_{39}^{\theta_1\theta_2} : [\mathbf{2}, E1, I+, J0^\theta, Y_1, X1^\theta]$
$P_{177} = e_{39}^{\theta_1\theta_2\sigma_0} : [\mathbf{1}, E2, I+, J0^\theta, Y1, X2^\theta]$
$P_{178} = e_{39}^{\theta_1\theta_3} : [\mathbf{2}, O, E2^\theta, J1, Y1, X1^\theta]$
$P_{179} = e_{39}^{\theta_1\sigma_0} : [\mathbf{1}, O, E1, J2^\theta, Y1, X2^\theta]$

Remark. There is a difference of notation between this paper and the former one. Our Ei^θ corresponds to θEi of the table 5 of [2], $J0$ to J, etc.

3.6 A Corollary

We consider the equivalence relation among the set $\mathcal{P}$ of these 180 points modulo the action of $Sp(2, \mathbb{Z})$. The plausible answer is the following.

Proposition-Conjecture. The cardinality of $Sp(2,\mathbb{Z})\backslash\mathcal{P}$ is at most 40. Moreover we hope this is exactly 40.

More precisely, we can split the $\mathcal{P}$ into two parts (disjoint): $\mathcal{P}_1$ and $\mathcal{P}_2$, both stable under the action of $Sp(2,\mathbb{Z})$. For the first the equivalence class under $Sp(2,\mathbb{Z})$ is exactly 12. For the remaining part, we know that it consists of 14 pairs of points P_i and ${P_i}^\theta$, which covers the set $\mathcal{P}_2$ under $Sp(2,\mathbb{Z})$. The problem here is that we cannot yet show that P_i and ${P_i}^\theta$ are not equivalent for each i.

3.7 *The Methods of Proof*

The problem of enumeration of 0-cells is formulated in a rather elementary language in abstract algebra. And the method is a computation by brutal force utilizing the "elimination method" (but advanced one) on computers. But the intermediate steps of machine computation require sometimes a huge resource in memory and in time (explosion!). So we might fail finally. Fortunately the workstation we used had a storage large enough for our purpose.

In the foundation of the polynomial computation, there is Gröbner basis (Buchberger). Heuristically speaking, it is an extension of the Euclid division algorithm to polynomials of many variables. This theory emerged around a few decades ago, and its derivatives and applications have been developed. There were many technical improvements to have really effective algorithm.

The important fact is we can compute the primary decomposition (of Emmy Noether) for an ideal I of $\mathbb{Q}[X_1,\cdots,X_n]$, etc. For our purpose, the system **Risa/Asir** developed by a group of people (those belonged to Fujitsu company formerly) including Anai, Noro, and Yokoyama is very useful (see the reference of our former paper [2]).

There is a package to judge whether an ideal I in $R = \mathbb{Q}[X_1,\cdots,X_n]$ has dimension 0, i.e., R/I a vector space of finite dimension, and also to get its primary decomposition. We also have the characteristic equations corresponding to the maximal ideals in I.

To have a list of 0-cells (modulo a conjecture) is a bridgehead for a further progress. But note that there exist many 1-cells connecting one 0-cells and points at infinity (the "ideal" 0-cells).

4 An Attempt Toward the Enumeration of 4-Cell

Somehow we have a reasonable candidate of the whole set of 0-cells in $\partial\mathcal{F}_2$. The next problem is enumerate the cells of other dimensions 1, 2, 3, 4. Probably the case of dimension 3 would be the most difficult case. Anyway we should start from the following.

Problem. How to find a candidate of the complete set of 4-dimensional cells?

A natural strategy is the following:

Set $\mathcal{S}[W_i]$ be the set of 180 points belonging to the wall W_j for each j. Then calculate the intersection $\mathcal{S}[W_i] \cap \mathcal{S}[W_j]$ for each pair (i, j) such that $i \neq j$.

4.1 *The Pairs* $\{W_i, W_j\}$ *with Empty* $\mathcal{S}[W_i] \cap \mathcal{S}[W_j]$

The following 146 intersections are empty (here $\mathcal{S}$ in the symbol $\mathcal{S}[W_i]$ is dropped):

$1:[R]\cap[R^\theta]$	$2:[R]\cap[E1^\theta]$	$3:[R]\cap[E2^\theta]$	$4:[R]\cap[I+^\theta]$	$5:[R]\cap[I-]$
$6:[R]\cap[I-^\theta]$	$7:[R]\cap[J0]$	$8:[R]\cap[J1]$	$9:[R]\cap[J2]$	$10:[R]\cap[Y3]$
$11:[R]\cap[X1]$	$12:[R]\cap[X2]$	$13:[R]\cap[X3^\theta]$	$14:[R^\theta]\cap[E1]$	$15:[R^\theta]\cap[E2]$
$16:[R^\theta]\cap[I+]$	$17:[R^\theta]\cap[I-]$	$18:[R^\theta]\cap[I-^\theta]$	$19:[R^\theta]\cap[J0^\theta]$	$20:[R^\theta]\cap[J1^\theta]$
$21:[R^\theta]\cap[J2^\theta]$	$22:[R^\theta]\cap[Y3]$	$23:[R^\theta]\cap[X1^\theta]$	$24:[R^\theta]\cap[X2^\theta]$	$25:[R^\theta]\cap[X3]$

$26:[O]\cap[J0]$	$27:[O]\cap[J0^\theta]$	$28:[E1]\cap[E1^\theta]$	$29:[E1]\cap[I+^\theta]$	$30:[E1]\cap[I-^\theta]$
$31:[E1]\cap[J0]$	$32:[E1]\cap[J1]$	$33:[E1]\cap[J1^\theta]$	$34:[E1]\cap[J2]$	$35:[E1]\cap[X1]$
$36:[E1]\cap[X3^\theta]$	$37:[E1^\theta]\cap[I+]$	$38:[E1^\theta]\cap[I-]$	$39:[E1^\theta]\cap[J0^\theta]$	$40:[E1^\theta]\cap[J1]$
$41:[E1^\theta]\cap[J1^\theta]$	$42:[E1^\theta]\cap[J2^\theta]$	$43:[E1^\theta]\cap[X1^\theta]$	$44:[E1^\theta]\cap[X3]$	$45:[E2]\cap[E2^\theta]$
$46:[E2]\cap[I+^\theta]$	$47:[E2]\cap[I-]$	$48:[E2]\cap[J0]$	$49:[E2]\cap[J1]$	$50:[E2]\cap[J2]$

$51:[E2]\cap[J2^\theta]$	$52:[E2]\cap[X2]$	$53:[E2]\cap[X3^\theta]$	$54:[E2^\theta]\cap[I+]$	$55:[E2^\theta]\cap[I-^\theta]$
$56:[E2^\theta]\cap[J0^\theta]$	$57:[E2^\theta]\cap[J1^\theta]$	$58:[E2^\theta]\cap[J2]$	$59:[E2^\theta]\cap[J2^\theta]$	$60:[E2^\theta]\cap[X2^\theta]$
$61:[E2^\theta]\cap[X3]$	$62:[I+]\cap[I+^\theta]$	$63:[I+]\cap[I-]$	$64:[I+]\cap[I-^\theta]$	$65:[I+]\cap[J0]$
$66:[I+]\cap[J1]$	$67:[I+]\cap[J1^\theta]$	$68:[I+]\cap[J2]$	$69:[I+]\cap[J2^\theta]$	$70:[I+]\cap[X1]$
$71:[I+]\cap[X2]$	$72:[I+]\cap[X3^\theta]$	$73:[I+^\theta]\cap[I-]$	$74:[I+^\theta]\cap[I-^\theta]$	$75:[I+^\theta]\cap[J0^\theta]$

Remark. The fact that $\mathcal{S}[W_i] \cap \mathcal{S}[W_j] = \phi$ does not imply that $[W_i] \cap [W_j] = \phi$, because the latter intersection might be a union of cells with "vertices at infinity"

$76:[I+^\theta]\cap[J1]$	$77:[I+^\theta]\cap[J1^\theta]$	$78[I+^\theta]\cap[J2]$	$79:[I+^\theta]\cap[J2^\theta]$	$80:[I+^\theta]\cap[X1^\theta]$
$81:[I+^\theta]\cap[X2^\theta]$	$82:[I+^\theta]\cap[X3]$	$83:[I-]\cap[I-^\theta]$	$84:[I-]\cap[J0]$	$85:[I-]\cap[J0^\theta]$
$86:[I-]\cap[J1]$	$87:[I-]\cap[J1^\theta]$	$88:[I-]\cap[J2]$	$89:[I-]\cap[J2^\theta]$	$90:[I-]\cap[X1]$
$91:[I-]\cap[X2^\theta]$	$92:[I-]\cap[X3]$	$93:[I-]\cap[X3^\theta]$	$94:[I-^\theta]\cap[J0]$	$95:[I-^\theta]\cap[J0^\theta]$
$96:[I-^\theta]\cap[J1]$	$97:[I-^\theta]\cap[J1^\theta]$	$98:[I-^\theta]\cap[J2]$	$99:[I-^\theta]\cap[J2^\theta]$	$100:[I-^\theta]\cap[X1^\theta]$

$101: [I-^{\theta}] \cap [X2]$	$102: [I-^{\theta}] \cap [X3]$	$103: [I-^{\theta}] \cap [X3^{\theta}]$	$104: [J0] \cap [J0^{\theta}]$	$105: [J0] \cap [J1]$
$106: [J0] \cap [J1^{\theta}]$	$107: [J0] \cap [J2]$	$108: [J0] \cap [J2^{\theta}]$	$109: [J0] \cap [Y3]$	$110: [J0] \cap [X1^{\theta}]$
$111: [J0] \cap [X2^{\theta}]$	$112: [J0] \cap [X3]$	$113: [J0^{\theta}] \cap [J1]$	$114: [J0^{\theta}] \cap [J1^{\theta}]$	$115: [J0^{\theta}] \cap [J2]$
$116: [J0^{\theta}] \cap [J2^{\theta}]$	$117; [J0^{\theta}] \cap [Y3]$	$118: [J0^{\theta}] \cap [X1]$	$119: [J0^{\theta}] \cap [X2]$	$120: [J0^{\theta}] \cap [X3^{\theta}]$
$121: [J1] \cap [J1^{\theta}]$	$122: [J1] \cap [J2]$	$123: [J1] \cap [J2^{\theta}]$	$124: [J1] \cap [Y3]$	$125: [J1] \cap [X1]$

$126: [J1] \cap [X2^{\theta}]$	$127: [J1] \cap [X3]$	$128: [J1^{\theta}] \cap [J2]$	$129: [J1^{\theta}] \cap [J2^{\theta}]$	$130: [J1^{\theta}] \cap [Y3]$
$131: [J1^{\theta}] \cap [X1^{\theta}]$	$132: [J1^{\theta}] \cap [X2]$	$133: [J1^{\theta}] \cap [X3^{\theta}]$	$134: [J2] \cap [J2^{\theta}]$	$135: [J2] \cap [Y3]$
$136: [J2] \cap [X1^{\theta}]$	$137: [J2] \cap [X2]$	$138: [J2] \cap [X3]$	$139: [J2^{\theta}] \cap [Y3]$	$140: [J2^{\theta}] \cap [X1]$
$141: [J2^{\theta}] \cap [X2^{\theta}]$	$142: [J2^{\theta}] \cap [X3^{\theta}]$	$143: [Y2] \cap [Y3]$	$144: [X1] \cap [X1^{\theta}]$	$145: [X2] \cap [X2^{\theta}]$
$146: [X3] \cap [X3^{\theta}]$				

(i.e., the ideal cells) of very low dimension. We cannot give examples for this case. We discuss similar problem in the case when $\#(\mathcal{S}[W_i] \cap \mathcal{S}[W_j]) = 1$ in the subsequent subsections.

4.2 *The Pairs with the Cardinality of the Intersection Is 1*

Here is the list of $\{W_i,\ W_j\}$ with $\#(\mathcal{S}[W_i] \cap \mathcal{S}[W_j]) = 1$:

1: $[\mathbf{2}] \cap [I-] = \{P_3 = e_0^{\theta_1}\}$	2: $[\mathbf{2}] \cap [I-^{\theta}] = \{P_2 = e_0^{\theta_2}\}$
3: $[O] \cap [I+] = \{P_4 = e_0^{\theta_1\theta_2}\}$	4: $[O] \cap [I+^{\theta}] = \{P_1 = e_0\}$
5: $[R] \cap [I+] = \{P_6 = e_1^{\theta_1\theta_2}\}$	6: $[R] \cap [J0^{\theta}] = \{P_6 = e_1^{\theta_1\theta_2}\}$
7: $[R] \cap [X1^{\theta}] = \{P_{28} = e_5^{\theta_1\theta_2}\}$	8: $[R] \cap [X2^{\theta}] = \{P_{28} = e_5^{\theta_1\theta_2}\}$
9: $[R^{\theta}] \cap [I+^{\theta}] = \{P_5 = e_1^{\theta_3}\}$	10: $[R^{\theta}] \cap [J0] = \{P_5 = e_1^{\theta_3}\}$
11: $[R^{\theta}] \cap [X1] = \{P_{27} = e_5\}$	12: $[R^{\theta}] \cap [X2] = \{P_{27} = e_5\}$
13: $[E1] \cap [E2^{\theta}] = \{P_3 = e_0^{\theta_1}\}$	14: $[E1^{\theta}] \cap [E2] = \{P_2 = e_0^{\theta_2}\}$
15: $[I+] \cap [Y3] = \{P_4 = e_0^{\theta_1\theta_2}\}$	16: $[I+^{\theta}] \cap [Y3] = \{P_1 = e_0\}$
17: $[I-] \cap [Y3] = \{P_3 = e_0^{\theta_1}\}$	18: $[I-^{\theta}] \cap [Y3] = \{P_2 = e_0^{\theta_2}\}$

Here

P_1: $e_0 = [\mathbf{1}; \mathbf{2}; O; E1^{\theta}; E2^{\theta}; I+^{\theta}; Y1; Y3; X1; X2]$ $\quad$ P_2: $e_0^{\theta_2} = [\mathbf{1}; \mathbf{2}; O; E_1^{\theta}; E_2; I-^{\theta}; Y1; Y3; X1; X2^{\theta}]$

P_3: $e_0^{\theta_1} = [\mathbf{1}; \mathbf{2}; O; E1; E2^{\theta}; I-; Y_1; Y_3; X_2; X1^{\theta}]$ $\quad$ P_4: $e_0^{\theta_1\theta_2} = [\mathbf{1}; \mathbf{2}; O; E1; E2; I+; Y1; Y3; X1^{\theta}; X2^{\theta}]$

P_5: $e_1^{\theta_3} = [\mathbf{1}; \mathbf{2}; R^{\theta}; E1^{\theta}; E2^{\theta}; I+^{\theta}; J0; Y1; Y2]$ $\quad$ P_6: $e_1^{\theta_1\theta_2} = [\mathbf{1}; \mathbf{2}; R; E1; E2; I+; J0^{\theta}; Y1; Y2]$

P_{27}: $e_5 = [R^{\theta}; E1^{\theta}; E2^{\theta}; Y1; Y2; X1; X2]$ $\quad$ P_{28}: $e_5^{\theta_1\theta_2} = [R; E1; E2; Y1; Y2; X1^{\theta}; X2^{\theta}]$

4.3 A Case Study with $\#(\mathcal{S}[W_i] \cap \mathcal{S}[W_j]) = 1$

Let us try to understand the case 5 : $[R] \cap [I+] = \{P_6 = e_1^{\theta_1\theta_2}\}$ in the above list. By definition $e_1^{\theta_1\theta_2} = \frac{1}{3}\begin{pmatrix} -1 & 1 \\ 1 & -1 \end{pmatrix} + i\frac{\sqrt{2}}{3}\begin{pmatrix} 2 & 1 \\ 1 & 2 \end{pmatrix}$, and we introduce the local parameters ξ_i, η_i at this point by

$$Z = e_1^{\theta_1\theta_2} + \begin{pmatrix} \xi_1 & \xi_0 \\ \xi_0 & \xi_2 \end{pmatrix} + \sqrt{-1}\begin{pmatrix} \eta_1 & \eta_0 \\ \eta_0 & \eta_2 \end{pmatrix}.$$

Then the defining equation of the wall $[R]$ is given by $|z_{11} + z_{22} - 2z_{12} + 1|^2 - 1 = 0$, hence by

$$-\frac{2}{3}(\xi_1 + \xi_2 - 2\xi_0) + \frac{4\sqrt{2}}{3}(\eta_1 + \eta_2 - 2\eta_0) + (\xi_1 + \xi_2 - 2\xi_0)^2 + (\eta_1 + \eta_2 - 2\eta_0)^2 = 0.$$

Similarly for the wall $[I+]$, the defining equation

$$|(z_1 + 1)(z_2 + 1) - z_{12}^2|^2 - 1 = 0$$

is equivalent to

$$\frac{4}{3}(\xi_1 + \xi_2 - \xi_0) + \frac{4}{3}\sqrt{2}(\eta_1 + \eta_2 - \eta_0) + \text{ higher degree term} = 0,$$

because

$$Re\{(z_{11} + 1)(z_{22} + 1) - z_{12}^2\} = -\frac{1}{3} + \frac{2}{3}(\xi_1 + \xi_2 - \xi_0) - \frac{\sqrt{2}}{3}(\eta_1 + \eta_2 - \eta_0)]$$
$$+ \xi_1\xi_2 - \xi_0^2 - (\eta_1\eta_2 - \eta_0)^2$$

and

$$Im\{(z_{11} + 1)(z_{22} + 1) - z_{12}^2\} = \frac{2}{3}\sqrt{2} + \frac{2\sqrt{2}}{3}(\xi_1 + \xi_2 - \xi_0) + \frac{2}{3}(\eta_1 + \eta_2 - \eta_0)$$
$$+ \xi_1\eta_2 + \xi_2\eta_1 - 2\xi_0\eta_0.$$

Here we assume that ξ_i, η_i are infinitesimally small and if we neglect the term of degree ≥ 2, then we have the system of two linear relations:

$$L_R : \quad -(\xi_1 + \xi_2 - 2\xi_0) + 2\sqrt{2}(\eta_1 + \eta_2 - 2\eta_0) = 0,$$

$$L_{I+} : \quad (\xi_1 + \xi_2 - \xi_0) + \sqrt{2}(\eta_1 + \eta_2 - \eta_0) = 0,$$

i.e., the defining equations of the tangent spaces of the walls $[R]$ and $[I+]$. Since these two relations are different, these walls intersect infinitesimally in dimension 4 at P_0. Therefore $P_6 = e_1^{\theta_1\theta_2}$ is not an isolated point in the intersection. This means that we have to suppose the existence of other 0-cells or higher dimensional cells at infinity.

We can push this way further, noting the fact that $e^{\theta_1\theta_2} \in$ [**1**, **2**, $E1$, $E2, J0^\theta$, $Y1, Y2$]. Obviously the linearized inequalities for [$Y1$] and [$Y2$] are given by

$$L_{Y1}: \quad \eta_2 - \eta_1 \geq 0, \qquad L_{Y2}: \quad \eta_1 - 2\eta_0 \geq 0,$$

respectively. The linearized inequalities for [**1**] and [**2**] are given by

$$L_{\mathbf{1}}: \quad 2\sqrt{2}\eta_1 \geq \xi_1, \qquad L_{\mathbf{2}}: \quad 2\sqrt{2}\eta_2 \geq \xi_2.$$

There remain three more linear inequalities corresponding to [$E1$], [$E2$], and [$J0^\theta$]:

$$L_{E1}: \quad 2\sqrt{2}(\eta_1 + \eta_2 - \eta_0) \geq (-\xi_1 + 2\xi_2 - 2\xi_0),$$

$$L_{E2}: \quad 2\sqrt{2}(\eta_1 + \eta_2 - \eta_0) \geq (2\xi_1 - \xi_2 + 2\xi_0),$$

and

$$L_{J0^\theta}: \quad 2\sqrt{2}(\eta_1 + \eta_2 - \eta_0) \geq (\xi_1 + \xi_2 - 4\xi_0),$$

respectively. The last three equations together with the equation L_{I+} is unified to a single equation:

$$2\sqrt{2}(\eta_1+\eta_2-\eta_0) \geq \sup\{-\xi_1+2\xi_2-2\xi_0,\ 2\xi_1-\xi_2+2\xi_0,\ -\xi_1-\xi_2+4\xi_0,\ -2(\xi_1+\xi_2-\xi_0)\}.$$

Together with L_R, we have nine linear inequalities defining a linear cone C_+ in $\mathbb{R}^6$.

Now we want to get this C_+ in the 4-dimensional linear space given by $L_R \cap L_{I+}$ explicitly. Firstly the solution of the system L_R and L_{I+} is given by

$$\sqrt{2}(\eta_1 + \eta_2) = -\frac{5}{2}(\xi_1 + \xi_2) + 3\xi_0, \quad \text{and} \quad \sqrt{2}\eta_0 = -\frac{1}{2}(\xi_1 + \xi_2) + 2\xi_0.$$

If we introduce a new variable $\delta = \sqrt{2}(\eta_2 - \eta_1)$, then we eliminate the variables η_0 and $\eta_1 + \eta_2$ and have to consider the cone C_+ inside the 4-dimensional real Euclid space with coordinates $(\xi_1, \xi_2, \xi_3, \delta) \in \mathbf{R}^4$ specified by the inequalities L_{Y1}, L_{Y2}, $L_{\mathbf{1}}$, $L_{\mathbf{2}}$, L_{E1}, L_{E2}, and L_{J0^θ}.

The equality $L_{Y1}: \sqrt{2}(\eta_2 - \eta_1) \geq 0$ is equivalent to $\delta \geq 0$. The inequality $\eta_1 - 2\eta_0 \geq 0$ is equivalent to $\frac{1}{2}\{\sqrt{2}(\eta_1 + \eta_2) - \delta\} - 2\sqrt{2}\eta_0 \geq 0$. Rewrite $\eta_1 + \eta_2$ and η_0 by ξ_i $(i = 1, 2, 3)$ to get

$$-\frac{5}{2}(\xi_1 + \xi_2) + 3\xi_0 + (-4)\{-\frac{1}{2}(\xi_1 + \xi_2) + 2\xi_9\}$$
$$= \frac{7}{2}(\xi_1 + \xi_2) - 5\xi_0 \geq \delta \geq 0.$$

In particular this implies

$$\frac{7}{10}(\xi_1 + \xi_2) \geq \xi_0.$$

The inequalities

$$L_1: \quad \sqrt{2}(\eta_1 + \eta_2) - \delta \geq \xi_1, \quad \text{and} \quad L_2: \quad \sqrt{2}(\eta_1 + \eta_2) + \delta \geq \xi_2$$

are equivalent to

$$-\frac{5}{2}(\xi_1 + \xi_2) + 3\xi_0 - \xi_1 \geq \delta, \quad \text{and} \quad -\frac{5}{2}(\xi_1 + \xi_2) + 3\xi_0 - \xi_2 \geq -\delta.$$

Moreover the sum of these two inequalities is

$$6\xi_0 - 6(\xi_1 + \xi_2) \geq 0,$$

hence we have $\xi_0 \geq \xi_1 + \xi_2$. Now this equality and $\frac{7}{10}(\xi_1 + \xi_2) \geq \xi_0$ imply that

$$(\frac{7}{10} - 1)(\xi_1 + \xi_2) \geq 0,$$

which means that $\xi_1 + \xi_2 \leq 0$.

The inequality L_{J0^θ} is equivalent to $-2(\xi_1 + \xi_2 - \xi_0) \geq \xi_1 + \xi_2 - 4\xi_0$, which in turn equivalent to $\xi_1 + \xi_2 \geq 2\xi_0$. Then we have $\xi_0 \geq \xi_1 + \xi_2 \geq 2\xi_0$, hence $\xi_0 \leq 0$.

Now we have the following claim:

Claim. $\xi_0 \geq \frac{5}{8}(\xi_1 + \xi_2)$.

Proof of Claim. The inequalities L_{E1} and L_{E2} are reduced to

$$-2(\xi_1 + \xi_2 - \xi_0) \geq -\xi_1 + 2\xi_2 - 2\xi_0, \quad \text{and} \quad -2(\xi_1 + \xi_2 - \xi_0) \geq -2\xi_1 - \xi_2 - 2\xi_0.$$

This is equivalent to the set of inequalities

$$0 \geq \xi_1 + 4\xi_2 - 4\xi_0, \text{ and } 0 \geq 4\xi_1 + \xi_2 - 4\xi_0.$$

In particular the sum of these two inequalities is $0 \geq 5(\xi_1 + \xi_2) - 8\xi_0$. □

The claim implies that

$$\frac{7}{10}(\xi_1 + \xi_2) \geq \xi_0 \geq \frac{5}{8}(\xi_1 + \xi_2).$$

Hence we have $\xi_1 + \xi_2 \geq 0$. Therefore, together with $\xi_1 + \xi_2 \leq 0$, we have $\xi_1 + \xi_2 = 0$ and also $\xi_0 = 0$ by the above claim and by the last sandwich formula.

We have $\xi_0 = 0$ and $\xi_1 + \xi_2 = 0$. Hence the inequalities

$$0 \geq \xi_1 + 4\xi_2 - 4\xi_0, \text{ and } 0 \geq 4\xi_1 + \xi_2 - 4\xi_0$$

in the proof of the above claim are reduced to $0 \geq 3\xi_2$ and $0 \geq 3\xi_1$. Therefore we have $\xi_1 = \xi_2 = 0$ too, and the only free variable is δ. Thus finally for the cone C we have

$$C_+ = \{(\xi_1, \xi_2, \xi_0, \delta) = (0, 0, 0, \delta) \mid \delta \geq 0\}.$$

Therefore around the point P_0 we have a one-dimensional cone as the infinitesimal intersection of $[R]$ and $[I+]$.

Conjecture. The actual intersection $[R] \cap [I+]$ would be a curve emanating from the point P_0 which should end at a point at infinity on the boundary of $\Gamma_2 \backslash \mathfrak{H}_2$.

There is the same problem for the remaining 17 cases $[W_i] \cap [W_j] = \{P_k\}$ in the above list.

4.4 The Pairs $\{W_i, W_j\}$ with $\#(\mathcal{S}[W_i] \cap \mathcal{S}[W_j]) = 2$

Here we give only the list:

$$\begin{array}{ll}
[\mathbf{2}] \cap [J2] = \{P_{86} = e_{18}^{\theta_3\theta_2},\ P_{146} = e_{30}^{\theta_3\theta_2}\}, & [\mathbf{2}] \cap [J2^\theta] = \{P_{87} = P_{86}^\theta = e_{18}^{\theta_1},\ P_{147} = P_{146}^\theta = e_{30}^{\theta_1}\}, \\
[O] \cap [I-] = \{P_3 = e_0^{\theta_1},\ P_{61} = e_{11}^{\theta_1}\}, & [O] \cap [I-^\theta] = \{P_2 = e_0^{\theta_2},\ P_{60} = e_{11}^{\theta_2}\}, \\
[E1] \cap [Y3] = \{P_3 = e_0^{\theta_1},\ P_4 = e_0^{\theta_1\theta_2}\}, & [E1^\theta] \cap [Y3] = \{P_1 = e_0,\ P_2 = e_0^{\theta_2}\}, \\
[E2] \cap [I-^\theta] = \{P_2 = e_0^{\theta_2},\ P_{118}\}, & [E2] \cap [Y3] = \{P_2 = e_0^{\theta_2},\ P_4 = e_0^{\theta_1\theta_2}\}, \\
[E2^\theta] \cap [I-] = \{P_3 = e_0^{\theta_1},\ P_{123}\}, & [E2^\theta] \cap [Y3] = \{P_1 = e_0,\ P_3 = e_0^{\theta_1}\}.
\end{array}$$

Here $P_1 = e_0$, $P_2 = e_0^{\theta_2}$, $P_3 = e_0^{\theta_1}$, and $P_4 = e_0^{\theta_1\theta_2}$ are already defined as above. Moreover

$$\begin{array}{ll}
P_{60} : e_{11}^{\theta_2} = [\mathbf{1}; O; I-^\theta; Y2; X1; X2^\theta], & P_{61} : e_{11}^{\theta_1} = [\mathbf{1}; O; I-; Y2; X2; X1^\theta], \\
P_{86} : e_{18}^{\theta_3\theta_2} = [\mathbf{1}; \mathbf{2}; E1^\theta; J2; Y1; X1; X3^\theta; X2^\theta], & P_{87} : e_{18}^{\theta_1} = [\mathbf{1}; \mathbf{2}; E1; J2^\theta; Y1; X3; X2; X1^\theta], \\
P_{118} : e_{25}^{\theta_2} = [O; E2; I-^\theta; Y1; Y2; X1; X2^\theta], & P_{123} : e_{25}^{\theta_1\theta_3} = [O; E2^\theta; I-; Y1; Y2; X2; X1^\theta], \\
P_{146} : e_{30}^{\theta_3\theta_2} = [\mathbf{1}; \mathbf{2}; R^\theta; E1^\theta; J2; Y1; X3^\theta], & P_{147} : e_{30}^{\theta_1} = [\mathbf{1}; \mathbf{2}; R; E1; J2^\theta; Y1; X3].
\end{array}$$

The cases when the cardinality of the intersection is large are skipped.

5 The Differential Geometry of the Symmetric Space $\mathfrak{H}_2$

We refer to Sakai [5], Chap. IV, Sect. 5 for general facts on the differential geometry.

5.1 The Cross Ratio and the Canonical Distance on $\mathfrak{H}_2 = Sp(2, \mathbf{R})/K$

Given two points $Z,\ W$ in $\mathfrak{H}_2$, their cross ratio is defined as

$$\mathcal{R}(Z,W) = (Z - W)(Z - \bar{W})^{-1}(\bar{Z} - \bar{W})(\bar{Z} - W)^{-1}$$

(Siegel [6]), which is diagonalized as

$$\operatorname{diag}(r_1, r_2) \quad (0 \le r_1 \le r_2 < 1).$$

Then the symplectic distance $\rho(Z, W)$ is given by

$$\rho(Z,W) := \{\sum_{i=1}^{2}(\log t_i)^2\}^{1/2} \quad \text{with } t_i = \frac{1 + r_i^{1/2}}{1 - r_i^{1/2}}.$$

The metric form ds^2 of this distance is a Kähler-Einstein metric invariant under the action of the real symplectic group of genus 2. Let $z_{ij} = x_{ij} + \sqrt{-1}y_{ij}$ for $(i,j) = (1,1),\ (1,2),\ (2,2)$, and $Y = \operatorname{Im}(Z)$, we have the explicit formula of ds^2:

$$\begin{aligned} ds^2 =& \det(Y)^{-2}[\sum_{i=1}^{2} y_{3-i}^2 dz_{ii} \otimes d\bar{z}_{ii} + y_{12}^2 \sum_{i=1}^{2} dz_{ii} \otimes d\bar{z}_{3-i3-i} \\ &+ 2(y_{11}y_{22} + y_{12}^2)dz_{12} \otimes d\bar{z}_{12} + \\ &- 2y_{12}y_{22}(dz_{11} \otimes d\bar{z}_{12} + +dz_{12} \otimes d\bar{z}_{11}) \\ &- 2y_{11}y_{12}(dz_{22} \otimes d\bar{z}_{12} + +dz_{12} \otimes d\bar{z}_{22})]. \end{aligned}$$

Then we can confirm the following immediately.

Fact. ds^2 is invariant under $\theta = \theta_1\theta_2\theta_3$ and σ_0. But neither of θ_i $(i = 1, 2, 3)$ keeps ds^2 invariant. We arrange the coordinates of the space $\mathfrak{H}_2$ in the order $y_{11}, y_{22}, y_{12}, x_{11}, x_{22}, x_{12}$. Then the matrix of the Riemann metric is given as a block of 3×3 matrices:

$$\det(Y)^{-2}\begin{pmatrix} G(Y) & \mathbf{0}_3 \\ \mathbf{0}_3 & G(Y) \end{pmatrix} \quad \text{with} \quad G(Y) = \begin{pmatrix} y_{22}^2 & y_{12}^2 & -2y_{12}y_{22} \\ y_{12}^2 & y_{11}^2 & -2y_{11}y_{12} \\ -2y_{12}y_{22} & -2y_{11}Y_{12} & 2(y_{11}y_{22} + y_{12}^2) \end{pmatrix}.$$

Here $\mathbf{0}_3$ is the zero matrix of 3×3.

5.2 Geodesic Curves

Let $Z(s)$ be a curve in $\mathfrak{H}_2$ parametrized by its length s. Then we know the equation of the geodesics.

Proposition 5.1 (Siegel [6]). *The system of the equations for a geodesic curve $Z(s)$ is given by*

$$\ddot{Z}(s) + i\dot{Z}(s)Y^{-1}(s)\dot{Z}(s) = 0.$$

Here we employ the Newtonian dots symbol for derivations.

If we spell out the above equation component-wise, we have

$$\ddot{z}_{11} + \frac{i}{\det(Y)}\{y_{22}\dot{z}_{11}^2 + y_{11}\dot{z}_{12}^2 - 2y_{12}\dot{z}_{11}\dot{z}_{12}\} = 0;$$

$$\ddot{z}_{22} + \frac{i}{\det(Y)}\{y_{22}\dot{z}_{12}^2 + y_{11}\dot{z}_{22}^2 - 2y_{12}\dot{z}_{12}\dot{z}_{22}\} = 0;$$

$$\ddot{z}_{12} + \frac{i}{\det(Y)}\{y_{22}\dot{z}_{11}\dot{z}_{12} + y_{11}\dot{z}_{12}\dot{z}_{22} - y_{12}(\dot{z}_{11}\dot{z}_{22} + \dot{z}_{12}^2)\} = 0.$$

Or more precisely component-wise, we have

$$(1) \quad \ddot{y}_{11} - \frac{1}{\det(Y)}\{y_{22}(\dot{y}_{11}^2 - \dot{x}_{11}^2) + +2y_{12}(\dot{y}_{11}\dot{y}_{12} - \dot{x}_{11}\dot{x}_{12}) + y_{11}(\dot{y}_{12}^2 - \dot{x}_{12}^2)\} = 0,$$

$$(2) \quad \ddot{y}_{22} - \frac{1}{\det(Y)}\{y_{11}(\dot{y}_{22}^2 - \dot{x}_{22}^2) + +2y_{12}(\dot{y}_{22}\dot{y}_{12} - \dot{x}_{22}\dot{x}_{12}) + y_{22}(\dot{y}_{12}^2 - \dot{x}_{12}^2)\} = 0,$$

$$(3) \quad \ddot{y}_{12} - \frac{1}{\det(Y)}\{y_{22}(\dot{y}_{11}\dot{y}_{12} - \dot{x}_{11}\dot{y}_{12}) + y_{11}(\dot{y}_{22}\dot{y}_{12} - \dot{x}_{22}\dot{x}_{12}) - y_{12}(\dot{y}_{11}\dot{y}_{22} - \dot{x}_{11}\dot{x}_{22} + \dot{y}_{12}^2 - \dot{x}_{12}^2)\} = 0,$$

$$(4) \quad \ddot{x}_{11} - \frac{1}{\det(Y)}\{2y_{22}\dot{y}_{11}\dot{x}_{11} - 2y_{12}(\dot{y}_{11}\dot{x}_{12} + \dot{y}_{12}\dot{x}_{11}) + 2y_{11}\dot{y}_{12}\dot{x}_{12}\} = 0,$$

$$(5) \quad \ddot{x}_{22} - \frac{1}{\det(Y)}\{2y_{11}\dot{y}_{22}\dot{x}_{22} - 2y_{12}(\dot{y}_{22}\dot{x}_{12} + \dot{y}_{12}\dot{x}_{22}) + 2y_{22}\dot{y}_{12}\dot{x}_{12}\} = 0,$$

$$(6) \quad \ddot{x}_{12} - \frac{1}{\det(Y)}\{y_{22}(\dot{y}_{11}\dot{x}_{12} + \dot{x}_{11}\dot{y}_{12}) + y_{11}(\dot{y}_{22}\dot{y}_{12} + \dot{x}_{22}\dot{x}_{12}) - y_{12}(\dot{y}_{11}\dot{x}_{22} + \dot{x}_{11}\dot{y}_{22} + 2\dot{y}_{12}\dot{x}_{12})\} = 0.$$

From the formula of the geodesic, we can obtain the Christoffel symbols immediately.

The Christoffel symbol We arrange the variable of the coordinates of $\mathfrak{H}_2$ as $y_{11}, y_{22}, y_{12}, x_{11}, x_{22}, x_{12}$, then the $126 = 6 \times 21$ Christoffel symbols are listed as follows:

$$\begin{array}{lllll}
\Gamma_{11}^1 = -\dfrac{y_{22}}{\det(Y)}, & \Gamma_{12}^3 = \dfrac{1}{2}\dfrac{y_{12}}{\det(Y)}, & \Gamma_{13}^1 = -\dfrac{y_{12}}{\det(Y)}, & \Gamma_{13}^3 = -\dfrac{1}{2}\dfrac{y_{22}}{\det(Y)}, & \Gamma_{14}^4 = -\dfrac{y_{22}}{\det(Y)}, \\
\Gamma_{15}^5 = -\dfrac{y_{11}}{\det(Y)}, & \Gamma_{15}^6 = \dfrac{1}{2}\dfrac{y_{12}}{\det(Y)}, & \Gamma_{16}^4 = -\dfrac{y_{12}}{\det(Y)}, & \Gamma_{16}^6 = -\dfrac{1}{2}\dfrac{y_{22}}{\det(Y)}, & \Gamma_{22}^2 = -\dfrac{y_{11}}{\det(Y)}, \\
\Gamma_{23}^2 = -\dfrac{y_{12}}{\det(Y)}, & \Gamma_{23}^3 = -\dfrac{1}{2}\dfrac{y_{11}}{\det(Y)}, & \Gamma_{24}^6 = \dfrac{1}{2}\dfrac{y_{12}}{\det(Y)}, & \Gamma_{26}^5 = -\dfrac{y_{12}}{\det(Y)}, & \Gamma_{26}^6 = -\dfrac{1}{2}\dfrac{y_{11}}{\det(Y)}, \\
\Gamma_{33}^1 = -\dfrac{y_{11}}{\det(Y)}, & \Gamma_{33}^2 = -\dfrac{y_{11}}{\det(Y)}, & \Gamma_{33}^3 = \dfrac{y_{11}}{\det(Y)}, & \Gamma_{34}^4 = -\dfrac{y_{11}}{\det(Y)}, & \Gamma_{34}^6 = -\dfrac{1}{2}\dfrac{y_{22}}{\det(Y)}, \\
\Gamma_{35}^5 = -\dfrac{y_{22}}{\det(Y)}, & \Gamma_{35}^6 = -\dfrac{1}{2}\dfrac{y_{11}}{\det(Y)}, & \Gamma_{36}^4 = -\dfrac{y_{11}}{\det(Y)}, & \Gamma_{36}^5 = \dfrac{y_{22}}{\det(Y)}, & \Gamma_{36}^6 = \dfrac{y_{12}}{\det(Y)}, \\
\Gamma_{44}^1 = \dfrac{y_{22}}{\det(Y)}, & \Gamma_{45}^3 = -\dfrac{1}{2}\dfrac{y_{12}}{\det(Y)}, & \Gamma_{46}^1 = \dfrac{y_{12}}{\det(Y)}, & \Gamma_{46}^3 = \dfrac{1}{2}\dfrac{y_{22}}{\det(Y)}, & \Gamma_{55}^2 = \dfrac{y_{11}}{\det(Y)}, \\
\Gamma_{56}^2 = \dfrac{y_{12}}{\det(Y)}, & \Gamma_{56}^3 = \dfrac{1}{2}\dfrac{y_{11}}{\det(Y)}, & \Gamma_{66}^1 = \dfrac{y_{11}}{\det(Y)}, & \Gamma_{66}^2 = \dfrac{y_{22}}{\det(Y)}, & \Gamma_{66}^3 = -\dfrac{y_{12}}{\det(Y)},
\end{array}$$

By this table, we can check the following fact immediately.

Proposition 5.2. *Any of* $[Xi]$ $(i = 1, 2, 3)$, $[Xi^{\theta}]$ $(i = 1, 2, 3)$ *is not totally geodesic.*

Proof. The case of $[X1]$: The tangent subspace $T_Z([X1])$ of $T_Z(\mathfrak{H}_2)$ is generated by five vector fields $\frac{\partial}{\partial y_{ij}}$ $(ij = 11,\ 22,\ 12)$ and $\frac{\partial}{\partial x_{12}}, \frac{\partial}{\partial x_{22}}$. Then the covariant derivations

$$\nabla_{(\frac{\partial}{\partial x_{12}})}(\frac{\partial}{\partial y_{11}}) = \frac{y_{12}}{\det(Y)}\frac{\partial}{\partial_{11}} - \frac{1}{2}\frac{y_{22}}{\det Y}\frac{\partial}{\partial x_{12}}$$

and

$$\nabla_{(\frac{\partial}{\partial x_{12}})}(\frac{\partial}{\partial y_{12}}) = -\frac{y_{11}}{\det(Y)}\frac{\partial}{\partial x_{11}} - \frac{y_{22}}{\det(Y)}\frac{\partial}{\partial x_{22}} + \frac{y_{12}}{\det(Y)}\frac{\partial}{\partial x_{12}}$$

in $T_Z(\mathfrak{H}_2)$ do not belong to $T_Z([X1])$. In other words, the second fundamental form of $[X1]$ in $\mathfrak{H}_2$ does not vanish, i.e., it is not a totally geodesic. We can handle the other cases $[X2]$, $[Xi^{\theta}]$ $(i = 1, 2)$ by the same way because of the reason of symmetry.

The case $[X3]$: In this case, the tangent subspace $T_Z([X3])$ is generated by five elements $\frac{\partial}{\partial y_{ii}}$ $(i = 1, 2)$, $\frac{\partial}{\partial y_{12}}$ and $\frac{\partial}{\partial x_{ii}}$ $(i = 1, 2)$. Among 15 covariant derivatives $\nabla_{(\frac{\partial}{\partial x_j})}(\frac{\partial}{\partial x_i})$ we find that the covariant derivation

$$\nabla_{(\frac{\partial}{\partial x_{22}})}(\frac{\partial}{\partial y_{11}}) = \frac{1}{2}\frac{y_{12}}{\det(Y)}\frac{\partial}{\partial x_{12}}$$

does not belong to the tangent subspace $T_Z([X3])$. □

Proposition 5.3. *The intersection* $[X1] \cap [X2]$ *of dimension 4 in* $\mathfrak{H}_2$ *is not totally geodesic.*

Proof. The tangent space at $Z \in [X1] \cap [X2]$ is generated by 4 elements $\frac{\partial}{\partial y_{ii}}$ $(i = 1, 2)$, $\frac{\partial}{\partial y_{12}}$ and $\frac{\partial}{\partial x_{12}}$.

Therefore it is enough to know either

$$\nabla_{(\frac{\partial}{\partial x_{12}})}(\frac{\partial}{\partial y_{11}}) = \frac{y_{12}}{\det(Y)} \frac{\partial}{\partial x_{11}} - \frac{1}{2} \frac{y_{22}}{\det(Y)} \frac{\partial}{\partial x_{12}}$$

or

$$\nabla_{(\frac{\partial}{\partial x_{12}})}(\frac{\partial}{\partial y_{22}}) = \frac{y_{12}}{\det(Y)} \frac{\partial}{\partial x_{22}} - \frac{1}{2} \frac{y_{11}}{\det(Y)} \frac{\partial}{\partial x_{12}}$$

does not belong to $T_Z([X1] \cap [X2])$. □

We can give some examples of totally geodesic submanifolds in $\mathfrak{H}_2$. Typical cases are obtained by the orbits of a point in $\mathfrak{H}_2$ under a reductive subgroups of $Sp(2, \mathbf{R})$. Some of them are discussed in the next section.

6 A Case Study of the Symplectic Distances Between 0-Cells in the 4-Cell $[X1] \cap [X2]$

A part of the results of this section is obtained by a collaboration with Takahiro Hayata. He made pictures of a number of i-cells $(i = 1, 2, 3)$ in the Euclidean metric of $\mathfrak{H}_2$.

6.1 *The Intersection* $[X1] \cap [X2]$

We consider the intersection of $[X1]$ and $[X2]$. Among 180 cells, the following 15 0-cells belong to both of $[X1]$ and $[X2]$:

$$e_0,\ e_5,\ e_{11},\ e_i,\ e_i^{\theta_3} \quad (i = 3, 7, 8, 9, 18, 25).$$

Here the coordinates of these points are given by

$$e_0 = \frac{1}{2}\begin{pmatrix} 1 & 0 \\ 0 & 1 \end{pmatrix} + i\frac{\sqrt{3}}{2}\begin{pmatrix} 1 & 0 \\ 0 & 1 \end{pmatrix},\ e_5 = \frac{1}{2}\begin{pmatrix} 1 & 0 \\ 0 & 1 \end{pmatrix} + i\frac{1}{2}\begin{pmatrix} 2 & 1 \\ 1 & 2 \end{pmatrix}, e_{11} = \frac{1}{2}\begin{pmatrix} 1 & 0 \\ 0 & 1 \end{pmatrix} + i\frac{1}{2}\begin{pmatrix} 2\omega_0 & \omega_0 \\ \omega_0 & \frac{1}{8}\omega_5 \end{pmatrix},$$

$$e_3 = \frac{1}{2}\begin{pmatrix} 1 & 1 \\ 1 & 1 \end{pmatrix} + i\frac{\sqrt{3}}{2}\begin{pmatrix} 1 & 0 \\ 0 & 1 \end{pmatrix}, \quad e_7 = \frac{1}{2}\begin{pmatrix} 1 & 1 \\ 1 & 1 \end{pmatrix} + i\frac{1}{2}\begin{pmatrix} 2\omega_0 & \omega_0 \\ \omega_0 & 3 - \omega_0 \end{pmatrix}$$

$$e_8 = \frac{1}{2}\begin{pmatrix}1 & 1\\ 1 & 1\end{pmatrix} + i\frac{\omega_3}{2}\begin{pmatrix}2 & 1\\ 1 & 2\end{pmatrix}, \quad e_9 = \frac{1}{2}\begin{pmatrix}1 & 1\\ 1 & 1\end{pmatrix} + i\frac{1}{2}\begin{pmatrix}2\omega_0 & \omega_0\\ \omega_0 & 2\omega_4\end{pmatrix}$$

$$e_{18} = \frac{1}{2}\begin{pmatrix}1 & 1\\ 1 & 1\end{pmatrix} + i\begin{pmatrix}\omega_0 & \omega_{21}\\ \omega_{21} & \omega_*\end{pmatrix}, \quad e_{25} = \frac{1}{2}\begin{pmatrix}1 & 1\\ 1 & 1\end{pmatrix} + i\frac{\omega_{19}}{2}\begin{pmatrix}2 & 1\\ 1 & 2\end{pmatrix}.$$

Here ω_i are defined by as found in table 2 of [2].

6.2 The Candidates of the Boundary Components

Observation.

(i) Each of the following walls has empty intersection in the set of 180 zero cells inside $\mathcal{X} := [X1] \cap [X2]$:

$$[R],\ [E1],\ [E2],\ [I+],\ [I-],\ [I-^{\theta}],\ [J0^{\theta}],\ [J1],\ [J1^{\theta}],\ [J2],\ [J2^{\theta}].$$

(ii) We have the following non-empty intersection with each wall $\in [X1],\ [X2],\ [X1^{\theta}],\ [X2^{\theta}]$:

$$\mathcal{X} \cap [X3] = \{e_3, e_8, e_9, e_{18}\}, \quad \mathcal{X} \cap [X3^{\theta}] = \{e_3^{\theta_3}, e_8^{\theta_3}, e_9^{\theta_3}, e_{18}^{\theta}\},$$

$$\mathcal{X} \cap [Y1] = \{e_0, e_3, e_3^{\theta_3}, e_5, e_8, e_8^{\theta_3}, e_{18}, e_{18}^{\theta_3}, e_{25}, e_{25}^{\theta_3}\},$$
$$\mathcal{X} \cap [Y2] = \{e_5, e_{11}, e_7, e_7^{\theta_3}, e_8, e_8^{\theta_3}, e_9, e_9^{\theta_3}, e_{18}, e_{25}, e_{25}^{\theta_3}\},$$

$$\mathcal{X} \cap [Y3] = \{e_0,\ e_3,\ e_3^{\theta_3}\},\ \mathcal{X} \cap [\mathbf{1}] = \{e_0,\ e_3,\ e_3^{\theta_3},\ e_7,\ e_7^{\theta_3},\ e_9,\ e_9^{\theta_3}, e_{11}, e_{18}, e_{18}^{\theta_3}\},$$

$$\mathcal{X} \cap [\mathbf{2}] = \{e_0,\ e_3,\ e_3^{\theta_3},\ e_{18},\ e_{18}^{\theta}\},\ \mathcal{X} \cap [\theta R] = \{e_5\},$$
$$\mathcal{X} \cap [O] = \{e_0, e_7, e_8, e_9, e_{18}, e_{25}\}$$

$$\mathcal{X} \cap [E1^{\theta}] = \{e_0, e_5, e_{11}, e_7^{\theta_3}, e_{25}, e_{25}^{\theta_3}\}, \quad \mathcal{X} \cap [E2^{\theta}] = \{e_0, e_5, e_{11}, e_7, e_{25}, e_{25}^{\theta_3}\}$$

$$\mathcal{X} \cap [I+^{\theta}] = \{e_0, e_7^{\theta_3}, e_8^{\theta_3}, e_9^{\theta_3}, e_{18}^{\theta_3}, e_{25}^{\theta_3}\}, \quad \mathcal{X} \cap [J0] = \{e_8^{\theta_3}, e_9^{\theta_3}, e_{18}^{\theta_3}\}$$

Combinatorial Chemistry We define the number of the local band between two zero cells e_i and e_j as the number of the walls $[W]$ such both of the zero cells are contained in $[X_1] \cap [X_2] \cap [W]$.

Here is the list of band number:

$$b(e_0,e_3)=4,\quad b(e_0,e_3^{\theta_3})=4,\quad b(e_0,e_{18})=4,\quad b(e_0,e_{18}^{\theta_3})=4,$$

$$b(e_0,e_{25})=4,\quad b(e_0,e_{25}^{\theta_3})=4,\quad b(e_0,e_5)=3,\quad b(e_0,e_{11})=3,$$

$$b(e_0,e_7)=3,\quad b(e_0,e_7^{\theta_3})=3,\quad b(e_0,e_8)=2,\quad b(e_0,e_8^{\theta_3})=2,$$

$$b(e_0,e_9)=2,\quad b(e_0,e_9^{\theta_3})=2.$$

These data of bands give heuristic information how strong two points are combined. This suggests two points with bigger band number belong to an intersection of higher codimension.

Example 0. For example, for $e_0,\ e_3$, and e_3^{θ}, we find that

$$\mathcal{X}\cap[Y1]\cap[Y3]\cap[\mathbf{1}]\cap[\mathbf{2}]=\{e_0,\ e_3,\ e_3^{\theta_3}\}.$$

The distances of these three points are given by $\rho(e_0,e_3)=\rho(e_0,e_3^{\theta_3})=\sqrt{2}\log(\frac{7+\sqrt{13}}{6})$: Here $r_i=\frac{1}{13}$ and $t_i=\frac{7+\sqrt{13}}{6}$ for $i=1,2$, and also by $\rho(e_3,e_3^{\theta_3})=\sqrt{2}\log 3$:

We note here that these three points belong to $[Y3], y_{12}=0$, and $[Y_1], y_{11}-y_{22}=0$.

Since $\sqrt{3}=1.732\dots<\frac{7+\sqrt{13}}{6}=1.7675918$, this means that e_0 is rather near to the middle point of the line segment $[e_3,e_3^{\theta}]$ but is a different point. Therefore the triangle $e_0e_3e_3^{\theta_3}$ might determine a totally geodesic surface segment inside the intersection

$$[X_1]\cap[X_2]\cap[Y_1]\cap[Y_3]\subset\left\{\begin{pmatrix}\frac{1}{2} & x_{12}\\ x_{12} & \frac{1}{2}\end{pmatrix}+i\begin{pmatrix}y & 0\\ 0 & y\end{pmatrix}\mid x_{12}\in[-\frac{1}{2},+\frac{1}{2}],\ y>0\right\}=:S$$

of dimension two. We confirm the fact that $e_0,\ e_3,\ e_3^{\theta_3}$ actually belong to a totally geodesic surface $\mathcal{S}$.

Here we define a map

$$\varphi:z=z+iy\in\mathfrak{H}_1\mapsto\begin{pmatrix}\frac{1}{2} & x\\ x & \frac{1}{2}\end{pmatrix}+iy\begin{pmatrix}1 & 0\\ 0 & 1\end{pmatrix},$$

from $\mathfrak{H}_1$ to S. Then for $z=x+iy$ and $w=u+iv\in\mathfrak{H}_1$, we compute $\mathcal{R}(\varphi(z),\varphi(w))$ to get

$$\mathcal{R}(\varphi(z),\varphi(w))=\{(y+v)^2+(x-u)^2\}^{-2}\times[\{(y^2-v^2)+(x-u)^2\}^2+4v^2(x-u)^2]\begin{pmatrix}1 & 0\\ 0 & 1\end{pmatrix}.$$

Note here

$$\{(y^2-v^2)+(x-u)^2\}^2+4v^2(x-u)^2=\{(y+v)^2+(x-u)^2\}\cdot\{(y-v)^2+(x-u)^2\}.$$

Then $r_1=r_2=\frac{(y-v)^2+(x-u)^2}{(y+v)^2+(x-u)^2}$. Hence we have

$$\rho(\varphi(z),\varphi(w))=\sqrt{2}\rho_1(z,w)$$

with the hyperbolic metric ρ_1 on $\mathfrak{H}_1$. We find that φ is an isometric map and $\varphi(i)=i\begin{pmatrix}1&0\\0&1\end{pmatrix}$ is the middle point of the geodesic segment $[e_3,e_3^{\theta_3}]$. We have $\rho(e_3,\varphi(i))=\frac{\sqrt{2}}{2}\log 3$ and $\rho(e_0,\varphi(i))=\frac{\sqrt{2}}{2}\log\frac{4}{3}$. Note that $\varphi(i\frac{\sqrt{3}}{2})=e_0$ is different from $\varphi(i)$. The triangle $[e_0,e_3,e_3^{\theta_3}]$ is totally geodesic and the image of a hyperbolic triangle in $\mathfrak{H}_1$ by φ is inside the 1-cells.

Ideal 1-cells From the 0-cell e_3, a geodesic line segment $\{\frac{1}{2}\begin{pmatrix}1&1\\1&1\end{pmatrix}+it\begin{pmatrix}1&0\\0&1\end{pmatrix}\mid t\geq\omega_0\}$ emanates, which is included in the one-dimensional intersection $[X_1]\cap[X_2]\cap[X_3]\cap[Y_1]\cap[Y_2]$. This infinite line segment should be considered to have the virtual or ideal point at infinity. There is a similar geodesic curve emanating from $e_3^{\theta_3}$ to infinity.

6.3 Distance Between Special Pairs of Points

We want to calculate $r_i^{1/2}$ for special pairs Z, W with $\mathrm{Re}(Z)=\mathrm{Re}(W)$. If $Z=U+iY$, $W=U+iV$, the invariance of the cross ratio implies $\mathcal{R}(Z,W)=\mathcal{R}(iY,iV)$. Therefore we may assume that $\mathrm{Re}(Z)=\mathrm{Re}(W)=0$. By definition,

$$r_1+r_2=\mathrm{tr}(\mathcal{R}(iY,iV)),\quad r_1r_2=\det(\mathcal{R}(iY,iV))\quad(r_1\leq r_2).$$

Since $\mathcal{R}(iY,iV)=(Y-V)(Y+V)^{-1}(Y-V)(Y+V)^{-1}$, we have

$$r_1r_2=\det(Y-V)^2\det(Y+V)^{-2}=\det(Y-V)^2\det(Y+V)^2/\det(Y+V)^4.$$

In particular, $r_1^{1/2}r_2^{1/2}=|\det(Y-V)/\det(V+V)$.

Moreover let $(Y+V)^*$ be the cofactor matrix of $Y+V$, and let $M:=(Y-V)(Y+V)^*$. Then

$$M=\begin{pmatrix}\det(Y)-\det(V)+y_{11}v_{22}-y_{22}v_{11}, & 2(y_{12}v_{11}-y_{11}v_{12})\\ 2(y_{12}v_{22}-y_{22}v_{12}) & \det(Y)-\det(V)-(v_{11}v_{22}-y_{22}v_{11})\end{pmatrix}$$

and

$$R(iY,iV) = \frac{1}{\det(Y+V)^2}M^2, \quad r_1r_2 = \frac{\det(M)^2}{\det(Y+V)^4}, \quad r_1+r_2 = \frac{\mathrm{tr}(M^2)}{\det(Y+V)^2}.$$

Formulae Let $Y = \begin{pmatrix} y_{11} & y_{12} \\ y_{12} & y_{22} \end{pmatrix}$, $V = \begin{pmatrix} v_{11} & v_{12} \\ v_{12} & v_{22} \end{pmatrix}$. Then

$$\det(M) = \{\det(Y) - \det(V)\}^2 - (y_{11}v_{22} - y_{22}v_{11})^2 - 4(y_{11}v_{12} - y_{12}v_{11})(y_{22}v_{12} - y_{12}v_{11})$$

and

$$\mathrm{tr}(M^2) = 2\big[\{\det(Y)\det(V)\}^2 + (y_{11}v_{22} - y_{22}v_{22})^2 + 4(y_{11}v_{12} - y_{12}v_{11})(y_{22}v_{12} - y_{12}v_{22}))\big]$$

In particular

$$r_2^{1/2} = \frac{1}{\det(Y+V)}\big[\{(y_{11}v_{22} - y_{22}v_{11})^2 + 4(y_{11}v_{12} - y_{12}v_{11})(y_{22}v_{12} - y_{12}v_{22})\}^{1/2}$$
$$+|\det(Y) - \det(V)|\big]$$

and

$$\pm r_1^{1/2} = \frac{1}{\det(Y+V)}\big[\{(y_{11}v_{22} - y_{22}v_{11})^2 + 4(y_{11}v_{12} - y_{12}v_{11})(y_{22}v_{12} - y_{12}v_{22})\}^{1/2}$$
$$-|\det(Y) - \det(V)|\big]$$

The proof of these formulae is done by direct computation, and note the equality $(r_2^{1/2} \pm r_1^{1/2})^2 = r_1 + r_2 \pm 2(r_1r_2)^{1/2}$.

6.4 Examples of Totally Geodesic Submanifolds

Example 1. The submanifold identified with $\mathbf{R}_{>0} \times \mathfrak{H}_1$

Our subspace $\mathfrak{H}_2^\theta := \{Z \in \mathfrak{H}_2 \mid \mathrm{Re}(Z) = 0\}$ consisting of points fixed under the involution θ is a homogeneous space under the action of $A \in GL(2,\mathbf{R})$: $Z \mapsto AZ \cdot {}^tA$, hence should be isomorphic with the product $\mathbf{R}_{>0} \times \mathfrak{H}_1$. We want to see the relation between the standard coordinates of this product structure and the matrix coordinates in $\mathfrak{H}_2$. Given iY, $iV \in \mathfrak{H}_2^\theta$, then we let $\lambda = \det(Y)^{1/2}$, $\mu = \det(V)^{1/2}$ and for $\lambda^{-1}Y$, $\mu^{-1}V$ we can find points $z = x + iy$ and $w = u + iv$ in $\mathfrak{H}_1$ such that

$$\lambda^{-1/2}Y = \begin{pmatrix} \frac{1}{y} & \frac{x}{y} \\ \frac{x}{y} & y + \frac{x^2}{y} \end{pmatrix}, \text{ and } \mu^{-1/2}V = \begin{pmatrix} \frac{1}{v} & \frac{u}{v} \\ \frac{u}{v} & v + \frac{u^2}{v} \end{pmatrix}.$$

Then we have the following formulae:

Formula

(i) $\det(Y+V) = \lambda\mu\left[\frac{(\lambda-\mu)^2}{\lambda\mu} + \frac{1}{yv}|z-\bar{w}|^2\right]$,
(ii) $\det(Y-V) = \lambda\mu\left[\frac{(\lambda-\mu)^2}{\lambda\mu} - \frac{1}{yv}|z-w|^2\right]$,
(iii) $\det(M) = (\lambda\mu)^2\left[\frac{(\lambda-\mu)^2}{\lambda\mu}\frac{(\lambda+\mu)^2}{\lambda\mu} - \frac{1}{(yv)^2}|z-w|^2|z-\bar{w}|^2\right]$,
(iv) $\mathrm{tr}(M^2) = 2(\lambda\mu)^2\left[\frac{(\lambda-\mu)^2}{\lambda\mu}\frac{(\lambda+\mu)^2}{\lambda\mu} + \frac{1}{(yv)^2}|z-w|^2|z-\bar{w}|^2\right]$.

Since $r_1 + r_2 = \mathrm{tr}(M^2)/\det(Y+Z)^2$ and $2r_1^{1/2}r_2^{1/2} = 2\det(M)/\det(Y+Z)^2$, for $(r_2^{1/2} \pm r_1^{1/2})^2$ we have two formulae which are both perfect squares:

$$\begin{cases} 4(\lambda\mu)^2(\frac{\lambda-\mu}{\sqrt{\lambda\mu}})^2(\frac{\lambda+\mu}{\sqrt{\lambda\mu}})^2/\det(Y+Z)^2, \\ 4(\lambda\mu)^2\frac{1}{yv}|z-w|^2|z-\bar{w}|^2/\det(Y+V)^2. \end{cases}$$

Now we can choose a real parameter ϕ_1 such that

$$\cosh(\phi_1) = \frac{|z-\bar{w}|}{2\sqrt{\lambda\mu}}, \text{ and } \sinh(\phi_1) = \frac{|z-w|}{2\sqrt{\lambda\mu}},$$

Then we have

$$\frac{|z-\bar{w}|^2}{yv} = 4\cosh^2(\phi_1) = 2\cosh(2\phi_1) + 2.$$

Similarly we choose a real parameter ϕ_2 such that

$$\frac{1}{2}\left(\sqrt{\frac{\lambda}{\mu}} - \sqrt{\frac{\mu}{\lambda}}\right) = \sinh(\phi_2), \text{ and } \frac{1}{2}\left(\sqrt{\frac{\lambda}{\mu}} - \sqrt{\frac{\mu}{\lambda}}\right) = \cosh(\phi_2).$$

Then

$$\left(\frac{\lambda-\mu}{\sqrt{\lambda\mu}}\right)^2 = 4\sinh^2\phi_2 = 2\cosh(2\phi_2) - 2.$$

Then we have $\det(Y+V) = 2\{\cosh(2\phi_1) + \cosh(2\phi_2)\}$. For numerators, we have

$$2\frac{|z-w||z-\bar{w}|}{yv} = 4\sinh(2\phi_1), \text{ and } 2\left(\frac{\lambda}{\mu} - \frac{\mu}{\lambda}\right) = 4\sinh(2\phi_2)).$$

Thus we have

$$r_2^{1/2} \pm r_1^{1/2} = \begin{cases} 2\dfrac{\sinh(2\phi_1)}{\cosh(2\phi_1)+\cosh(2\phi_2)}, \text{ or} \\ 2\dfrac{\sinh(2\phi_2)}{\cosh(2\phi_1)+\cosh(2\phi_2)}. \end{cases}$$

Now let us introduce the other set of parameters ψ_1, ψ_2 by $2\phi_1 = \psi_1 + \psi_2$, $2\phi_2 = \psi_1 - \psi_2$; then by the addition formula for the hyperbolic triangle functions, the right-hand sides are equal to either of $\tanh(\psi_1) \pm \tanh(\psi_2)$. This means that $r_i^{1/2} = \tanh(\psi_i)$ or $r_i^{1/2} = \tanh(\psi_{3-i})$ for the pair $r_1^{1/2}, r_2^{1/2}$. Hence $\log t_i = 2\psi_i$ or $= 2\psi_{3-i}$.

Thus the distance between iY and iV is given as $2\sqrt{\psi_1^2 + \psi_2^2}$.

An obvious conclusion of the formula above is that $\mathfrak{H}_2^\theta$ is a totally geodesic submanifold in $\mathfrak{H}_2$. The images under some element of $Sp(2, \mathbf{R})$ are also totally geodesic, because they are an isometry of $\mathfrak{H}_2$. In particular, the intersections $[X1] \cap [X2] \cap [X3]$ and $[X1] \cap [X2] \cap [X3^\theta]$ are both totally geodesic. The former contains e_3, e_{18}, e_8, and e_9, and the latter $e_3^{\theta_3}$, $e_{18}^{\theta_3}$, $e_8^{\theta_3}$, e_9^{θ}.

Remark. Note that $\mathfrak{H}_2^\theta$ is also a Lagrangian submanifold in $\mathfrak{H}_2$ with respect to the Kähler 2-form associated to the invariant metric form.

Example 2. A totally geodesic surface Let us consider the 2-dimensional subspace of $\left\{Z \in \mathfrak{H}_2 \mid \mathrm{Re}(Z) = \frac{1}{2}\begin{pmatrix} 1 & 1 \\ 1 & 1 \end{pmatrix}\right\}$ given by

$$\left\{\frac{1}{2}\begin{pmatrix} 1 & 1 \\ 1 & 1 \end{pmatrix} + i\begin{pmatrix} y_{11} & y_{12} \\ y_{12} & y_{11} \end{pmatrix} \mid y_{11} > 0,\ y_{11}^2 - y_{12}^2 > 0\right\}$$

Let $A = e^{\theta_1}\begin{pmatrix} \cosh(\theta_2) & \sinh(\theta_2) \\ \sinh(\theta_2) & \cosh(\theta_2) \end{pmatrix}$ and put $V = AY \cdot {}^tA$. Then we have $r_i^{1/2} = \dfrac{\sinh(\theta_2) \pm \sinh(\theta_1)}{\cosh(\theta_1) + \cosh(\theta_2)}$. Here rewrite $2\theta_2 = \psi_1 + \psi_2$, $2\theta_1 = \psi_1 - \psi_2$. Then $r_i^{1/2} = \pm\tanh(p_j) = (i, j = 1, 2)$, and $t_i = e^{2\psi_j}$ and $\log t_i = 2\psi_j$. Therefore $\rho(iY, iV) = 2\sqrt{\psi_1^2 + \psi_2^2}$.

Some cells of lower dimension are automatically totally geodesic, when they are orbit of some reductive subgroup of $Sp(2, \mathbf{R})$, and the restriction of the Killing form of $Sp(2, \mathbf{R})$ to this subgroup is non-degenerate.

6.5 A Non-irreducible Intersection $[X1] \cap [X2] \cap [E2^\theta] \cap [E1^\theta]$ of Codimension 4

The contents of this subsection was noticed by Takahiro Hayata, though he worked with the Euclid metric on $\mathfrak{H}_2$.

The subset $[X1]\cap[X2]\cap[E2^\theta]\cap[E1^\theta]$ is defined by the relations $x_{11}=x_{22}=\frac{1}{2}$ and

$$f_6(Z):|(z_{11}-1)z_{22}-z_{12}^2|^2-1=0,$$

$$f_8(Z):|z_{11}(z_{22}-1)-z_{12}^2|^2-1=0,$$

corresponding to $W_6:[E1^\theta]$ and $W_8:[E2^\theta]$. Since $x_{11}=x_{22}=\frac{1}{2}$, this means

$$\{y_{11}y_{22}+\frac{1}{4}+x_{12}^2-y_{12}^2\}^2+(\frac{1}{2}y_{11}-\frac{1}{2}y_{22}-2x_{12}y_{12})^2-1=0,$$

$$\{y_{11}y_{22}+\frac{1}{4}+x_{12}^2-y_{12}^2\}^2+(-\frac{1}{2}y_{11}+\frac{1}{2}y_{22}-2x_{12}y_{12})^2-1=0.$$

Subtracting the second equation from the first to get

$$(\frac{1}{2}y_{11}-\frac{1}{2}y_{22}-2x_{12}y_{12})^2-(-\frac{1}{2}y_{11}+\frac{1}{2}y_{22}-2x_{12}y_{12})^2=4x_{12}y_{12}(y_{11}-y_{22})=0.$$

Therefore we have three irreducible components $S_1:x_{12}=0$, $S_2:y_{12}=0$, or $S_3:y_{11}-y_{22}=0$. Namely, under the condition $x_{11}=x_{22}=\frac{1}{2}$,

$$S_1:\quad x_{12}=0,\text{ and }\{\det(Y)+\frac{1}{4}\}^2+\frac{1}{4}(y_{11}-y_{22})^2-1=0,$$

$$S_2:\quad y_{12}=0,\text{ and }(x_{12}^2+\frac{1}{4}+y_{11}y_{22})^2+\frac{1}{4}(y_{11}-y_{22})^2=1,$$

$$S_3:\quad y_{11}=y_{22},\text{ and }(x_{12}^2+\frac{1}{4}+y_{11}^2-y_{12}^2)^2+4x_{12}^2y_{12}^2=1.$$

Then we have $e_0\in S_1\cap S_2\cap S_3$, $e_3\in S_2\cap S_3$, $e_5\in S_1\cap S_3$, and $e_{11}\in S_1$. Moreover if we set

$$\tilde{S}_1:=\left\{\begin{pmatrix}\frac{1}{2}&0\\0&\frac{1}{2}\end{pmatrix}+i\begin{pmatrix}y_{11}&y_{12}\\y_{12}&y_{11}\end{pmatrix}\mid y_{11}>0,\ y_{11}^2-y_{12}^2>0\right\},$$

then S_1 is an open subset of this totally geodesic surface (cf. Example 2 of the previous subsection). $[X1]\cap[X2]\cap[E2^\theta]\cap[E1^\theta]$ contains also e_{25} and $e_{25}^{\theta_3}$. Since e_{25} and $e_{25}^{\theta_3}$ are both outside of S_1 and S_2, these points belong to S_3.

Fact. Here $e_0,\ e_5$ in the curve $S_1\cap S_3$ which is a geodesic in $\tilde{S}_1$.

Guess. The geodesic triangle $e_0,\ e_{25},\ e_{25}^{\theta_3}$ would not be contained in the surface S_3.

6.6 *Some Computations of the Distance Inside* $[X1] \cap [X2]$

Here are some examples of the distance between two points. Say, we have $\rho(e_0, e_5) = \frac{\sqrt{2}}{2} \log 3$: Here $r_i = (2 - \sqrt{3})^2$ and $t_i = \sqrt{3}$ for both $i = 1,\ 2$.

The numerical data of the distances from e_0 are given as:
$\rho(e_0, e_5) = 0.776836\ldots$, $\rho(e_0, e_{11}) = 0.701250\ldots$, $\rho(e_0, e_{25}) = 0.803123\ldots$, $\rho(e_0, e_7) = 0.703282\ldots$, $\rho(e_0, e_9) = 1.068227\ldots$, $\rho(e_0, e_8) = 1.131457\ldots$.

We note here that 10 points

$$e_8,\ e_9,\ e_{25},\ e_7,\ e_5,\ e_{11},\ e_{25}^{\theta_3},\ e_7^{\theta_3},\ e_8^{\theta_3},\ e_9^{\theta_3}$$

all belong to the intersection $[X_1] \cap [X_2] \cap [Y_2]$ of dimension 3. Among them the five points

$$e_8,\ e_{25},\ e_5,\ e_{25}^{\theta_3},\ e_8^{\theta_3}$$

belong to the intersection

$$[X_1] \cap [X_2] \cap [Y_2] \cap [Y_1] \subset \left\{ \begin{pmatrix} \frac{1}{2} & x_{12} \\ x_{12} & \frac{1}{2} \end{pmatrix} + i \begin{pmatrix} y & \frac{1}{2}y \\ \frac{1}{2}y & y \end{pmatrix} \mid x_{12} \in [-\frac{1}{2}, +\frac{1}{2}],\ y > 0 \right\}$$

of dimension 2, and the remaining five

$$e_9,\ e_7,\ e_{11},\ e_7^{\theta_3},\ e_9^{\theta_3}$$

to the intersection

$$[X_1] \cap [X_2] \cap [Y_2] \cap [\mathbf{1}] \subset \left\{ \begin{pmatrix} \frac{1}{2} & x_{12} \\ x_{12} & \frac{1}{2} \end{pmatrix} + i \begin{pmatrix} \omega_0 & \frac{1}{2}\omega_0 \\ \frac{1}{2}\omega_0 & y_{22} \end{pmatrix} \mid x_{12} \in [-\frac{1}{2}, +\frac{1}{2}],\ y_{22} > \frac{1}{4}\omega_0 \right\}$$

of dimension 2. Heuristically speaking these 10 points take an arrangement of a "ladder":

$$\begin{array}{ccccccccc} e_8 & -- & e_{25} & -- & e_5 & -- & e_{25}^{\theta_3} & -- & e_8^{\theta_3} \\ | & & | & & | & & | & & | \\ e_9 & -- & e_7 & -- & e_{11} & -- & e_7^{\theta_3} & -- & e_9^{\theta_3} \end{array}$$

In spite of the fact that all of the above eight 0-cells belong to the 3-dimensional intersection $[X1] \cap [X2] \cap [Y2]$, they are not geodesically coplanar. But each small square belongs to the same 2-dimensional intersection. The whole ladder is a hyperbolic "origami."

6.6.1 Some Data of the Distance Around the Ladder Above

Here are tables of the distances between two among the above 8 points, which bring simple observations below.

The distances from e_5:

$\rho(e_5, e_{11}) = 0.198254\ldots,\quad \rho(e_5, e_{25}) = 0.250495\ldots, \rho(e_5, e_8) = 0.992858\ldots,$
$\rho(e_5, e_7) = 0.305867\ldots, \rho(e_5, e_9) = 1.002317\ldots.$

The distances from e_{11}:

$\rho(e_{11}, e_5) = 0.198254\ldots,\quad \rho(e_{11}, e_7) = 0.233865\ldots, \rho(e_{11}, e_9) = 0.951749\ldots,$
$\rho(e_{11}, e_8) = 0.991124\ldots, \rho(e_{11}, e_{25}) = 0.314231\ldots.$

The distances from e_{25}:

$\rho(e_{25}, e_5) = 0.250495\ldots,\quad \rho(e_{25}, e_{11}) = 0.314231\ldots, \rho(e_{25}, e_7) = 0.190781\ldots,$
$\rho(e_{25}, e_9) = 0.800095\ldots, \rho(e_{25}, e_8) = 0.778154\ldots.$

The distances from e_8:

$\rho(e_8, e_9) = 0.2638130\ldots,\quad \rho(e_8, e_7) = 0.7840507\ldots,\quad \rho(e_8, e_{11}) = 0.991124\ldots,$
$\rho(e_8, e_5) = 0.992858\ldots, \rho(e_8, e_{25}) = 0.778154\ldots,$

The distances from e_9:

$$\rho(e_9, e_7) = 0.7451460845226703$$

The distances from e_{18}:

$$\rho(e_{18}, e_8) = 0.504750\ldots,\quad \rho(e_{18}, e_9) = 0.437330\ldots,\quad \rho(e_{18}, e_{25}) = 0.859732\ldots,$$
$$\rho(e_{18}, e_7) = 0.818413\ldots, \rho(e_{18}, e_5) = 1.033202\ldots,\quad \rho(e_{18}, e_{11}) = 0.993451\ldots.$$

Observation 1. We have

$$\rho(e_8, e_8^{\theta_3}) = 1.788066\ldots, \rho(e_5, e_8) = 0.992858\ldots, \text{ and}$$
$$2\rho(e_5, e_8) - \rho(e_8, e_8^{\theta_3}) = 2 \times 0.992858\ldots - 1.788066\ldots = 0.197650\cdots > 0.$$

Hence the 3 points $e_5, e_8, e_8^{\theta_3}$ are never colinear. Similarly

$$\rho(e_{25}, e_{25}^{\theta_3}) = 0.499661\ldots, \rho(e_{25}, e_5) = 0.250453\ldots, \text{ and}$$
$$2\rho(e_{25}, e_5) - \rho(e_{25}, e_{25}^{\theta_3}) = 0.250453\ldots \times 2 - 0.499661\ldots$$
$$= 0.001906 - 0.000661 = 0.0013\ldots > 0.$$

Then $e_5, e_{25}, e_{25}^{\theta_3}$ are not colinear.

Remark. The precision of decimal computations of the distance might be not precise enough here.

Observation 2. We have

$$\rho(e_7, e_7^{\theta_3}) = 0.46644836434646214 \text{ and } \rho(e_{11}, e_7) = 0.23386505331938626, \text{i.e.,}$$

$$2\rho(e_{11}, e_7) - \rho(e_7, e_7^{\theta_3}) = 0.0012\cdots > 0.$$

This means that e_{11} is quite near to the middle point between e_7 and $e_7^{\theta_3}$ but different. Therefore the three points $e_{11}, e_7, e_7^{\theta_3}$ are coplanar, but not colinear.

Remark. We can analyze the intersection $[X1^\theta] \cap [X2^\theta]$ which is diffeomorphic to $[X1] \cap [X2]$ by the map $\theta_2\theta_3$. Similarly the intersections $[X1] \cap [X2^\theta]$ and $[X1^\theta \cap X2]$ are also diffeomorphic. But they are combinatorially different from $[X1] \cap [X2]$.

Observation 3. We have

$$\mathcal{S}[X1^\theta] \cap \mathcal{S}[X2^\theta] = \{e_i^{\theta_1\theta_2}\ (i = 0, 5, 11),\ \ e_j^{\theta_1\theta_2},\ e_j^\theta\ (j = 3, 7, 8, 9, 18, 25)\}$$

with cardinality 15, and

$$\mathcal{S}[X1] \cap \mathcal{S}[X2^\theta] = \{e_0^{\theta_2},\ e_{11}^{\theta_2},\ e_i^\theta,\ e_i^{\theta_2\theta_3}\ (i = 3, 7, 8, 9, 18, 25)\}$$

with cardinality 14.

References

1. Gottschling, Erhard, Explizite Bestimmung der Randflächen des Fundamentalbereiches des der Modulgruppe zweiten Grades. Math. Ann. **138**, 103–124 (1959)
2. T. Hayata, T. Oda, T. Yatouda, Zero cells of the Siegel-Gottschling fundamental domain of genus 2. Experimental Math. **21**(3), 266–279 (2012)
3. J.-P. Labesse, J. Schwermer (eds), *Cohomology of Arithmetic Groups and Automorphic Forms*, Proceedings of Conference held in Luminy/Marseille, May 1989, Lecure Notes im Math. vol. 1447 (Springer, NewYork)
4. T. Oda, in *Geometry and Analysis of Automorphic Forms of Several Variables*, Cohomology Of Siegel Modular Varieties Of Genus 2 and Corresponding Automorphic Forms (World Scientific, Singapore, 2011) pp. 211–253
5. S. Takashi, *Riemannian Geometry*, Translations of Mathematical Monographs, vol. 149 (American Mathematical Society, Providence 1992)
6. C.L. Siegel, Symplectic geometry. Amer. J. Math. **65**, 1–86 (1943); the book of Academic Press with the same title (1964)

Nonvanishing of L-Functions Associated to Cusp Forms of Half-Integral Weight

B. Ramakrishnan and Karam Deo Shankhadhar

Mathematics Subject Classification (2010): Primary 11F37, 11F66; Secondary 11F25, 11F30.

1 Introduction

In [4], W. Kohnen proved that, given a real number t_0 and a positive real number ϵ, for all k large enough the sum of the functions $L^*(f, s)$ with f running over a basis of (properly normalized) Hecke eigenforms of weight k does not vanish on the line segment $\operatorname{Im} s = t_0$, $(k-1)/2 < \operatorname{Re}(s) < k/2-\epsilon, k/2+\epsilon < \operatorname{Re}(s) < (k+1)/2$. As a consequence, he proved that for any such point s, for k large enough there exists a Hecke eigenform of weight k on $SL_2(\mathbb{Z})$ such that the corresponding L-function value at s is nonzero. Using similar methods, in [8], A. Raghuram generalized Kohnen's result for the average of L-functions over a basis of newforms (of integral weight) of level N with primitive character modulo N. In this article, we extend Kohnen's method to forms of half-integral weight. As a consequence, we show that for any given point s inside the critical strip $k/2 - 1/4 < \operatorname{Re}(s) < k/2 + 3/4$, there exists a Hecke eigen cusp form f of half-integral weight $k + 1/2$ on $\Gamma_0(4N)$ with

B. Ramakrishnan (✉)
Harish-Chandra Research Institute, Chhatnag Road, Jhunsi, Allahabad 211 019, India
e-mail: ramki@hri.res.in

K.D. Shankhadhar
Harish-Chandra Research Institute, Chhatnag Road, Jhunsi, Allahabad 211 019, India

The Institute of Mathematical Sciences, IV cross road, CIT Campus,
Taramani, Chennai 600 113 (India)
e-mail: karam@imsc.res.in

B. Heim et al. (eds.), *Automorphic Forms*, Springer Proceedings
in Mathematics & Statistics 115, DOI 10.1007/978-3-319-11352-4_16

character ψ such that the corresponding L-function value at s is nonzero and the first Fourier coefficient of f is nonzero. It should be noted that the normalization of Fourier coefficients of forms of half-integral weight is still an open question. Also, contrary to the result of Kohnen in the case of integral weight modular forms on $SL_2(\mathbb{Z})$, we get the nonvanishing result inside the critical strip including the central line (see Remark 4.1). Our results are obtained for N sufficiently large if k is fixed and vice versa. In particular when $N = 1$, for sufficiently large k, either f is a newform in the full space or f is a Hecke eigenform in the Kohnen plus space.

2 Notations and Main Theorems

Let $N \geq 1$, $k \geq 3$ be integers and ψ be an even Dirichlet character modulo $4N$. Let $S_{k+1/2}(4N, \psi)$ be the space of cusp forms of weight $k + 1/2$, level $4N$ with character ψ ([2, 10]). Let $L(f, s)$ be the L-function associated to the cusp form $f \in S_{k+1/2}(4N, \psi)$ defined by $L(f, s) = \sum_{n\geq 1} a_f(n)n^{-s}$, where $a_f(n)$ denotes the n-th Fourier coefficient of f. Then by [6, Proposition 1], the completed L-function defined by $L^*(f, s) := (2\pi)^{-s}(\sqrt{4N})^s\Gamma(s)L(f, s)$ has the following functional equation

$$L^*(f|H_{4N}, k + 1/2 - s) = L^*(f, s), \tag{1}$$

where H_{4N} is the Fricke involution on $S_{k+1/2}(4N, \psi)$ defined by

$$f|H_{4N}(z) = i^{k+1/2}(4N)^{-k/2-1/4}z^{-k-1/2}f(-1/4Nz).$$

For $f, g \in S_{k+1/2}(4N, \psi)$, let $\langle f, g\rangle$ denote the Petersson scalar product of f and g. It is known that the space $S_{k+1/2}(4N, \psi)$ has an orthogonal basis of Hecke eigenforms with respect to all Hecke operators $T(p^2)$, $p \nmid 2N$. Let $\{f_1, f_2, \ldots, f_d\}$ be such an orthogonal basis of Hecke eigenforms, where d is the dimension of the space $S_{k+1/2}(4N, \psi)$ (see, for example, [10]). Let K be the operator defined by $f|K(z) = \overline{f(-\overline{z})}$. Since $KH_{4N} = H_{4N}K$ on $S_{k+1/2}(4N, \psi)$, we have $f|(KH_{4N})^2 = f$. Also, the operators K and H_{4N} commute with the Hecke operators $T(p^2)$, $p \nmid 2N$. Therefore, for the basis elements f_j, $1 \leq j \leq d$, we may assume that $f_j|KH_{4N} = \lambda_{f_j} f_j$, where $\lambda_{f_j} = \pm 1$.

We now state the main results of this article.

Theorem 2.1. *Let $N \geq 1$ be a fixed integer. Let $\{f_1, f_2, \ldots, f_d\}$ be an orthogonal basis as above. Let $r_0 \in \mathbb{R}$. Then there exists a constant $C = C(r_0)$ depending only on r_0 such that for $k > C$, the function*

$$\sum_{j=1}^{d} \frac{L^*(f_j, s)}{\langle f_j, f_j\rangle}\lambda_{f_j}a_{f_j}(1)$$

does not vanish for any point $s = \sigma + ir_0$ with $k/2 - 1/4 < \sigma < k/2 + 3/4$.

Theorem 2.2. *Let $k \geq 3$ be a fixed integer. Let $\{f_1, f_2, \ldots, f_d\}$ be an orthogonal basis as above. Let $r_0 \in \mathbb{R}$. Then there exists a constant $C' = C'(r_0)$ depending only on r_0 such that for $N > C'$, the function*

$$\sum_{j=1}^{d} \frac{L^*(f_j, s)}{\langle f_j, f_j \rangle} \lambda_{f_j} a_{f_j}(1)$$

does not vanish for any point $s = \sigma + i r_0$ with $k/2 - 1/4 < \sigma < k/2 + 3/4$.

The following corollary is an easy consequence of the above two theorems.

Corollary 2.3. *Let s_0 be a point inside the critical strip $k/2 - 1/4 < \mathrm{Re}(s_0) < k/2 + 3/4$. If either k or N is suitably large, then there exists a Hecke eigenform f belonging to $S_{k+1/2}(4N, \psi)$ such that $L(f, s_0) \neq 0$ and $a_f(1) \neq 0$.*

Remark 2.4. Though we consider Hecke eigenforms of half-integral weight, the L-function corresponding to such a Hecke eigenform does not have an Euler product.

3 Proof

The proof is on the same lines as that of Kohnen [4] and so we give only a sketch. First, let us recall the Poincaré series in $S_{k+1/2}(4N, \psi)$. We define the n-th Poincaré series in $S_{k+1/2}(4N, \psi)$ as follows.

$$P_{n,k+1/2,4N,\psi}(z) = \frac{1}{2} \sum_{\substack{(c,d)\in\mathbb{Z}^2 \\ (c,d)=1, 4N|c}} \overline{\psi}(d) \left(\frac{c}{d}\right) \left(\frac{-4}{d}\right)^{k+1/2} (cz+d)^{-(k+1/2)} e\left(n \frac{a_0 z + b_0}{cz+d}\right), \tag{2}$$

where in the summation above, for each coprime pair (c, d) and $4N|c$, we make a fixed choice of $(a_0, b_0) \in \mathbb{Z}^2$ with $a_0 d - b_0 c = 1$ and $e(x)$ stands for $e^{2\pi i x}$. We have the following characterization of the Poincaré series.

$$\langle f, P_{n,k+1/2,4N,\psi} \rangle = \frac{\Gamma(k - 1/2)}{i_{4N} (4\pi n)^{k-1/2}} a_f(n), \qquad f \in S_{k+1/2}(4N, \psi), \tag{3}$$

where i_{4N} is the index of $\Gamma_0(4N)$ in $SL_2(\mathbb{Z})$.
Next, let us define the kernel function for the special values of the L-function associated to a cusp form of half-integral weight. A similar function for forms of integral weight was considered by Kohnen [4]. Let $z \in \mathcal{H}$ and $s \in \mathbb{C}$ with $1 < \sigma < k - 1/2, \sigma = \mathrm{Re}(s)$. Define

$$R_{s;k,N,\psi}(z) = \gamma_k(s) {\sum}' \overline{\psi}(d) \left(\frac{c}{d}\right) \left(\frac{-4}{d}\right)^{k+1/2} (cz+d)^{-(k+1/2)} \left(\frac{az+b}{cz+d}\right)^{-s}, \tag{4}$$

where

$$\gamma_k(s) = \frac{1}{2}\, e^{\pi i s/2}\, \Gamma(s)\Gamma(k+1/2-s), \tag{5}$$

and the sum $\sum'$ varies over all matrices $\begin{pmatrix} a & b \\ c & d \end{pmatrix} \in \Gamma_0(4N)$. The condition $1 < \sigma < k - 1/2$ ensures that the above series converges absolutely and uniformly on compact subsets of $\mathcal{H}$ and hence it represents an analytic function on $\mathcal{H}$. The function $R_{s,\psi}(z) := R_{s;k,N,\psi}(z)$ is a cusp form in $S_{k+1/2}(4N, \psi)$.
For a given $c, d \in \mathbb{Z}$ with $\gcd(c,d) = 1$ and $4N|c$, we choose a_0, b_0 such that $a_0 d - b_0 c = 1$. Then any other solution a, b of $ad - bc = 1$ is given by $a = a_0 + nc$ and $b = b_0 + nd$ for some $n \in \mathbb{Z}$. Hence,

$$R_{s,\psi}(z) = \gamma_k(s) \sum_{\substack{(c,d)=1\\ 4N|c}} \sum_{n\in\mathbb{Z}} \overline{\psi}(d) \left(\frac{c}{d}\right)\left(\frac{-4}{d}\right)^{k+1/2} (cz+d)^{-k-1/2} \left(\frac{a_0 z + b_0}{cz+d} + n\right)^{-s}. \tag{6}$$

Using Lipschitz's formula

$$\sum_{n=-\infty}^{\infty} (z+n)^{-s} = \frac{e^{-\pi i s/2}(2\pi)^s}{\Gamma(s)} \sum_{n\geq 1} n^{s-1} e(nz) \qquad (z \in \mathcal{H}, \sigma > 1), \tag{7}$$

we get

$$\begin{aligned} R_{s,\psi}(z) &= \gamma_k(s) \sum_{\substack{(c,d)=1\\ 4N|c}} \overline{\psi}(d) \left(\frac{c}{d}\right)\left(\frac{-4}{d}\right)^{k+1/2} (cz+d)^{-k-1/2} \frac{e^{-\pi i s/2}(2\pi)^s}{\Gamma(s)} \\ &\qquad \times \sum_{n\geq 1} n^{s-1} e\left(n \frac{a_0 z + b_0}{cz+d}\right) \\ &= (2\pi)^s \Gamma(k+1/2-s) \sum_{n\geq 1} n^{s-1} \\ &\qquad \times \frac{1}{2} \sum_{\substack{(c,d)=1\\ 4N|c}} \overline{\psi}(d) \left(\frac{c}{d}\right)\left(\frac{-4}{d}\right)^{k+1/2} (cz+d)^{-k-1/2} e\left(n \frac{a_0 z + b_0}{cz+d}\right). \end{aligned} \tag{8}$$

Here, we have used the absolute convergence of the above sum in the region $1 < \sigma < k - \beta - 1/2$ and so the interchange of summations is allowed, where $\beta = k/2 - 1/28$ is the exponent of the estimate for the Fourier coefficients of a cusp form of weight $k + 1/2$ on $\Gamma_0(4N)$ obtained by Iwaniec [1]. Thus, for $1 < \sigma < k - \beta - 1/2$,

$$R_{s,\psi}(z) = (2\pi)^s \Gamma(k+1/2-s) \sum_{n\geq 1} n^{s-1} P_{n,k+1/2,4N,\psi}(z). \tag{9}$$

Using Eq. (3) with the last equation, we get for $1 < \sigma < k - \beta - 1/2$,

$$\langle f, R_{\overline{s},\psi}\rangle = \frac{\pi\Gamma(k-1/2)}{i_{4N}2^{k-3/2}(4N)^{k/2+1/4-s/2}}L^*(f, k+1/2-s),$$

for all $f \in S_{k+1/2}(4N,\psi)$. Using this, we have

$$R_{s,\psi} = \frac{2^{-2k+1+s}\pi\Gamma(k-1/2)}{i_{4N}N^{k/2+1/4-s/2}}\sum_{j=1}^{d}\frac{L^*(f_j|K, k+1/2-s)}{\langle f_j, f_j\rangle}f_j, \tag{10}$$

where the sum varies over the orthogonal basis $\{f_j\}$. Using $f_j|KH_{4N} = \lambda_{f_j}f_j$ together with the functional equation (1), we get for $1 < \sigma < k - \beta - 1/2$,

$$R_{s,\psi} = \frac{2^{-2k+1+s}\pi\Gamma(k-1/2)}{i_{4N}N^{k/2+1/4-s/2}}\sum_{j=1}^{d}\frac{L^*(f_j, s)}{\langle f_j, f_j\rangle}\lambda_{f_j}f_j. \tag{11}$$

This equality has been established for $1 < \sigma < k-\beta-1/2$. Since the right-hand side is an entire function of s, this gives an analytic continuation of the kernel function $R_{s,\psi}$ for all $s \in \mathbb{C}$.

Next, we need the Fourier expansion of the function $R_{s,\psi}$. In an earlier version of [6], the Fourier expansion of the function $R_{s,\psi}(z)$ was derived, which we present here. Let

$$R_{s,\psi}(z) = \sum_{n\geq 1} a_{s,\psi}(n)e^{2\pi i n z}$$

be the Fourier expansion of $R_{s,\psi}$, where the Fourier coefficients $a_{s,\psi}(n)$ are given by

$$\begin{aligned} a_{s,\psi}(n) &= (2\pi)^s\Gamma(k+1/2-s)n^{s-1} + e^{\pi i s/2}(-2\pi i)^{k+1/2}n^{k-1/2} \\ &\times \sum_{\substack{(a,c)\in\mathbb{Z}^2, ac\neq 0\\ \gcd(a,c)=1, 4N|c}} \psi(a)\left(\frac{c}{a}\right)\left(\frac{-4}{a}\right)^{k+1/2} c^{s-k-1/2}a^{-s}e^{2\pi i n a'/c} \\ &\quad {}_1f_1(s, k+1/2; -2\pi i n/ac), \end{aligned} \tag{12}$$

where a' is an integer which is the inverse of a modulo c and

$${}_1f_1(\alpha,\beta;z) = \frac{\Gamma(\alpha)\Gamma(\beta-\alpha)}{\Gamma(\beta)}{}_1F_1(\alpha,\beta;z). \tag{13}$$

Here ${}_1F_1(\alpha,\beta;z)$ is the Kummer's degenerate hypergeometric function.

We now give the proof of our theorems. Assume that $\sum_j \frac{L^*(f_j,s)}{\langle f_j,f_j\rangle}\lambda_{f_j}a_{f_j}(1) = 0$, for s as in the theorem. This implies that (for the values of s) the first Fourier coefficient of $R_{s,\psi}$ is zero. Dividing by $(2\pi)^s\Gamma(k+1/2-s)$, we obtain

$$1 + e^{\pi i s/2} \frac{(2\pi)^{k+1/2-s}}{i^{k+1/2}\Gamma(k+1/2-s)} \sum_{\substack{a,c\in\mathbb{Z},ac\neq 0\\(a,c)=1,4N|c}} \psi(a)\left(\frac{c}{a}\right)\left(\frac{-4}{a}\right)^{k+1/2} \frac{c^{s-k-1/2}}{a^s}$$
$$\times e^{2\pi i a'/c}\,{}_1f_1(s,k+1/2;-2\pi i/ac) = 0. \tag{14}$$

In particular, let $s = k/2 + 1/4 - \delta + ir_0$, where $0 \leq \delta < 1/2$. Now, one has

$$|{}_1f_1(s,k+1/2;-2\pi i/ac)| \leq 1$$

(see [4]). Taking the absolute value in (14) and using the above estimate, we get

$$1 \leq A(r_0) \frac{\pi^{k/2+1/4+\delta}}{|\Gamma(k/2+1/4+\delta-ir_0)|} \frac{1}{(2N)^{k/2+1/4+\delta}} \left(\sum_{\substack{a,c\in\mathbb{Z},ac\neq 0,\\(a,2Nc)=1}} \frac{1}{a^{k/2+1/4-\delta}\cdot c^{k/2+1/4+\delta}} \right)$$

$$\leq A(r_0)\, B \frac{\pi^{k/2+1/4+\delta}}{|\Gamma(k/2+1/4+\delta-ir_0)|} \frac{1}{(2N)^{k/2+1/4+\delta}},$$

where $A(r_0)$ is a constant depending only on r_0 and $B > 0$ is an absolute constant. To prove Theorem 2.1, we fix N and allow k tend to infinity and for the proof of Theorem 2.2, we fix k and allow N tend to infinity. In either case, the right-hand side goes to zero (for fixed N one should use the Stirling's approximation), a contradiction. This completes the proof.

4 Remarks

Remark 4.1. In [4], the nonvanishing result was obtained for s inside the critical strip with the condition that $\mathrm{Re}(s) \neq k/2$, the center of the critical strip. However, since the level of the modular forms considered in this paper is greater than 1, we need not assume this condition and our results are valid for all s inside the critical strip $k/2 - 1/4 < \mathrm{Re}(s) < k/2 + 3/4$. We also remark that the same is true in [8], since the level M is greater than 1. The reason is as follows. When $M = 1$ (i.e., when one considers the case of forms of integral weight on $SL_2(\mathbb{Z})$), while deriving the Fourier expansion of the function R_s, the term corresponding to $ac = 0$ has two contributions ($c = 0$ and $a = 0$). Therefore, in the estimation of the first Fourier coefficient, there is an extra term on the right-hand side (see [4, p. 189, Eq.(10)]). Due to the appearance of this extra term, in order to get a contradiction, the central values have to be omitted. Since the case $a = 0$ does not arise for the levels $M > 1$, we do not get the extra term on the right-hand side in the estimation. Therefore, this gives the advantage of considering all the values of s inside the critical strip. In particular, one obtains nonvanishing results for forms at the center of the critical strip when the level is greater than 1.

Remark 4.2. The average sum in Theorem 2.1 (and also in Theorem 2.2) contains an extra factor λ_f (and of course the first Fourier coefficient), which does not appear in Kohnen's result. In the case of level 1, we have the functional equation $L^*(f, k-s) = (-1)^{k/2}L^*(f, s)$, whereas when the level M is greater than 1, we have a different functional equation in the sense that on the one side we have $L^*(f, s)$ and on the other side we have $L^*(f|H_M, k-s)$ and therefore, in the final form of the functional equation, the root number will depend on the function, especially the eigenvalue of f under the Fricke involution H_M. (In the case of half-integral weight, the Fricke involution is $H_M = H_{4N}$ in our notation.) Therefore, we will have an extra factor, which we call λ_f. Note that the extra factor corresponding to the eigenvalues under H_M also appears in Raghuram's results (see [8]). Since normalization of Fourier coefficients is not known in the case of half-integral weight, we also have the first Fourier coefficients appearing in the average sum.

Remark 4.3. Let us consider the space $S_{k+1/2}(4N, \psi)$, where ψ is an even primitive Dirichlet character modulo $4N$. Then it is known from the work of Serre and Stark [9] that the space $S_{k+1/2}(4N, \psi)$ is the space of newforms. Hence, the orthogonal basis consists of newforms. In this case, the Hecke eigenform f in Corollary 2.3 will be a newform of level $4N$.

Remark 4.4. Let us consider the case $N = 1$. That is, we consider the space $S_{k+1/2}(4)$. Let $\{f_1, f_2, \ldots, f_{d_1}\}$ be an orthogonal basis of $S^{new}_{k+1/2}(4)$ which are newforms (see [7]). Let $\{g_1, g_2, \ldots, g_{d_2}\}$ be an orthogonal basis of $S^{+}_{k+1/2}(4)$ (see [3]), which are Hecke eigenforms such that the set $\{g_1 \pm g_1|W(4), \ldots, g_{d_2} \pm g_{d_2}|W(4)\}$ forms an orthogonal basis of $S^{old}_{k+1/2}(4)$. Here, $d_1 + 2d_2 = d$ is the dimension of the space $S_{k+1/2}(4)$ and $W(4)$ is the Atkin-Lehner W-operator for the prime $p = 2$ on $S_{k+1/2}(4)$. Thus, an orthogonal basis of Hecke eigenforms for the space $S_{k+1/2}(4)$ is given as follows:

$$\{f_1, f_2, \ldots, f_{d_1}, g_1 \pm g_1|W(4), g_2 \pm g_2|W(4), \ldots, g_{d_2} \pm g_{d_2}|W(4)\}.$$

In this case, we get the following result as a consequence of Theorem 2.1. Let s_0 be a point inside the critical strip $k/2 - 1/4 < \mathrm{Re}(s_0) < k/2 + 3/4$. If k is suitably large, then there exists a j, with $1 \le j \le d_1$ or $1 \le j \le d_2$ such that

$$L(f_j, s_0) \neq 0, a_{f_j}(1) \neq 0 \text{ or } L(g_j \pm g_j|W(4), s_0) \neq 0, a_{g_j}(1) \pm 2^{-k}a_{g_j}(4) \neq 0. \tag{15}$$

For the last assertion in the above equation (15), we use the fact that $W(4) = 2^{-k}U(4)$ on $S^{+}_{k+1/2}(4)$, where $U(4)$ is the Hecke operator for $p = 2$ on $S_{k+1/2}(4)$. Note that $W(4) = H_4$ on $S_{k+1/2}(4)$ and therefore, in the second case of (15), for a j with $1 \le j \le d_2$, it follows from the functional equation that either $L(g_j, s_0) \neq 0$ or $L(g_j, k+1/2-s_0) \neq 0$. Hence, for any given point s inside the critical strip, our theorem gives (for sufficiently large k) the existence of a newform f in $S^{new}_{k+1/2}(4)$ such that $L(f, s) \neq 0$ or a Hecke eigenform g in the plus space $S^{+}_{k+1/2}(4)$ such that $L(g,s) \neq 0$ or $L(g,k+1/2-s) \neq 0$. Correspondingly, we also get the nonvanishing of the first Fourier coefficient (if it is a newform in $S^{new}_{k+1/2}(4)$) or the first or the 4th Fourier coefficient (if it is a Hecke eigenform in $S^{+}_{k+1/2}(4)$).

4.1 Additional Remarks

In view of a recent result of Kumar [5], we make the following additional remarks.

Let N be any positive integer and let $E(k+1/2, 4N)$ denote the set of all Hecke eigenforms $h \in S_{k+1/2}(4N)$ (the vector space of cusp forms of weight $k+1/2$ on $\Gamma_0(4N)$ with trivial character) such that the L-value $L(h, k/2+1/4) \neq 0$. In [5, §4], the following results are obtained.

Theorem 4.5. *For $h \in E(k+1/2, 4N)$ one has*

$$\frac{L'(h,k/2+1/4)}{L(h,k/2+1/4)} = -\Psi(k/2+1/4) + \log(\pi),$$

where Ψ is the logarithmic derivative of the gamma function Γ. Further, for such an h, $L'(h,k/2+1/4) \neq 0$ and the real number

$$\exp\left(\frac{L'(h,k/2+1/4)}{L(h,k/2+1/4} + \Psi(k/2+1/4)\right)$$

is transcendental. Moreover, considering the quotient $\frac{L'(h,k/2+1/4)}{L(h,k/2+1/4)}$ as a function of k one deduces that the function $\frac{L'(h,k/2+1/4)}{L(h,k/2+1/4)} + \Psi(k/2+1/4)$ is independent of k and the function $\frac{L'(h,k/2+1/4)}{L(h,k/2+1/4)} \to -\infty$ as $k \to \infty$.

Proposition 4.6. *If $\frac{L'(h_0,k_0/2+1/4)}{L(h_0,k_0/2+1/4)}$ is algebraic (resp. transcendental) for some $h_0 \in E(k_0+1/2, 4N)$, then $\frac{L'(h,k/2+1/4)}{L(h,k/2+1/4)}$ is algebraic (resp. transcendental) for all $h \in E(k+1/2, 4N)$ and for all $k \in \mathbb{N}$ with $k \equiv k_0 \pmod 2$.*

Note: In [5, Sect. 4], the above results are proved for the case $N = 1$ and was remarked that a similar method will lead to the results for N square-free. In fact following the same arguments one can get the above results for any positive integer N.

The above results (Theorem 4.5 and Proposition 4.6) are about the properties of the functions in the set $E(k+1/2, 4N)$ under the assumption that it is a nonempty set. It is to be noted that the main results of our present work guarantees the existence of an element in the set $E(k+1/2, 4N)$. In fact, by taking $r_0 = 0$ and $\sigma = k/2+1/4$ in Theorem 2.1 and Theorem 2.2, we get the following two corollaries.

Corollary 4.7. *There exists a constant C_1 such that for any $k > C_1$ and any N, the set $E(k+1/2, 4N)$ is nonempty.*

Corollary 4.8. *There exists a constant $C_2 > 0$ such that for any $N > C_2$ and any $k \geq 3$, the set $E(k+1/2, 4N)$ is nonempty.*

Acknowledgements The authors thank the referee for making valuable suggestions. The first author enjoyed the conference and he thanks the organizers for the excellent organization and also for their warm hospitality.

References

1. H. Iwaniec, Fourier coefficients of modular forms of half-integral weight. Invent. Math. **87**, 385–401 (1987)
2. N. Koblitz, *Introduction to Elliptic Curves and Modular Forms* (Springer, NewYork, 1997)
3. W. Kohnen, Modular forms of half integral weight on $\Gamma_0(4)$. Math. Ann. **248**, 249–266 (1980)
4. W. Kohnen, Non-vanishing of Hecke L-functions associated to cusp forms inside the critical strip. J. Number Theory **67**, 182–189 (1997)
5. N. Kumar, Non-vanishing of derivatives of certain modular L-functions. Funct. Approx. Comment. Math. **51**(1), 121–132 (2014)
6. M. Manickam, V. Kumar Murty, B. Ramakrishnan, Twisted averages of L-functions, Preprint 2012.
7. M. Manickam, B. Ramakrishnan, T.C. Vasudevan, On the theory of newforms of half-integral weight. J. Number Theory **34**, 210–224 (1990)
8. A. Raghuram, *Non-vanishing of L-functions of cusp forms inside the critical strip*, Number theory, Ramanujan Math. Soc. Lect. Notes Ser., **1** (Ramanujan Math. Soc., Mysore, 2005) pp. 97–105
9. J.-P. Serre, H. M. Stark, *Modular Forms of Weight* $1/2$, Modular functions of one variable, VI (Proc. Second Internat. Conf., Univ. Bonn, Bonn, 1976), Lecture Notes in Math., vol. 627 (Springer, Berlin, 1977) pp. 27–67
10. G. Shimura, On Modular forms of half integral weight. Annals of Math. **97**, 440–481 (1973)

Hermitian Modular Forms of Degree 2 over the Eisenstein Integers

Martin Woitalla

1 Hermitian Modular Group and the Attached Half-Space

In this section we introduce the Hermitian modular group with respect to the Eisenstein integers. For more details we refer to [7] where the theory is analogously developed for the Gaussian integers. Let $l \in \mathbb{N}$ and $R \subset \mathbb{C}$ be a subring of the complex numbers and let $I_l \in \mathrm{Mat}(l, R)$ be the identity of the matrix ring. We extend the usual complex conjugation componentwise to $\mathrm{Mat}(l, R)$ and define

$$A^* := \overline{A}^{\mathrm{t}} \quad \text{for all} \quad A \in \mathrm{Mat}(l, R)\,.$$

For the rest of this text we will concentrate ourselves on the case where R is the ring of integers of Eisenstein's number field $\mathbb{Q}(\sqrt{-3})$, which we will denote by $\mathfrak{o}$. This ring is given by

$$\mathfrak{o} := \mathbb{Z} + \mathbb{Z}\omega \subset \mathbb{C} \quad \text{where} \quad \omega := \frac{1 + i\sqrt{3}}{2}.$$

We denote by

$$\mathrm{N} : \mathbb{Q}(\sqrt{-3}) \to \mathbb{Q} \quad , \quad \mathrm{N}(a) := a\bar{a}$$

the norm of this number field. Then $\mathfrak{o}$ equipped with N forms a Euclidian ring. For any $n \in \mathbb{N}$ we consider the matrix

$$J := \begin{pmatrix} 0 & I_n \\ -I_n & 0 \end{pmatrix} \in \mathrm{Mat}(2n, \mathbb{C})\,.$$

M. Woitalla (✉)
Lehrstuhl A für Matheamtik, RWTH University, Templergraben 55, 52062 Aachen, Germany
e-mail: Martin.Woitalla@matha.rwth-aachen.de

B. Heim et al. (eds.), *Automorphic Forms*, Springer Proceedings
in Mathematics & Statistics 115, DOI 10.1007/978-3-319-11352-4_17

We have a right group action of $\mathrm{GL}(n, \mathbb{C})$ on $\mathrm{Mat}(n, \mathbb{C})$ which is given by

$$(M, A) \mapsto A[M] := M^* A M$$

The (special) *unitary group of signature* (n, n) is defined as

$$\begin{aligned} \mathrm{U}(n,n) :=& \quad \{M \in \mathrm{GL}(2n, \mathbb{C}) \mid J[M] = J\} \\ =& \quad \{M^* \in \mathrm{GL}(2n, \mathbb{C}) \mid J[M] = J\} \end{aligned}$$

respectively

$$\mathrm{SU}(n,n) = \{M \in \mathrm{SL}(2n, \mathbb{C}) \mid J[M] = J\}$$

and $\mathrm{U}(n,n)$ hence equals the stabiliser of the Hermitian form represented by the matrix $\mathrm{i}J$. The corresponding modular group of degree n with respect to $\mathfrak{o}$ is given by

$$\Gamma_n := \mathrm{SU}(n,n) \cap \mathrm{Mat}(2n, \mathfrak{o})\,.$$

It is a well-known fact that Γ_n is generated by the elements

$$J \quad , \quad \begin{pmatrix} I & H \\ 0 & I \end{pmatrix} \quad , \quad \begin{pmatrix} U^* & 0 \\ 0 & U^{-1} \end{pmatrix}$$

where H runs through the lattice of integral Hermitian matrices $\mathrm{Her}(n, \mathfrak{o})$ and U runs through the group $\mathrm{SL}_n(\mathfrak{o})$. We will also refer to these elements as *canonical generators*. We now consider the action of the unitary group on the half-space by fractional linear transformations. The *Hermitian half-space* $\mathcal{H}_n$ is a natural generalisation of the usual upper half-plane in $\mathbb{C}$ and consists of all matrices

$$Z \in \mathrm{Mat}(n, \mathbb{C}) \quad \text{with} \quad \frac{1}{2\mathrm{i}}(Z - Z^*) > 0\,.$$

The special unitary group $\mathrm{SU}(n,n)$ acts on $\mathcal{H}_n$ as a group of biholomorphic automorphisms via

$$Z \mapsto M\langle Z\rangle := (AZ+B)(CZ+D)^{-1} \quad , \quad Z \in \mathcal{H}_n\,,\ M = \begin{pmatrix} A & B \\ C & D \end{pmatrix} \in \mathrm{SU}(n,n)\,.$$

From now on we always assume $M \in \mathrm{SU}(n,n)$ to be in the *block decomposition* from above and n is always assumed to be in $\mathbb{N}, n \geq 2$. The case $n = 1$ is excluded from the definition since one needs an additional boundedness requirement there. For any $f : \mathcal{H}_n \to \mathbb{C}$, $k \in \mathbb{Z}$ and $M \in \mathrm{SU}(n,n)$ we define the *slash operator* as

$$f \mid_k M : \mathcal{H}_n \to \mathbb{C} \quad , \quad Z \mapsto \det(CZ + D)^{-k} f(M\langle Z\rangle)$$

Using this operator modular forms can be defined.

Definition 1. Let $k, n \in \mathbb{Z}$ and $n > 1$ and let $\Lambda \leq \Gamma_n$ be a subgroup of finite index. A *modular form of weight k with respect to* Λ is a function $f : \mathcal{H}_n \to \mathbb{C}$ with the following properties:

(M1) f is holomorphic
(M2) $f \mid_k M = f$ for all $M \in \Lambda$.

We denote the space of modular forms of weight k by $[\Lambda, k]$.

Consider $f \in [\Gamma_n, k]$. In view of the canonical generators of Γ_n condition (M2) implies $f(Z+H) = f(Z)$ for all $Z \in \mathcal{H}_n$ and $H \in \mathrm{Her}(n, \mathfrak{o})$ so f is periodic and therefore possesses a Fourier expansion. Due to Koecher's principle this expansion has the form

$$f(Z) = \sum_{\substack{T \in \mathrm{Her}(n,\mathbb{C}),\, T \geq 0 \\ \forall g \in \mathfrak{o}^n :\, g^* T g \in \mathbb{Z}}} \alpha(T)\, e^{2\pi i\, \mathrm{tr}(TZ)} \quad , \quad Z \in \mathcal{H}_n. \tag{1}$$

Note again that we excluded the case $n = 1$. The following theorem goes back to Hel Braun and guarantees that the vector space of modular forms of fixed weight is always finite dimensional. For more details we refer to [1] and [2].

Theorem 1 (H. Braun). *Let* $f \in [\Gamma_n, k]$ *with Fourier expansion as in (1). Then the following statements hold:*

(a) $f \equiv 0$, *whenever* $k < 0$.
(b) There exists a constant $c_n \in \mathbb{R}$ *such that the following is true for all* $k \geq 0$*:*
We have $f \equiv 0$, *whenever* $\alpha(T) = 0$ *for all* T *with* $\mathrm{trace}(T) \leq \frac{k}{4\pi} c_n$.

Since for any $C > 0$ there are only finitely many T in the Fourier expansion of f satisfying $\mathrm{trace}(T) \leq C$, we conclude $\dim[\Gamma_n, k] < \infty$ for all $k \in \mathbb{Z}$. This fact has the remarkable consequence that two modular forms can be identified by comparing finitely many Fourier coefficients. We will make use of this property in Sect. 4.

2 The Graded Ring of Degree 2

In this section we give a description of the graded ring of modular forms of degree 2 with respect to the full modular group. The results presented here can be found in [3] and [4]. By a theorem of U. Hirzebruch the group $\mathrm{Bih}(\mathcal{H}_2)$ is generated by

$$Z \mapsto M\langle Z\rangle\,,\; M \in \mathrm{SU}(2,2) \quad \text{and} \quad \mathrm{I}_{\mathrm{tr}} : Z \mapsto Z^{\mathrm{tr}}$$

For a proof confer, e.g., [7], p. 50.

Definition 2. A modular form $f \in [\Gamma_2, k]$ is called (skew-)symmetric if

$$f \circ \mathrm{I}_{\mathrm{tr}} = f \quad \text{resp.} \quad f \circ \mathrm{I}_{\mathrm{tr}} = -f\,.$$

We use the notation $[\Gamma_2, k]^{\mathrm{sym}}$ and $[\Gamma_2, k]^{\mathrm{skew}}$ for the corresponding subspaces.

Example 1. Examples of symmetric modular forms are given by Eisenstein series of Siegel type. Let $k > 4$ be even. Then for every $Z \in \mathcal{H}_2$ we define

$$E_k(Z) := \sum_{M:\Gamma_{2,0}\backslash\Gamma_2} 1|_k\, M(Z)$$

where $\Gamma_{2,0}$ is the subgroup whose C-block equals zero. Moreover we have $E_k \in [\Gamma_2, k]^{\mathrm{sym}}$. The "missing" Eisenstein series E_4 can be defined as a Maaß lift. It is also possible to choose a theta series

$$\Theta(Z, S) = \sum_{G\in\mathrm{Mat}(2,4,\mathfrak{o})} e^{\pi i\,\mathrm{trace}(ZGSG^*)}$$

where S is a positive definite even 4×4 matrix with $\det(S) = 16/9$; for more details, confer [4].

Remark 1. The group of abelian characters of $\mathrm{U}(n,n) \cap \mathrm{Mat}(4,\mathfrak{o})$ is generated by det and hence is isomorphic to C_3. The modular group Γ_2 has index 3 in $\mathrm{U}(n,n) \cap \mathrm{Mat}(4,\mathfrak{o})$ and representatives of the cosets are given by $I_4, \omega I_4, \omega^2 I_4$. As multiples of I have to act trivial there is no difference when considering modular forms for Γ_2 or the bigger group with character $\det^l$ where $l \in \mathbb{Z}$.

We introduce two submanifolds of $\mathcal{H}_2$:

$$H_2(\mathbb{R}) = \{Z \in \mathcal{H}_2\,;\, Z = Z^{\mathrm{tr}}\} \quad , \quad \mathcal{H}_3 = \left\{\begin{pmatrix} \tau & z \\ w & \tau' \end{pmatrix} \in \mathcal{H}_2\,;\, z = -w\right\}$$

For the description of the graded ring we need the following proposition.

Proposition 1. *There exist cuspidal Borcherds products*

$$\phi_9 \in [\Gamma_2, 9]^{\mathrm{skew}} \quad \textit{and} \quad \phi_{45} \in [\Gamma_2, 45]^{\mathrm{sym}}$$

where ϕ_9 is a Maaß lift and ϕ_{45} is not. The zeros of ϕ_9 resp. ϕ_{45} are all of first order and given by

$$\bigcup_{M\in\Gamma_2} M\langle H_2(\mathbb{R})\rangle \quad \textit{and} \quad \bigcup_{M\in\Gamma_2} M\langle \mathcal{H}_3\rangle$$

This leads us to the main result of the paragraph.

Theorem 2.

(a) The graded subring $\bigoplus_{k\in 2\mathbb{Z}} [\Gamma_2, k]^{sym}$ is generated by

$$E_4, E_6, E_{10}, E_{12} \quad \textit{and} \quad \phi_9^2\,.$$

The modular forms $E_4, E_6, E_{10}, E_{12}, \phi_9$ are algebraically independent.

(b) The graded ring $\bigoplus_{k\in\mathbb{Z}}[\Gamma_2, k]$ *is generated by*

$$E_4, E_6, \phi_9, E_{10}, E_{12} \quad \text{and} \quad \phi_{45}.$$

There exists a unique polynomial $p \in \mathbb{C}[X_1, \ldots, X_5]$ *satisfying*

$$\phi_{45}^2 = p(E_4, E_6, \phi_9, E_{10}, E_{12}).$$

Moreover we have

$$\bigoplus_{k\in\mathbb{Z}}[\Gamma_2, k] \cong \mathbb{C}[X_1, \ldots, X_6]/(X_6^2 - p).$$

3 Theta Constants

Following [4] we use theta series to construct modular forms. We shall study the transformation behaviour under canonical generators and obtain a representation of Γ_2.

Definition 3. Let $p \in \mathfrak{o}^2$. For any $Z \in \mathcal{H}_2(\mathbb{C})$ we define the theta series of characteristic p as

$$\Theta_p(Z) := \sum_{g\in\frac{i}{\sqrt{3}}p+\mathfrak{o}^2} e^{2\pi i\, g^* Z g}.$$

This construction yields a holomorphic function $\Theta_p : \mathcal{H}_2 \to \mathbb{C}$. Let us now consider the following relation on $\mathfrak{o}^2$:

$$p \sim q :\Longleftrightarrow \quad q \equiv \pm p \pmod{i\sqrt{3}}.$$

One easily checks that $(\mathfrak{o}^2, \sim)$ is an equivalence relation. The equivalence classes are represented by $p_0, \ldots, p_4$ where

$$p_0 = \begin{pmatrix}0\\0\end{pmatrix}, p_1 = \begin{pmatrix}1\\0\end{pmatrix}, p_2 = \begin{pmatrix}0\\1\end{pmatrix}, p_3 = \begin{pmatrix}1\\-1\end{pmatrix}, p_4 = \begin{pmatrix}1\\1\end{pmatrix}.$$

We show that the map

$$\Theta : \mathfrak{o}^2 \to \mathrm{Hol}(\mathcal{H}_2, \mathbb{C}) \quad , \quad p \mapsto \Theta_p$$

separates the equivalence classes of $(\mathfrak{o}^2, \sim)$.

Lemma 1. *One has*

$$p \sim q \iff \Theta_p = \Theta_q.$$

Therefore we set $\vartheta_j := \Theta_{p_j}$ *for* $j = 0, \ldots, 4$. *The set* $\mathfrak{B}_\vartheta = \{\vartheta_0, \ldots, \vartheta_4\} \subset \mathbb{C}^{\mathcal{H}_2}$ *is linearly independent.*

Proof. The claim follows by a simple calculation and by regarding the Fourier expansion of the theta series since the sets of matrices appearing in the expansion of the representatives are mutually disjoint. □

In the next step we describe the transformation of Θ_p under canonical generators.

Lemma 2. *Let $p \in \{p_0, \ldots, p_4\}$. Then we have the identities*

$$\Theta_p\big|_1 \begin{pmatrix} U^* & 0 \\ 0 & U^{-1} \end{pmatrix} = \Theta_{Up} \qquad \text{for all} \qquad U \in \mathrm{SL}_2(\mathfrak{o})$$

$$\Theta_p\big|_1 \begin{pmatrix} I & H \\ 0 & I \end{pmatrix} = \nu_p(H)\,\Theta_p \qquad \text{for all} \qquad H \in \mathrm{Her}(2, \mathfrak{o})\,.$$

The quantity $\nu_p(S)$ is a root of $X^3 - 1 \in \mathbb{Q}[X]$. Moreover the theta constants have the following behaviour under transformations of J:

$$\begin{aligned}
3\,\vartheta_0|_1\,J &= -\vartheta_0 - \vartheta_1 - \vartheta_2 - \vartheta_3 - \vartheta_4 \\
3\,\vartheta_1|_1\,J &= -2\vartheta_0 + \vartheta_1 - 2\vartheta_2 + \vartheta_3 + \vartheta_4 \\
3\,\vartheta_2|_1\,J &= -2\vartheta_0 - 2\vartheta_1 + \vartheta_2 + \vartheta_3 + \vartheta_4 \\
3\,\vartheta_3|_1\,J &= -2\vartheta_0 + \vartheta_1 + \vartheta_2 + \vartheta_3 - 2\vartheta_4 \\
3\,\vartheta_4|_1\,J &= -2\vartheta_0 + \vartheta_1 + \vartheta_2 - 2\vartheta_3 + \vartheta_4
\end{aligned}$$

Proof. The behaviour under generators of the first two types is deduced by direct calculations, the last one follows by applying the theta transformation formula. □

We now define $\mathcal{V} = \langle \mathfrak{B}_\vartheta \rangle_{\mathbb{C}}$ and for every $M \in \Gamma_2$ we consider the map

$$\mathcal{V} \to \mathcal{V} \quad , \quad \vartheta_j \mapsto \vartheta_j\big|_1 M \quad \text{for} \quad j = 0, \ldots 4 \tag{2}$$

Because of the last lemma this is a well-defined automorphism of $\mathcal{V}$.

Definition 4. We denote by $\Psi(M)$ the transpose of the representation matrix of (2) w.r.t. $\mathfrak{B}_\vartheta$.

Let us consider the main congruence subgroup of level $i\sqrt{3}$

$$\Gamma_2(i\sqrt{3}) := \left\{ M \in \Gamma_2 \mid M \equiv \pm I_4 \pmod{i\sqrt{3}} \right\}.$$

By regarding the block decomposition of unitary matrices one observes that for every $M \in \Gamma_2(i\sqrt{3})$ and $p \in \{p_0, \ldots, p_4\}$ the identity

$$\Theta_p\big|_1 M = \Theta_{A^* p}$$

holds. In view of Lemma 1 and by virtue of the equation

$$\Theta_{p_j}(Z^{\mathrm{tr}}) = \Theta_{\overline{p_j}}(Z) = \Theta_{p_j}(Z)$$

we obtain

Theorem 3. *We have* $\vartheta_j \in [\Gamma_2(i\sqrt{3}), 1]^{\mathrm{sym}}$ *for* $j = 0, \ldots, 4$.

4 Congruence Subgroups

Theorem 3 gives rise to a 5-dimensional representation of Γ_2. Since the factor group $\Gamma_2/\Gamma_2(i\sqrt{3})$ is isomorphic to the finite simple group $\mathrm{PSp}(2, \mathbb{F}_3)$ we can immediately derive the following.

Proposition 2. *There exists a unique 5-dimensional representation given by*

$$\Psi : \Gamma_2 \to \mathrm{GL}(5, \mathbb{C}) \quad , \quad M \mapsto \Psi(M).$$

The kernel of Ψ *is given by*

$$\ker(\Psi) = \Gamma_2(i\sqrt{3}).$$

Moreover $\vartheta_0, \ldots, \vartheta_4$ are algebraically independent and from the fact that the ring of symmetric forms with respect to the full modular group is contained in $\mathbb{C}[\vartheta_0, \ldots, \vartheta_4]$ follows

$$\bigoplus_{k=0}^{\infty} [\Gamma_2(i\sqrt{3}), k]^{\mathrm{sym}} = \mathbb{C}[\vartheta_0, \ldots, \vartheta_4], \tag{3}$$

see also [5]. Let Γ now be a congruence subgroup of level $i\sqrt{3}$, i.e., $\Gamma(i\sqrt{3}) \subset \Gamma \subset \Gamma_2$. Then we can define an action of Γ on $\mathbb{C}[\vartheta_0, \ldots, \vartheta_4]$ by using our representation Ψ: Let $M \in \Gamma$ and $f \in \mathbb{C}[\vartheta_0, \ldots, \vartheta_4]$. One defines

$$\begin{aligned} M \cdot f(\vartheta_0, \ldots, \vartheta_4) :=& \ f\left(\vartheta_0|_1 M^{-1}, \ldots, \vartheta_4|_1 M^{-1}\right) \\ =& \ f\left(\Psi(M)^{-1}.(\vartheta_0, \ldots, \vartheta_4)\right) \end{aligned}$$

and sets

$$\mathbb{C}[\vartheta_0, \ldots, \vartheta_4]^{\Gamma} = \{f \in \mathbb{C}[\vartheta_0, \ldots, \vartheta_4] \mid M \cdot f = f \text{ for all } M \in \Gamma\}.$$

This is the ring of invariants of Γ. We have the following connection with the graded ring of modular forms for Γ.

Corollary 1. *Let $\Gamma \subset \Gamma_2$ be a congruence subgroup of level $i\sqrt{3}$. Then we have*

$$\bigoplus_{k=0}^{\infty}[\Gamma,k]^{\text{sym}} = \mathbb{C}[\vartheta_0,\dots,\vartheta_4]^{\Gamma}.$$

Proof. This follows from (3) and the definition of Ψ. □

We want to determine the graded rings for some of the most important congruence subgroups of level $i\sqrt{3}$. Here we fix the notation for these groups

$$\Gamma_S(i\sqrt{3}) := \left\{M \in \Gamma_2;\ M \equiv \begin{pmatrix} I_2 & * \\ 0_{2,2} & I_2 \end{pmatrix} \pmod{i\sqrt{3}}\right\}$$

$$\Gamma_0(i\sqrt{3}) := \left\{M \in \Gamma_2;\ M \equiv \begin{pmatrix} A & B \\ 0_{2,2} & D \end{pmatrix} \pmod{i\sqrt{3}}\right\}$$

$$\Gamma_0^{\nabla}(i\sqrt{3}) := \left\{M \in \Gamma_2;\ M \equiv \begin{pmatrix} * & * & * & * \\ * & * & * & * \\ 0 & 0 & \pm 1 & * \\ 0 & 0 & 0 & \pm 1 \end{pmatrix} \pmod{i\sqrt{3}}\right\}$$

$$\Gamma_{2,1}(i\sqrt{3}) := \left\{M \in \Gamma_2;\ M \equiv \begin{pmatrix} * & * & * & * \\ * & * & * & * \\ * & * & * & * \\ 0 & 0 & 0 & * \end{pmatrix} \pmod{i\sqrt{3}}\right\}.$$

Remark 2. As already mentioned above Ψ can be considered as a representation of the finite group $\mathrm{Sp}(2,\mathbb{F}_3)$. This group is of order 51.840 and contains 162 subgroups up to conjugation. Table 1 contains the indices of the four subgroups.

Now let Γ be a congruence subgroup as above. We first compute a Hironaka decomposition of $\mathcal{R} = \mathcal{R}(\Gamma) = \mathbb{C}[\vartheta_0,\dots,\vartheta_4]^{\Gamma}$. The next step is to compute the coordinates of the generators of $\bigoplus_{k=0}^{\infty}[\Gamma_2,k]^{\text{sym}}$ in order to find more natural primary and secondary invariants. For both steps MAGMA was used.

Let $f_1,\dots,f_r \in \mathbb{C}[\vartheta_0,\dots,\vartheta_4]^{\Gamma}$ be the computed primary invariants. Then we set $\mathcal{A} = \mathbb{C}[f_1,\dots,f_r]$. The following theorem describes the structure of the graded rings for each of the four congruence subgroups.

Table 1 Index of congruence groups

Γ	$\Gamma_S(i\sqrt{3})$	$\Gamma_0^{\nabla}(i\sqrt{3})$	$\Gamma_0(i\sqrt{3})$	$\Gamma_{2,1}(i\sqrt{3})$
$[\Gamma_2 : \Gamma]$	1.920	320	40	40

Theorem 4.

(a) The graded ring with respect to $\Gamma_S(i\sqrt{3})$ is a free $\mathcal{A}$-module of rank 3 and has a Hironaka decomposition of the form

$$\mathcal{R} = \mathcal{A} + E_4\mathcal{A} + E_4^2\mathcal{A}$$

with primary invariants equal to

$$\vartheta_0 \quad , \quad \vartheta_1^3 \quad , \quad \vartheta_2^3 \quad , \quad \vartheta_3^3 \quad , \quad \vartheta_4^3$$

(b) Let $\Gamma = \Gamma_0^\nabla(i\sqrt{3})$. Then there exists a modular form h_9 of weight 9 such that

$$\mathcal{R} = \mathcal{A} + E_4\,\mathcal{A} + E_4^2\,\mathcal{A} + h_9\left(\mathcal{A} + E_4\,\mathcal{A} + E_4^2\,\mathcal{A}\right)$$

and with primary invariants equal to

$$\vartheta_0 \quad , \quad \vartheta_1^3 \quad , \quad \vartheta_2^3 + \vartheta_3^3 + \vartheta_4^3, \quad , \quad \vartheta_2^6 + \vartheta_3^6 + \vartheta_4^6 \quad , \quad \vartheta_2^9 + \vartheta_3^9 + \vartheta_4^9$$

(c) Let $\Gamma = \Gamma_0(i\sqrt{3})$. Then there exist modular forms $h_{19}, h_{21}, h_{27} \in \mathcal{R}$ (the subscript indicates the weight) and a Hironaka decomposition is given by

$$\mathcal{R} = \mathcal{A} + E_4\,\mathcal{A} + E_{10}\,\mathcal{A} + E_{12}\,\mathcal{A} + h_{19}\,\mathcal{A} + h_{21}\,\mathcal{A} + h_{27}\,\mathcal{A} + h_{19}E_{12}\,\mathcal{A}.$$

with primary invariants equal to

$$\vartheta_0^2 \quad , \quad \vartheta_1\,\vartheta_2\,\vartheta_3\,\vartheta_4 \quad , \quad \vartheta_1^6 + \vartheta_2^6 + \vartheta_3^6 + \vartheta_4^6 \quad , \qquad \vartheta_1^{18} + \vartheta_2^{18} + \vartheta_3^{18} + \vartheta_4^{18} \quad ,$$
$$\vartheta_1^3\,\vartheta_2^3 + \vartheta_1^3\,\vartheta_3^3 + \vartheta_1^3\,\vartheta_4^3 + \vartheta_2^3\,\vartheta_3^3 + \vartheta_2^3\,\vartheta_4^3 + \vartheta_3^3\,\vartheta_4^3$$

(d) Let $\Gamma = \Gamma_{2,1}(i\sqrt{3})$. There exist modular forms $h_6, h_{12}, f_{12}, g_{12}, h_{18} \in \mathcal{R}$ (the subscript indicates the weight) such that a Hironaka decomposition is given by

$$\mathcal{R} = \mathcal{A} + E_6\,\mathcal{A} + E_{10}\mathcal{A} + E_{12}\,\mathcal{A} + E_6^2\,\mathcal{A} + \phi_9^2\,\mathcal{A} + E_6^3\,\mathcal{A} + E_6E_{12}\,\mathcal{A}$$
$$+ E_{10}^2\,\mathcal{A} + \phi_9^2E_6\,\mathcal{A} + E_{12}^2\,\mathcal{A} + \phi_9^2E_{12}\,\mathcal{A} + h_6\,\mathcal{A} + h_{12}\,\mathcal{A} + g_{12}\,\mathcal{A} + h_{18}\,\mathcal{A}.$$

In this case the primary invariants equal

$$E_4 \quad , \quad \vartheta_0^4 + 8\,\vartheta_0\,\vartheta_1^3 \quad , \quad \vartheta_2^6 - 10\,\vartheta_2^3\,\vartheta_3^3 - 10\,\vartheta_2^3\,\vartheta_4^3 + \vartheta_3^6 - 10\,\vartheta_3^3\,\vartheta_4^3 + \vartheta_4^6,$$
$$\vartheta_2^6\,\vartheta_3^3 - \vartheta_2^6\,\vartheta_4^3 - \vartheta_2^3\,\vartheta_3^6 + \vartheta_2^3\,\vartheta_4^6 + \vartheta_3^6\,\vartheta_4^3 - \vartheta_3^3\,\vartheta_4^6 \quad , \quad f_{12}\,.$$

Remark 3. Finally we want to mention an application to Siegel modular forms of level 3. The restrictions of the five theta constants to $H_2(\mathbb{R})$ correspond to the five modular forms of level 3 and weight one in [6]. However, it was shown in [6] that these forms do not generate the full graded ring of level 3.

Acknowledgements We would like to thank the organizers of the conference and the workshop as well as the referee for valuable suggestions.

References

1. Braun, H., Hermitian modular functions I. Ann. Math. **50**, 827–855 (1949)
2. Braun, H., Der Basissatz für hermitische Modulformen. Abh. Math. Sem. Univ. Hamburg **19**, 134–148 (1955)
3. T. Dern, Hermitesche Modulformen zweiten Grades, PhD thesis, Aachen, 2001
4. T. Dern, A. Krieg, Graded Rings of Hermitian modular forms of degree 2. Manuscripta Math. **110**, 251–272 (2003)
5. E. Freitag, R. Salvati Manni, Hermitean modular forms and the Burkhardt quartic, Manuscr. Math. **119**, 57–59 (2006)
6. E. Freitag, R. Salvati Manni, The Burkhardt group and modular forms. Transf. Groups **9**, 25–45 (2004)
7. A. Krieg, *Modular Forms on Half-Spaces of Quaternions* (Springer, NewYork, 1985)

MIX
Papier aus verantwortungsvollen Quellen
Paper from responsible sources
FSC® C105338

If you have any concerns about our products,
you can contact us on
ProductSafety@springernature.com

In case Publisher is established outside the EU,
the EU authorized representative is:
Springer Nature Customer Service Center GmbH
Europaplatz 3, 69115 Heidelberg, Germany

Printed by Libri Plureos GmbH
in Hamburg, Germany